ENVIRONMENTAL GOVERNANCE FOR SUSTAINABLE DEVELOPMENT

This book studies the role of the United Nations Environment Programme (UNEP) as an advocate for greater environmental responsibility and analyses the major achievements and outcomes of two landmark conferences – Stockholm (1972) and Rio (1992) – which set the agenda for the future role of the UNEP. It discusses the UNEP's evolution, objectives and the problems of differing perspectives within, its ability to deal with environmental challenges, its skill in successfully carrying out the mandate and contributing to the pursuit of environmental security.

The book also looks at five developing countries of South Asia, namely India, Bangladesh, Nepal, Pakistan and Sri Lanka, to study the role of the South Asia Co-operative Environment Programme (SACEP), which plays an active role in the management of environmental issues and constitutes an important landmark in regional cooperation in South Asia. The author evaluates the contributions of National Conservation Strategies not only in creating environmental awareness but also in strengthening environmental governance architecture by integrating Millennium Development Goals and Sustainable Development Goals into the development planning of these South Asian countries under study.

Drawing on in-depth research and interviews, this book will be of interest to students, teachers, researchers, policymakers and strategic analysts working in the fields of environment studies, sustainable development, environmental science and policy, environmental law and governance, geography, politics and international affairs.

Uma Nabhi, Assistant Professor, Department of Political Science, Maitreyi College, University of Delhi, India.

'The book comprehensively examines the UN programmes and policies on environmental governance and how have they given impetus to the idea of sustainable development as well as economic well-being of the South Asian countries. It is a must-read book for those who are interested in working on environmental issues in South Asia in general and India in particular.'

Sangit Kumar Ragi, *Professor, University of Delhi, India*

'At a very critical temporal juncture when the vengeful demi gods of neoliberal capitalism are reaching the stage of self-actualization, regarding the apocalyptical consequences of the age of high mass consumption. This book engages in the task of epistemic emendation of the literature on the critique of neoliberal institutionalism and global cosmopolitanism.'

Sanjeev Kumar H.M., *Professor, University of Delhi, India*

ENVIRONMENTAL GOVERNANCE FOR SUSTAINABLE DEVELOPMENT

South Asian Perspectives

Uma Nabhi

LONDON AND NEW YORK

Designed cover image: FrankyDeMeyer/Getty Images

First published 2023
by Routledge
4 Park Square, Milton Park, Abingdon, Oxon OX14 4RN

and by Routledge
605 Third Avenue, New York, NY 10158

Routledge is an imprint of the Taylor & Francis Group, an informa business

© 2023 Uma Nabhi

The right of Uma Nabhi to be identified as author of this work has been asserted in accordance with sections 77 and 78 of the Copyright, Designs and Patents Act 1988.

All rights reserved. No part of this book may be reprinted or reproduced or utilised in any form or by any electronic, mechanical, or other means, now known or hereafter invented, including photocopying and recording, or in any information storage or retrieval system, without permission in writing from the publishers.

Trademark notice: Product or corporate names may be trademarks or registered trademarks, and are used only for identification and explanation without intent to infringe.

British Library Cataloguing-in-Publication Data
A catalogue record for this book is available from the British Library

Library of Congress Cataloging-in-Publication Data
Names: Nabhi, Uma, author.
Title: Environmental governance for sustainable development : South Asian perspectives / Uma Nabhi.
Description: First Edition. | New York : Informa Law from Routledge, 2023. | Includes bibliographical references and index.
Identifiers: LCCN 2022043990 (print) | LCCN 2022043991 (ebook) | ISBN 9781032294582 (Hardback) | ISBN 9781032440675 (Paperback) | ISBN 9781003370246 (eBook)
Subjects: LCSH: United Nations Environment Programme. | Environmental monitoring. | Environmental risk assessment | Environmental policy—International cooperation. | Environmental responsibility—Developing countries. | South Asia Co-operative Environment Programme. | Environmental policy—Asia, South—Case studies. | Environmental management—International cooperation. | Sustainable development—Developing countries—International cooperation. | Communication in the environmental sciences—International cooperation.
Classification: LCC GE170 .N32 2023 (print) | LCC GE170 (ebook) | DDC 363.7/0526—dc23/eng20221229
LC record available at https://lccn.loc.gov/2022043990
LC ebook record available at https://lccn.loc.gov/2022043991

ISBN: 978-1-032-29458-2 (hbk)
ISBN: 978-1-032-44067-5 (pbk)
ISBN: 978-1-003-37024-6 (ebk)

DOI: 10.4324/9781003370246

Typeset in Bembo
by Apex CoVantage, LLC

CONTENTS

Preface *viii*
List of Abbreviations *xvi*

1 Economic Development and Environment: An Overview 1

2 Evolving Role of UNEP at 50 23

3 India and UNEP 76

4 Environmental Governance and Sustainable Development in South Asia: A Study of Bangladesh, Nepal, Pakistan and Sri Lanka 105

5 UNEP: An Appraisal 163

6 Conclusion 192

Bibliography *202*
Index *259*

PREFACE

The planet Earth is poised between two conflicting trends. On the one hand, a wasteful and invasive consumer society, coupled with increasing population growth, is threatening to destroy the very resources on which human life is based. At the same time, society is engaged in a struggle against time to reverse these trends and introduce suitable practices that will ensure the welfare of future generations.

The emergence of environmental challenges on the agenda of world politics has arisen as a response to environmental accidents and disasters, to increases in scientific knowledge and to activism and lobbying by non-governmental organizations and grassroots movements. Environmental issues encompass such concerns as ozone depletion, climate change, deforestation loss of biodiversity, desertification, management of toxic and hazardous wastes, pollution of the oceans and waters, acid rain, air pollution land degradation and the depletion of resources. The reason for concern is twofold. First, scientific evidence supports the proposition that ecological damage is increasing at a faster rate because of human activity. Second, environmental challenges have the potential for severe and irreversible impact not only on the ecosystem but also on the economic and social development of people and states. Environmental degradation is claimed as global when the common heritage of human kind is at stake. This invokes the idea of shared global resources to be managed in the common interest of all. Forests are viewed as a common heritage and a global environment concern because the destruction of forests and the lungs of the world can impact the global climate system. In a similar way loss of biodiversity is viewed as global because the world is deprived of potentially valuable resources of pharmaceutical and agricultural genetic resources. The impact of the loss of one component of the ecosystem has effects on the global ecosystem as a whole. These environmental issues require collective action. They thus fall within the realm of multilateral decision-making fora, the foremost of which is the United Nations.

The United Nations is viewed as the only institution within which it is possible to strengthen global environmental governance and achieve sustainable development and environment security at a global level. For the environment, this is a time of dynamic change where policy tools are continuously being developed and adapted to new issues and challenges. As the United Nations is mandated to play a central role on environmental matters, the United Nations Environment Programme was created by the United Nations to address environmental issues.

Since UNEP's establishment, a substantive shift in environmental priorities and the policies to address them has been happening. Although much remains to be done in the traditional pollution-related problems, the policies needed are increasingly being refined and implemented. In recent years, a set of environmental challenges have come to the top of the agendas, including climate change, biodiversity and the sustainable management of forests, oceans, freshwater and land resources.

In this context, the book deals with the evolution, growth and role of United Nations Environment Programme, which represents environmental governance at the global level, as an advocate for greater environmental responsibility and as a vehicle for new approaches to deal with pollution and conservation of natural resources. The scope of the book is expanded to include the period from 1972 to 2022 (UNEP at 50), and in this process, the action plans and outcomes of the two conferences, namely, the Stockholm and Rio Conference, which are considered extremely important as they set the agenda for the future role of UNEP, are highlighted. The book further attempts to probe various environmental governance issues and challenges raised by the five developing countries of South Asia, namely India, Bangladesh, Nepal, Pakistan and Sri Lanka and the significant and active role of the South Asia Co-operative Environment Programme (SACEP), in the management of environmental issues and challenges in the South Asian region.

In this context, the study attempts to examine two interrelated hypotheses. The *main hypothesis* centres on the proposition that while there is an urgent need for developing countries to improve environmental conditions, the economically advanced countries pursue an agenda that caters basically to their own interests, which is reflected in their efforts to dominate the United Nations Environment Programme. The United Nations Environment Programme, whose objectives are undoubtedly laudable and of extreme significance, is subjected to diverse pressures exerted by the developed countries on the one hand and the developing countries on the other, which is reflected in the implementation of its policies that occupies a significant portion of the book (Chapters 2 and 5).

The *secondary, related hypothesis* is South Asia-specific. The South Asian states, with their huge populations, poor economic conditions and inadequate infrastructural facilities, face immense environmental and region-specific challenges. While they need to have external aid in managing these problems, particularly from agencies like the UNEP and SACEP, their problems have to be addressed keeping in view the perspectives and problems of developing countries. In spite of financial constraints, UNEP is playing a significant role in managing and improving the environmental problems in the South Asian region through the South-Asia

Co-operative Environmental Programme, which is the main focus of an important part of the book (Chapter 4).

In response to the environmental challenges in the South Asian region, significant initiatives have been launched at the national, sub-regional and regional levels, and emphasis is placed on regional environmental cooperation.

In addition to this, the study addressed several important issues:

- One was related to UNEP's instrumental role in the adoption and implementation of international environmental conventions and treaties aimed at preserving the ozone layer, conserving biological diversity, coping with climate change, protecting the oceans and seas, controlling the movement of toxic wastes and controlling the trade in endangered wildlife species.
- Second, it attempts to explore the compulsions, constraints and challenges faced by UNEP in the implementation of its policies.
- Third, the book tries to evaluate how far UNEP has been successful as an indispensable contribution to environmental governance and to what extent it was able to solve the long-term environmental problems having immense consequences for the economic well-being and security of nations throughout the world.

The book further evaluates 50 years of the role of UNEP as a leader, catalyst and coordinator, focusing the study on its role as a bridge between science and policy, between governments and non-governmental activists and between environment and development. The book tries to examine the measures taken by UNEP that led to the process of developing an international legal regime for the environment.

This book also underlines the determining role of UNEP, with the help of the national governments, in the establishment of an environmental governance architecture through environmental ministries, agencies and policies in developing countries; helps them to participate fully in the process of international environmental cooperation and governance; and strengthens their capacities to deal with their own domestic environment and development-related issues.

This book is structured into *six chapters*.

The world scenario of today is entirely different from what existed immediately after the Second World War. In this emerging global order, a striking new perception of environmental interdependence has come to the forefront between the countries of the industrialized North and developing countries of the South, making possible accommodation between the two an issue of major significance on the international agenda of the 1990s. And this environmental interdependence based on pragmatic approach between and among nations has become a crucial factor in finding solutions to the common problems of mankind, particularly with regard to environment and development.

At the biosphere conference of 1968, environment was for the first time listed on the world agenda. Subsequently, the Stockholm Conference, or the United Nations Conference on the Human Environment (UNCHE), held in 1972, proclaimed the right of individuals to a healthy environment. This conference marked the beginning of the institutionalization of global environmental concerns within the United Nations system. The conference, whose theme was 'Only One Earth', was greeted as evidence of the commitment of governments to address these concerns.

Tensions between the developed industrialized countries and the developing countries surfaced and shaped the outcomes of the conference, and these differences have become a defining theme of 'global environmental politics and the pursuit of sustainable development and environmental security' since then. The Stockholm Conference created a declaration and a set of principles dealing with such matters as the responsibility of cooperative action to preserve the earth's ecology, the responsibilities of states for transboundary pollution and the sovereignty of states over their national rights to development. One of the major issues identified during the conference proceedings was the close connection between the environmental and the development agendas. One of the key issues negotiated at the conference was embodied in principle 21, which coupled state responsibility for transboundary pollution with the inalienable right to the exploitation of natural resources demanded by the Southern governments. The conference made several significant recommendations, the foremost among which was setting up of the United Nations Environment Programme in Nairobi.

The United Nations Environment Programme was established with the purpose of monitoring and coordinating environmental understanding. The developed countries were reluctant to agree to a new institution that would require increased funding. UNEP's mandate was to monitor, coordinate and catalyse. It has suffered from several constraints, including a paucity of funding, lack of political support from governments and geographical isolation within the UN system. This study examines how far the UNEP has been able to deal with these challenges and how successful it has been in carrying out its mandate and contributing to the pursuit of environmental governance and sustainable development.

In 1983, the process of globalizing the environmental movement was strengthened, when Gro Harlem Brundtland, the prime minister of Norway, submitted the report entitled 'Our Common Future' in 1987. The Brundtland Commission placed the concept of 'Sustainable Development' firmly onto the UN agenda. The report is a mix of progressive reforms on debt, reduction of military spending, socio-economic justice and conventional solutions.

The second main UN Conference on the Environment was known as the United Nations Conference on Environment and Development or 'the Earth Summit', held in Rio de Janeiro in June 1992. This conference shows the evidence of the commitment of the UN and the international community to the cause of environmental security and sustainable development.

The Rio Conference resulted in the Rio declaration, Agenda 21, a statement on forest principles' a treaty on climate change and a treaty on biodiversity. It also led to the formulation of an earth charter as a statement of ethics to guide environmental governance into the twenty-first century. The Commission on Sustainable

Development is mandated to ensure the effective follow-up of the conference to enhance international cooperation and to examine the progress of implementation of Agenda 21. However, major differences have emerged in the perceptions of the economically advanced countries of the North and those of the developing countries of the South on various environmental issues. In this context, the first chapter, 'Economic Development and Environment: An Overview', provides a conceptual basis and a brief background of the environment in the UN agenda. It begins with a brief overview of the concepts related to economic development and environment and deals with the emergence and evolution of the concept of sustainable development, and in the process the significance of Brundtland Report is explained. The chapter attempts to further study the links between the Bretton Woods Institutions and the UN, addressing the basic issue of 'who is going to govern the environmental agenda' – whether the Bretton Woods Institutions represented by World Bank, the International Monetary Fund and others or the UN system and UNEP. The chapter in its updated section further deals with the significance of the Millennium Development Goals and Sustainable Development Goals in the evolution of sustainable development agenda and environmental governance and tries to explore the manner in which environmental issues have been pursued within the United Nations and UNEP.

The second chapter, 'UNEP at fifty', deals with the evolution, growth and role of the UN Environment Programme covering the period from 1972 to 2022. This chapter deals with two crucial aspects: first, the stellar role of leading environmentalists and environmental movements in the evolution of the UNEP, and, second, the emergence of significant differences in the perspectives of the advanced developed countries of the North and the developing countries of the South. Underlining the role played by environmental movements in awakening global consciousness, it examines the linkages between British environmentalism, American environmentalism and new environmentalism in the evolution of the United Nations Environment Programme. A study of the major issues raised by both developed and developing countries during the proceedings of Stockholm and Rio Conferences, which were important milestones on the road to the establishment of the UNEP, is discussed. The chapter traces the developments leading to the creation of the United Nations Environment Programme, with a focus on its major institutional and functional components, namely environmental assessment, environmental management and supporting measures. This chapter also examines the main hypothesis that in spite of its laudable agenda, UNEP is being pressurized and dominated by differing viewpoints of developed and developing countries, which is reflected in the implementation of its agenda. Finally, it endeavours to highlight the achievements and constraints/challenges faced with regard to implementation of environmental negotiations/agreements that contributed to the growth of UNEP. While dealing with all these major environmental issues, from the perspective of North-South divide, critical viewpoints of leading environmental activists were highlighted. As a whole, the chapter concludes by observing that strong political and financial commitments are required, as support for a strengthened UNEP, which had demonstrated skill and showed the commitment to the cause of environmental security that guides environmental governance into the twenty-first century, in spite of its limited mandate and lack of funding.

The third chapter, 'UNEP and India', as the title suggests, is India-specific. It begins with a study of the impact of the cultural traditions and thoughts that had provided the ideological underpinning to the present environmental movements in India. This was essential (1) to underline the extant environmental consciousness within the Indian ethos and (2) to provide the background to existing environmental activism in India. While it is outside the scope of the chapter to deal with all the environmental movements, the main movements are studied in the context of the discourse. These included the 'Chipko Movement', which had demonstrated the workability of a powerful model of people's participation, a model that can be emulated and replicated; 'Silent Valley Movement' in Kerala, as one of the most important milestones in the shaping of public opinion as well as the formulation of official policy which had been halted on environmental grounds, which offered a unique foresight of peoples' participation; and 'Narmada Bachao Andolan', led by Baba Amte, Medha Patkar and others, which focused on two major dams under construction that submerged almost as much area as it was meant to irrigate. The movement intensified its resistance on central government by peaceful means; 'Tehri Project', led by Sunder Lal Bahuguna, prevented the further development of the project. This was undertaken to emphasize the extant environmental impulses within India.

The second part of the chapter tries to link the efforts of the UNEP in the management of various environmental issues in India which included desertification, deforestation, aquatic ecosystem, multipurpose valley projects, biological diversity, climate change, natural resource management, urbanization, human health and welfare, energy, industry and transportation. In the management of these environmental issues the tools and methodologies of environmental planning were primarily addressed through the training, strategies and guidelines of UNEP. It further examines how UNEP is guiding Indian initiatives and programmes.

The fourth chapter attempts to explore environmental governance and sustainable development challenges in four South Asian states, namely Bangladesh, Nepal, Pakistan and Sri Lanka. A significant component of this analysis was the role played by South Asia Co-operative Environment Programme, which constituted an important landmark in regional cooperation in South Asia. The chapter begins with an environmental perspective of the South Asian region, where the study underlines the interlinkages between economic growth, population, natural disasters, forests, biodiversity, renewable freshwater resources, oil pollution, air pollution, land and food issues.

A special focus of the chapter is the role of National Conservation Strategies in not only creating environmental awareness but creating an environmental governance architecture to achieve sustainable development in all South Asian countries. This chapter underlines the importance of regional and sub-regional cooperative efforts in the management of environmental challenges and issues that culminated in the establishment of a South Asian Co-operative Environment Programme in the South Asian region, where eight countries, namely Afghanistan, Bangladesh, Bhutan, India, Maldives, Nepal, Pakistan and Sri Lanka, are brought together solely on the subject of environment. In the updated section, 'An assessment of SACEP (1982–2022)', post-2015 development agenda that includes 17 Sustainable Development Goals (SDGs) to end poverty, fight inequality and injustice, and to tackle climate change by 2030 has been assessed and examined. Further, this chapter underlines the growing feelings among the nations that the environmental challenges and initiatives which are specific to South Asian region, that is, biodiversity conservation, waste management, climate change and regional seas programmes, can be taken up and solved through regional and sub-regional cooperative efforts, with the conviction that environment is one area where nations will have to come together and resolve all their differences to save the common heritage.

The fifth chapter attempts a critical appraisal of the United Nations Environment Programme during the period from Stockholm (1972) to post-Rio negotiations covering the ground till 2022. It tries to detail the challenges and constraints faced by UNEP in the implementation of its policies, while taking stock of its achievements.

'An appraisal of UNEP' underlined the 'North-South' debate in which environment and development appeared to represent conflicting values. While development-related priorities played a dominating role for developing countries of the South, environment-related issues played a significant role for developed countries of the North. The priority areas put forward by the developing countries of the South were human settlements, land and water resources, desertification, trade, the transfer of technology, oceans, conservation of nature, wildlife, genetic resources and energy. Whereas the significant environmental issues addressed by the developed countries of the North were global warming, conservation of biodiversity, ozone depletion and deforestation. In the process, the chapter analyses the first three sessions of the Governing Council which were formative and also the first three sessions of the Environment Assembly (EA replaced Governing Council in 2012 (The United Nations Environment Assembly replaced Governing Council in 2012 as the world's highest-level decision -making body on the environment, with a universal membership of all 193 member states)) that set the course of the UNEP to the present date.

It was observed that UNEP faced four major handicaps in the management of local, regional and international environmental matters, which were elaborated and explained as financial, managerial, political and constitutional. Despite these handicaps, the Regional Seas Programme was widely regarded as UNEP's most effective undertaking in its first decade of operation which was regional rather than global in nature. A special focus of this chapter is the views and opinions expressed by leading environmental activists as well as the officials conducting UNEP programmes on the role and relevance of UNEP (environmental activists and UNEP officials were interviewed). This chapter also exclusively deals with the challenges and reveals the constraints faced by UNEP in the implementation of environmental agendas

(1972–2022). The major challenges and constraints faced by UNEP are underlined as finances, internal management, its location in Nairobi, the nature of UNEP's relationship with other UN agencies, incompatibility between its role as conceptualizer of environmental policy and its commitment to action and effective leadership within UNEP. The evaluation at the same time notes the significant role played by UNEP in planning a new way forward to ensure the survival of life on earth and to correct the flaws that created environmental degradation and protect the very systems that support life on earth. It is observed that the prime challenge before UNEP is not only to promote and implement its environmental agenda but also to integrate it strategically with the goals of economic development and social well-being. It further observes that a strong regional architecture is central to UNEP's ability to manage and advance its global environmental agenda and governance.

On the whole, the analysis presents a detailed study of UNEP's role as the leading environmental authority that promoted the coherent implementation of the environmental dimension of sustainable development. The analysis further observes the need for the global community to embark on major structural reforms. UNEP as activator of the environmental action plans had given the international environmental movement a universality, legitimacy and acceptability in the developing countries which under the circumstances could hardly have been obtained.

In the concluding chapter the major objectives arising out of the study and findings of the main chapters are recapitulated in a broader perspective. The study concludes in an optimistic tone by emphasizing on a strong, effective and committed UNEP as essential not only to the global community as its environmental conscience but also to the success of the United Nations.

ABBREVIATIONS

ACC	Administrative Commission on Co-ordination
AIJ	Activities Implemented Jointly.
APELL	Programme on Awareness and Preparedness for Emergencies at the Local Level
ASTRA	Application of Science and Technology to Rural Areas
AWARD	Association of Voluntary Agencies for Rural Development
BRAC	Bangladesh Rural Advancement Committee
BWI	Bretton Woods Institutions
CBD	Convention on Biological Diversity
CBP	Control of Blindness Programme
CCD	Convention to Combat Desertification
CEA	Central Environmental Authority (Sri Lanka)
CEE	Centre for Environment and Education (Gujarat)
CFCs	Chloroflurocarbons
CGA	Central Ganga Authority
CITES	Convention on International Trade in Endangered Species of Wild Fauna and Flora
CNG	Compressed Natural Gas
CPCB	The Central Pollution Control Board
CPR	Committee on Permanent Representatives
CPRE	Council for the Preservation of Rural England
CSD	Commission on Sustainable Development
CSIR	Council for Scientific and Industrial Research
DAP	Draft Action Programme
DD	Development Decade
DGSM	Dasholi Gram Swarajya Mandal
DoEF	Department of Environment and Forest

Abbreviations

DOEM	Designated Officers on Environmental Matters
DSCWM	Department of Soil Conservation and Watershed Management
EC	European Community
ECB	Environmental Coordination Board
ECG	Environmental Core Group (Nepal)
ECLA	UN Commission for Latin America
ECOSOC	Economic and Social Council
EFCTC	European Fluro Carbon Technical Committee
EJA	Environment Impact Assessment
ELC	Environment Liaison Centre
EMG	Environmental Management Group
EPC	Environment Protection Council (Nepal)
ESCAP	Economic and Social Council for Asia and Pacific
EUAD	Environment and Urban Affairs Division
FAO	Food and Agricultural Organization (UN)
FCCC	Framework Convention on Climate Change
GA	General Assembly
GAP	Ganga Action Plan
GAR	General Assembly Resolution
GC	Governing Council
GDP	Gross Domestic Product
GEF	Global Environment Facility
GEMS	Global Environmental Monitoring System
GEO	Global Environmental Organization
GEO	Global Environment Outlook
GNP	Gross National Product
GoP	Government of Pakistan
IBRD	International Bank of Reconstruction and Development
ICAR	Indian Council of Agricultural Research
ICC	International Conservation Conference
IES	Indian Environment Society
IETC	International Environmental Technology Centre
IFSD	Institutional Framework for Sustainable Development
IGOs	Inter-Governmental Organizations
IGPF	Inter-Governmental Panel on Forests
IIED	International Institute for Environment and Development
ILO	International Labor Organization
IMF	International Monetary Fund
IMM	Intergovernmental Meetings of Ministers
INC	Inter-governmental Negotiating Committee
INFOTERA	International Referral System
IPCC	Inter-government Panel on Climate Change
IPBES	Intergovernmental Platform on Biodiversity and Ecosystem Services
IRPTC	The International Register of Potentially Toxic Chemicals

ITTO	International Tropical Timber Organization
IUCN	International Union for Conservation of Nature
IWC	Inland Waterways Commission
MEPA	Ministry of Environment and Parliamentary Affairs (Sri Lanka)
MINAR	Monitoring of Indian National Aquatic Resources
MTS	Medium Term Strategy
NACC	North American Conservation Congress
NBRI	National Botanical Research Institute
NCA	National Commission on Agriculture (Pakistan)
NCS	National Conservation Strategy
NEA	National Environment Act
NEAP	National Environmental Action Programme
NEPAP	National Environment Policy and Action Plan
NESC	National Environmental Steering Committee
NFCP	National Filaria Control Programme
NGOs	Non-Governmental Organizations
NLC	National Leprosy Control
NMCP	National Malaria Control Programme
NPC	National Planning Commission
NT	National Trust
NTMWD	National Technology Mission on Wasteland Development
NWDB	National Wasteland Development Board
OECD	Organization for Economic Co-operation and Development
OHP	Operational Hydrology Programme
OPEC	Oil and Petroleum Exporting Countries
PACD	Plan of Action to Combat Desertification
PCED	Peoples Commission on Environment and Development
PEPAC	Environmental Planning and Architectural Consensus (Pakistan)
Prep Com	Preparatory Committee
SAARC	South Asian Association for Regional Co-operation
SACEP	South Asian Co-operative Environment Programme
SC	Sierra Club
SC	Security Council
SPB	The Royal Society for the Protection of Birds
SPCS	Sarhad Provincial Conservation Strategy
STAP	Scientific and Technical Advisory Panel
SUN	The Sustainable United Nations Team
SUNFED	The Special UN Fund for Economic Development
SWMTEP	System-Wide Medium Term Environmental Programme
TERI	Tata Energy and Research Institute
TNCs	Trans National Corporations
UK	United Kingdom
UN	United Nations
UNCCEE	UN Collaborating Centre on Energy and Environment

UNCED	United Nations Conference on Environment and Development
UNCHE	The United Nations Conference on Human Environment
UNCRD	United Nations Centre for Regional Development
UNCTAD	United Nations Conference on Trade and Development
UNIDO	United Nations Industrial and Development Organization
UNDP	United Nations Development Programme
UNEP	United Nations Environment Programme
UNESCO	United Nations Education Scientific and Cultural Organization
UNGASS	UN General Assembly Special Session
USA	United States of America
WB	World Bank
WCC	World Climate Conference
WCED	World Commission on Environment and Development
WCIP	World Climate Impact Studies Programme
WCISP	World Climate Impact Studies Programme
WCS	World Conservation of Nature
WEO	World Environment Organization
WHO	World Health Organization
WMO	World Meteorological Organization
WRI	World Resources Institute
WTO	World Trade Organization
WWF	World Wide Fund for Nature

1
ECONOMIC DEVELOPMENT AND ENVIRONMENT

An Overview

Introduction

The world scenario of today is entirely different from what existed immediately after the Second World War. In this emerging global order, a striking new perception of the environmental interdependence has come to the forefront between the countries of the industrialized North and developing countries of the South, making possible accommodation between the two, an issue of major significance on the international agenda of the 1990s. And this environmental interdependence based on pragmatic approach between and among nations has become a crucial factor in finding solutions to the common problems of mankind, particularly with regard to environment and development. Humanity is confronted with a perpetuation of disparities between and within nations with worsening of poverty, hunger, ill health, illiteracy and the deterioration of the ecosystems on which the humanity depends for its well-being. However, an integration of environment and development concerns will lead to improved living standards, better protected and managed ecosystems and a safer, more prosperous future in a global partnership for sustainable development.[1]

At the biosphere conference of 1968, environment was for the first time listed on the world agenda.[2] Subsequently, the Stockholm Conference or the United Nations Conference on the Human Environment[3] was held in 1972, which proclaimed the right of individuals to a healthy environment, emphasizing the responsibility to protect and ameliorate the environment for posterity. This conference marked the beginning of the institutionalization of global environmental concerns within the United Nations system. The conference, whose theme was 'Only One Earth',[4] was greeted as evidence of the commitment of governments to address these concerns.

2 Economic Development and Environment

Tensions between the developed industrialized countries and the developing countries surfaced and shaped the outcomes of the conference, and these differences have become a defining theme of 'global environmental politics and the pursuit of environmental security' since then. The Stockholm Conference created a declaration and a set of principles dealing with such matters as the responsibility of cooperative action to preserve the earth's ecology, the responsibilities of states for transboundary pollution and the sovereignty of states over their national rights to development. One of the major issues identified during the conference proceedings was the close connection between the environment and the development agendas. One of the key issues negotiated at the conference was embodied in principle 21, which coupled state responsibility for transboundary pollution with the inalienable right to the exploitation of natural resources demanded by the Southern governments. The conference made several significant recommendations, the foremost among which was setting up of the United Nations Environment Programme[5] in Nairobi.

The United Nations Environment Programme was established with the purpose of monitoring and coordinating environmental understanding. The developed countries were reluctant to agree to a new institution that would require increased funding; UNEP's mandate was to monitor, coordinate and catalyse. It has suffered from several constraints, including a paucity of funding, lack of political support from governments and geographical isolation within the UN system. This study examines how far the UNEP has been able to deal with these challenges and how successful it has been in carrying out its mandate and contributing to the pursuit of environmental security.

In 1983, the process of globalizing the environmental movement was strengthened, when Gro Harlem Brundtland, the prime minister of Norway, submitted the report entitled 'Our Common Future' in 1987.[6] The Brundtland Commission placed the concept of 'sustainable development' firmly into the lexicon of environmental politics and onto the UN agenda. The report is a mix of progressive reforms on debt, reduction of military spending, socio-economic justice and conventional solutions.

The second main UN Conference on the Environment was known as the United Nations Conference on Environment and Development or 'the Earth Summit', held in Rio de Janeiro in June 1992. This conference shows the evidence of the commitment of the UN and the international community to the cause of environmental governance.[7]

The Rio Conference resulted in the Rio declaration, Agenda 21, a statement on forest principles' a treaty on climate change and a treaty on biodiversity. It also led to the formulation of an earth charter as a statement of ethics to guide environmental governance into the twenty-first century.[8] The Commission on Sustainable Development is mandated to ensure the effective follow-up of the conference to enhance international cooperation and to examine the progress of implementation of Agenda 21.[9] However, major differences have emerged in the perceptions of the economically advanced countries of the North and those of the developing countries of the South on various environmental issues. In this context, this chapter provides a conceptual basis and a brief background of the environment in the UN agenda. The chapter begins with a brief overview of the concepts related to economic development and environment and deals with the emergence and evolution of the concept of sustainable development, and in the process the significance of Brundtland Report is highlighted. The chapter attempts to further study the links between the Bretton Woods Institutions and the UN, addressing the basic issue of who is going to govern the environmental agenda, whether the Bretton Woods institutions like World Bank, the International Monetary Fund and others or the UN system, including UNEP. While explaining the role of 'Bretton Woods Institutions and the UN', the chapter examines how the Bretton Woods Institutions, which were originally conceived by Keynes as contributors to a world welfare system, are later manoeuvred to counter the growing strength of the developing countries forums, like UNEP and UN system, which are pressing for a more equitable role in defining the direction of the world economy and environment. The chapter, in its updated section, further deals with the significance of the Millennium Development Goals and Sustainable Development Goals in the evolution of sustainable development agenda and environmental governance.

Conventional wisdom on the pursuit of environmental preservation suggests two things: (1) it can be achieved through global governance, and (2) the United Nations can provide this improved governance. However, the United Nations has grappled with an issue which simply was not a matter for international concern 50 years ago and which was not incorporated in the UN's mandate.

In the recent past, a number of well-publicized environmental emergencies ranging from oil spills and toxic chemical releases to nuclear accidents and the deliberate human modification of the natural environment and their cumulative consequences have extended far beyond the borders of the countries and have furthered the imperative for international cooperation. Improved international coordination on environmental issues is the most important prerequisite of the trend towards a more integrated approach. Bilateral and multilateral environmental agreements have proved to be powerful instruments of change. Thus, Ian Johnson noted,

> Man, as both creature and moulder of his environment, has transformed his environment in countless ways. Both aspects, the natural and the man-made, are essential to his well-being and to the enjoyment of basic human rights, particularly, the right to life.[10]

These global environmental threats have become a major concern since all of humanity stands to be greatly affected. They can only be addressed effectively by measures which place environmental concerns at the centre of the development policies of all countries. The United Nations Development Programme, in its Human Development Report, 1994, noted that environmental problems respect no national border and their grim consequences travel the world.[11] Emphasizing on a successful cooperative effort to save the environment, Al Gore talks about the feasibility and desirability of establishing a world government, and '[s]ince the inclusion of the developing countries is imperative for such a strategy, he calls for a 'Global Marshall Plan',[12] which aims at concentrating all efforts on stabilizing the world population, developing environmentally sound technologies, modifying the economic rules of the game, concluding collective treaties and launching an information campaign for the citizens of the globe. To have a better understanding of sustainable development, it is important to understand the linkages between economic development and the concept of environment.

Economic Development

Economic development is defined as a sustainable increase in living standards that encompass material consumption, education, health and environmental protection.[13] In a broader sense, development includes equality of opportunity, political freedom and civil liberties. The challenge of development is to improve the quality of life, especially in the developing countries, and therefore it encompasses better education, higher standards of health and nutrition, less poverty, a cleaner environment, more equality of opportunity, greater individual freedom and a richer cultural life. Further, development is defined as growth plus change, and change in social, cultural, psychological and economic, qualitative as well as quantitative.[14] According to Amartya Sen 'development [is] defined as a process of expanding the capabilities of the people'. The French sociologist Raymond Aron[15] has different interpretations of the concept of development. In the first place, it is explained as the long-term statistical trend of economic growth. Alternately, it is understood as the contrast between rich and poor countries. Both these views are incorporated in the 'stages of growth' theory propagated by W.W. Rostow, in which development is regarded as a linear path along which all countries travel. At various stages in history, advanced countries reached the stage of 'take off'; poor countries will approach it in due course. If they do not, it is because they lack capital, technology, expertise and motivation. A third interpretation of the process, and one which applies only to the West, is explained as the gap between affluent capitalist countries and the Soviet bloc.

The disenchantment with GNP and the growth process is now becoming increasingly widespread. The well-known Pakistani economist Mahbub ul Haqq articulated it well when he says that '[d]eveloping countries have no choice but to turn inwards and adopt a different style more consistent with their own poverty'.[16] According to him, development goals are concerned with 'elimination of malnutrition, disease, illiteracy, squalor, unemployment and inequality'. They are not merely concerned with how much is produced but what is produced and how it is distributed; this requires a redefinition of economic and social objectives. Within this country, a person who has been arguing for such an approach and has been drawing attention to the potential of alternative technology is Professor Amulya Kumar N. Reddy, head of the cell for the Application of Science and Technology to Rural Areas (ASTRA) at the Indian Institute of Science at Bangalore.[17] In his scheme of things, genuine development is the index of how much energy a person consumes rather than his per capita income, which comprises both calories in the shape of food and other external forms like fuel, cooking and lighting. To him energy is a component of a development strategy which does three things: satisfies basic human needs, starting with the needs of the neediest; is directed towards self-reliance; and is in harmony with the environment. In other words, it views development from the base upwards, instead of expecting the benefits to trickle downwards.

The United Nation's preoccupation with development is rooted in the sharp division of its membership between rich and poor nations, a division that has frequently been characterized as a leading long-term threat to environment, peace and security.

Environment

The concern with the deteriorating quality of life has led over the years to explore the relationship between the environment and development. At one level, environment refers to the preservation of scenic natural landscape or dwindling wildlife species; at another, it comprises of industrial pollution or the threat to a citizen's amenities caused by the building of a road or a major factory. The *Encyclopaedia of Social Sciences* defines the concept of environment

> as a metaphor for enduring contradictions in human condition: the power of domination, yet the obligation of responsibility; the drive for betterment, tempered by the sensitivity of humility; the management of nature to improve the chances of survival, yet the universal appeal of sustainable development; the individualism of consumerism and the social solidarity of global citizenship.[18]

In his book *Temples or Tombs*, Darryl de Monte defines the concept of environment as the source of all natural resources: energy, land water, atmosphere and minerals.[19] Differing viewpoints have emerged with regard to pollution and the exhaustion of natural resources. The classic in this genre was Rachel Carson's *Silent Spring*, in 1962, which showed the interconnectedness between minute levels of pesticides, which could become concentrated in food chains and thus can lead to a serious environmental problem.[20]

An extreme ecological stand was taken by the 'Limits to Growth' study produced by the 'Club of Rome' exactly a decade later, which predicted that unless technology changed its current course, the world was in danger of running out of its resources. And it set the actual date of collapse in the latter half of the twenty-first century.[21] To avert doomsday, the 'Club of Rome', whose membership is restricted to rich Western countries, advocated a voluntary reduction in consumption levels particularly, of fossil fuels and a 'zero growth' economic policy. The club looked at the issue from the perspective of the affluent West and neglected to point out the vast disparity in consumption of resources between the first and third worlds. Developing countries also feared 'no growth' advocacy was a neo-imperialist plot to keep them backward, thereby perpetuating the unjust status quo.

Since the 1960s, interest in environment and environmentalism has become a powerful movement with widespread popular support and with a much broader scope of interest. The concern for acid rain became the first major environmental problem in the early 1970s. The debate that took place in the early 1970s was essentially based on the air and water pollution, but gradually, it was realized that environmental degradation is not only caused by industrialization but by poverty and lack of development. The problems of acid rain, ozone depletion and greenhouse effect had a much higher profile than other environmental issues, but now, environment has become concerned with all aspects of the natural environment: land, water, minerals, all living organisms and life processes, the atmosphere and climate, the polar icecaps and remote ocean deeps, and even outer space. It shifted its focus from the natural environment to the interrelationships with human conditions and well-being and with the status of international economic cooperation.

A strong sense prevailed in the international community that environmental issues would remain high on the agenda during the 1990s and beyond. In this context the linkage between economic development and environment is vital. Much debate has occurred around these two crucial issues. The momentum has its source in the convergence of several factors. One is related to ozone layer depletion and global climate change and their potential to inflict severe costs on human societies that have galvanized public and policy attention. Environmental degradation in the developing countries, combined with the prospect of the environmental costs of future economic development, has added urgency to the international aspects of the problem. These linkages between environment and economic development have stimulated widespread interest in a new paradigm of environmentally sensitive economic development and have led to the emergence of the concept of sustainable development. According to the UNDP Human Development Report 1994, this new paradigm of development, that is, sustainable development, 'puts people at the centre of development, regards economic growth as a means and not an end, protects the life opportunities of future generations as well as the present generations and respects the natural systems on which all life depends'.[22]

The Brundtland Report was crucial in defining the concept of sustainable development. An attempt is now made to critically study the concept of sustainable development with an overview of the key themes that have emerged in the discourse. Three significant aspects of the issue are: (1) the Brundtland Report and perspectives on sustainable development; (2) evolution of the objectives of sustainable development through the two landmark conferences of Stockholm and Rio, which were foundational; and (3) emergence of Millennium Development Goals (2000–2015) and Sustainable Development Goals (2015–2030).

The Brundtland Report and Perspectives on Sustainable Development

The emergence of the environment into the realm of high politics has begun in 1987 with the publication of a United Nations report. The report of the Commission on Environment and Development 'Our Common Future' became an international best seller, which is also known as Brundtland Report. Much of this report is devoted to spelling out in considerable detail the changes in attitudes and behaviour that are needed, in all areas and in all aspects of our societies, to promote and achieve 'sustainable development', which is the report's central recipe for addressing the problem of the interaction between environment and development. The Brundtland Report is of key importance because written after three years of intensive work throughout the world by a distinguished commission under the chairmanship of Gro Harlem Brundtland, the former prime minister of Norway, it gives the best and most thorough analysis on the implications and management of that interaction so as to 'build a future that is more prosperous, more just, more equal and secure'.[23]

Economic development involves changes in ecosystems. Sustainable development does not imply the maintenance of a fixed state but rather, as the Brundtland Report puts it, 'a process of the direction of investments, the orientation of technological development and institutional changes are made consistent with future as well as present needs'.[24] The report had changed the terms of international debate by convincingly demonstrating that environmental degradation is a 'survival issue' for the developing countries. 'Our Common Future' had succeeded in integrating the issues of poverty, hunger, debt and economic growth into environmental issues.

As defined in the Brundtland report 'sustainable development' is economic development that seeks to meet the needs and aspirations of the present without compromising the ability to meet those of the future. It probes various important concepts – in particular equity and the protection of the environmental resource base – so that it can support growth over the long term. On this basis sustainable development envisages 'a new era of growth in which developing countries play a large role and reap large benefits as a means of solving the problems of poverty and under-development'.[25]

However, the ultimate purpose of both environmental and developmental policies is the enhancement of the quality of life, beginning with the satisfaction of basic human needs. In recognition of this, the emergence of terms like 'the new development' or 'alternative styles of development' suggests a way of looking at the purposes of development, in which environmental considerations play a central role.[26]

Evolution of the Objectives of Sustainable Development

Two international events mark the evolution of sustainable development over the past decades: the Stockholm Conference of 1972 and the Rio Conference

8 Economic Development and Environment

of 1992. These two conferences are significant because they represent the formal institutionalization of the public's demand that governments address the growing environmental crises, while also marking the beginning of new periods of political activity.

The evolution of the objectives of sustainable development was driven by two political forces that shaped the Stockholm Conference: they are public pressures exerted by a diverse but growing environmental movement and the tensions between North and South. The environmental movement was galvanized by a multiplicity of environmental crises. Among them were localized disasters such as the mass chemical contamination in Bhopal, India, and the Chernobyl nuclear accident in the Ukraine. Some environmental crises spread across regions of the world, such as acidification in the United States and Northern Europe and destruction of tropical rain forests in Brazil, Asia and Central Africa. Other environmental problems acquired global proportions, such as stratospheric ozone depletion and greenhouse gas accumulation in the atmosphere. Public demand for government action created a number of new international environmental conventions,[27] compelled multilateral and bilateral development agencies to adopt new standards of behaviour, and created new international financial incentives.[28]

'Silent Spring' (1962),[29] 'The Population Bomb' (1970)[30] and the 'Limits to Growth' (1972)[31] captured the general anxieties of the public in industrialized countries by expressing doomsday scenarios caused by shrinking resource base, spreading pollution and ever-expanding populations.

The environmental agenda of the industrialized societies collided head-on with the political perspectives and priorities of the developing world.[32] In contrast to industrialization problems of the North, developing countries identified the issue of poverty alleviation as their most urgent challenge to arresting environmental degradation. They highlighted the relation between impoverishment and the degradation of natural resources through soil erosion, deforestation, desertification and diminishing water sources. The development-cum-environment agenda of the South accepted at Founex and Stockholm fared badly in the real-world politics of North-South power relations in the post-Stockholm period.[33]

These two international pressures, growing public demand and conflicting North-South development perspectives, found their way into two seminal formulations of sustainable development in the 1980s. The first formulation was recorded in the World Conservation Strategy (WCS), published by the International Union for Conservation of Nature and Natural Resources (IUCN), UNEP and the World Wildlife Fund (WWF) in 1980.[34] The WCS marked a significant departure from previous approaches to the development environment nexus in that it tried to establish a 'global framework for conservation' and affirmed the compatibility of promoting development objectives while 'achieving conservation'.[35] The strategy set forth a carefully organized set of requirements and action priorities for national governments that would help guide them in using their natural resource bases to promote human welfare while respecting the 'carrying capacities of ecosystems'.[36]

WCS was received as a compelling visionary statement that was not able to mobilize international political support and consequently proved unable to generate enduring practical and programmatic influence. But the WCS provided a strengthened conceptual basis for the second formulation of sustainable development offered by the World Commission on Environment and Development (WCED), also known as the Brundtland Commission, in its 1987 report, 'Our Common Future'.[37]

David Reed, in his book *Structural Adjustment, Environment and Sustainable Development*, analysing the impact of the Brundtland Commission's contributions to establishing sustainable development as the standard for international development, had divided it into three parts. *First*, WCED effectively established the present generations' responsibility for safeguarding future generation's development options and opportunities by protecting the planet's environment and natural resources. *Second*, it placed the alleviation of poverty in developing countries as the central axis around which global sustainability would revolve. *Third*, it recast the pursuit of sustainability in the context of the international economy by recognizing the need to reorder patterns of international trade and flows of capital.

One of the criticisms levelled against the Brundtland Commission Report lay not in its analytical contributions but in its prescriptions. The report and the proposals sent the message to governments and development agencies that 'growth as usual' policies would remain the lynchpin for promoting sustainable development practices. In this regard, the commission asserted that a 3–4 per cent growth rate in the industrialized nations was the pivotal economic requirement on which poverty alleviation in the developing world depended.[38]

The Brundtland Commission achieved remarkable success in establishing sustainable development as the standard against which the behaviour of governments and international institutions would measure their policies and activities.

The United Nations Conference on Environment and Development held in 1992 at Rio de Janeiro, also known as the Earth Summit, was the culmination of an intense period of awareness raising among policy-makers at the highest level.[39] The objectives of sustainable development emerging from the Earth Summit, while giving greater attention to changing North-South economic relations, did not alter the basic 'growth as usual' articulated a few years earlier by the Brundtland Commission. This attitude was particularly evident in the approach of northern countries that portrayed 'growth as usual' and technological innovations as the strategic pillars of sustainable development.

In the words of David Reed,

> [T]he evolution of sustainable development as the new development paradigm has been driven by public pressure and the North-South political struggle, the conceptual basis of sustainability has been enriched over the past two decades by contributions of a new intellectual order.[40]

The hallmark of that intellectual effort is reflected in the works of Nicholas Georgescu Rogen, K. William Kapp, Kenneth Boulding and Herman Daly, among many, which have posed fundamental challenges to the post-war development enterprise.[41]

Economic, Social and Environmental Dimensions of Sustainable Development

The concept of sustainable development is based on the definition given in 'Caring for the Earth', a joint publication of the IUCN, UNEP and WWF.[42] In this, sustainable development is explained as people-centred, its aim is to improve the quality of human life, and it is conservation based that is conditioned by the need to respect nature's capacity to provide resources and life-supporting services. In this perspective sustainable development means 'improving the quality of human life while living within the carrying capacity of supporting ecosystems'.[43] This definition analyses three basic components – the economic, the social and the environmental – that constitute the basis of sustainable development.

The economic component of sustainability requires that societies pursue economic growth paths that generate an increase in true income, not short-term policies that lead to long-term impoverishment and where manmade capital, human capital and natural capital are substitutable and remain complementary. It requires a differentiated approach to growth so that many developing areas need to increase their productive capacity and at the same time the industrialized societies need to reduce their consumption of natural resources and use them more efficiently.[44]

The social dimension of sustainable development is built on the premise that equity and interdependence are basic requirements of an acceptable quality of life, which is the ultimate aim of development. For a development path to be sustainable over a long period, wealth, resources and opportunities must be shared in such a manner that all citizens have access to minimum standards of security, human rights and social benefits, such as food, health, education, shelter and opportunities for self-development. Moreover, it, further, demands the active political participation of all social sectors and the accountability of governments to the broader public in making basic social policy.

The environmental dimension of sustainable development is predicated on maintaining the long-term integrity and productivity of the planet's life support systems and environmental infrastructure. Environmental sustainability requires the use of environmental goods and services in such a way as to not diminish the productivity of nature.

The United Nations Environment Programme was at the forefront of the effort to articulate and popularize the concept. UNEP's concept of sustainable development encompasses these features:

1. Help for the very poor, because they are left with no options but to destroy their environment;
2. The idea of self-reliant development, within natural resource constraints;
3. Cost-effective development using non-traditional economic criteria;
4. The issues of health, appropriate technology, food, self-reliance, clean water and shelter for all;
5. The notion that people-centred initiatives are needed.[45]

Summarizing the debate regarding Sustainable development, the rapporteurs Jacob, Gardner and Munro[46] (1987) articulated the objectives of sustainable development like this:

> Sustainable development seeks to respond to five broad requirements: (1) integration of conservation and development, (2) satisfaction of basic human needs, (3) achievement of equity and social justice, (4) provision of social self-determination and cultural diversity and (5) maintenance of ecological integrity.[47]

In contrast, the currently popular definition of sustainable development – the one adopted by the World Commission on Environment and Development – is quite brief: 'Sustainable development is development that meets the needs of the present without compromising the ability of future generations to meet their own needs.'[48]

While the WCED's statement of the fundamental objectives of sustainable development is brief, the commission is much more elaborate about the operational objectives of sustainable development. It states that 'the critical objectives' which follow from the concept of sustainable development are as follows:

1 Reviving growth;
2 Changing the quality of life;
3 Meeting essential needs for jobs, food, energy, water, and sanitation;
4 Ensuring a sustainable level of population;
5 Conserving and enhancing the resource base;
6 Reorienting technology and managing risk;
7 Merging environment and economics in decision-making;
8 Reorienting international economic relations;
9 Making development more participatory.[49]

Besides articulating these objectives, the three biggest challenges presented by the Brundtland Report are highlighted as follows:

1 Institutional challenges;
2 Energy challenges;
3 Population challenges.

The *first challenge is institutional*, and the report concludes that improved environmental protection and management is so crucially important a factor that it can no longer safely be left to weak, underfunded environmental institutions or ministries. The central agencies of government and the major sectoral ministries (environmental governance) play key roles in national decision-making, and they together with their client industries 'have the greatest influence on the form, character and distribution of the impacts of economic activity on the environmental resource base'[50] and determine whether that base is enhanced or degraded. The report is quite clear that all entities, whether public or private, whose activities have a potential effect on the environment must share the task and must be made both responsible and accountable for ensuring that all their activities contribute to sustainable development. Being responsible and accountable means that issues of various kinds must be fully considered before economic, financial and fiscal decisions are taken and must be fully integrated with so that what is done is both economically rational and environmentally desirable. The *second major challenge* emphasized in the Brundtland Report is *energy*. Development without energy is impossible. As the Brundtland Report says, 'A safe and sustainable energy pathway is crucial to sustainable development, we have not yet found it.'[51] The report looks to economic growth that is less energy-intensive, improved efficiency in the production and use of energy and the development of 'sustainable forms of renewable energy' as the main hopes for the future.

The Brundtland Report is blunt on *the third challenge*, *population*, and says that 'present rates of population growth cannot continue'.[52] The report is more explicit when it says that in many parts of the world, the population is growing at rates that cannot be sustained by available environmental resources at rates that are outstripping any reasonable expectations of improvements in housing, health care, food security or energy supplies.[53]

Sustainable development can only be pursued if population size and growth are in harmony with the changing productive potential of the ecosystems. There is a moral imperative in the idea of stewardship to protect future generations. As the 1990 White Paper, 'This Common Inheritance', says, 'we have a moral duty to look after our planet and to hand it on in good order to future generations'.[54]

Bretton Woods Institutions and the UN

Bretton Woods Institutions were originally conceived by Keynes and others as contributing to a world welfare system in the same way as was done within national economies following the Great Depression.[55] They were meant to be under the broad UN system, but are not accountable to it. The Bretton Woods Institutions were manoeuvred to counter the growing strength of Third World forums, which were pressing for a more equitable role in defining the direction of the world economy. Recognizing the possibility of unrest and radical movements in much of the Third World, in the wake of democratic and anti-colonial consciousness, the World Bank, under Robert McNamara, pushed for a global anti-poverty programme. The programme did not address the structural issues within and between nations which legitimated and generated poverty; instead, it concentrated on making the poor more productive.[56] The Bretton Woods Institutions have also demonstrated that they are incapable of managing the international economic system to the mutual benefit of all their members. Whatever accountability exists is primarily due to the systematic work of actors in civil society building global networks of concerned people. They continue to violate a wide cross section of universal, human rights as well as a number of ethical norms.[57] They have become largely unaccountable.

On the other hand, the UN system, compared to the BWI, has a much wider representation and potentially more plural structures of arbitration and negotiation. However, despite a series of well-intentioned covenant and treaties binding states, internal limitations as well as inability to translate this representation into binding legal regimes with effective mechanisms of implementation and enforcement has diminished its influence over the years. And yet, throughout the UN system, as well as in a number of allied institutions, work continues to further its original vision.

The promotion of the economic development of the non-industrial, underdeveloped and largely colonized areas of the world, including masses of Africa and Asia, the Pacific, Caribbean islands and parts of South America, was the original mandate of the North.[58] The Bretton Woods Institutions, founded in 1944 (one year ahead of the UN), began with missions quite distinct from their latter-day involvement with North-South relations. The International Monetary Fund (IMF) was conceived by the Bretton Woods founders as the guardian of global liquidity, a function that it was supposed to fulfil by monitoring member countries' maintenance of stable exchange rates and providing a temporary facility on which they could draw to overcome cyclical balance-of-payments difficulties. The International Bank for Reconstruction and Development was set up to assist in the reconstruction of war-torn economies by lending them at manageable rates of interest.[59] The IMF was deeply involved in stabilizing Third World economies with balance-of-payments difficulties. As for the World Bank, it had evolved as the prime multilateral agency for aid and development.

In the case of the World Bank, a turning point was the debate triggered by the 1951 report of a group of experts entitled 'Measures for the Economic Development of Under-Developed Countries', which proposed making grant-in-aid available to Third World countries.[60] Using this as a springboard, Third World countries at the General Assembly tried to push through resolutions establishing SUNFED, the Special UN Fund for Economic Development, which would be controlled not by the North but by the UN. The establishment of the UN Special Fund, later renamed the UN Development Programme, served as the channel of smaller quantities of mainly technical aid to developing countries.[61]

The 1990s have seen a shift among the developing countries and the former Soviet bloc to democratic regimes. This has happened at the same time as there has been a move away from the UN towards the Bretton Woods Institutions for discussions on finance, trade, environment and even poverty. The result of this is the developed countries now have a greater control of the agenda through Bretton Woods Institutions.[62] In the book *Earth Summit 2002: A New Deal*, Felix Dodds,[63] while analysing 'Reforming the International Institutions', emphasizes on the need for integration of the Bretton Woods Institutions within the UN system. According to him one of the key questions that needs to be addressed is: Are they (BWI) committed to 'democratic governance and transparency within the UN system'? Further, while analysing the unequal decision-making structures in the form of BWIs, he notes that the richer countries who contribute more to these institutions, have more influence.

ECOSOC has been conducting a series of 'dialogues' with the Bretton Woods Institutions, and the hope is that they will start to bring the institutions into a more focused and accountable relationship with the United Nations agencies and programmes, including UNEP.

Agenda for Development

As the United Nations approached its 50th year, the General Assembly, in considering Secretary-General Boutrous-Boutrous Ghali's 'An Agenda for Development',[64] embarked on the most detailed debate on development which links and includes environmental protection as one of the dimensions of development. The report made several important recommendations. *First*, it said that development should be recognized as the foremost and most far-reaching task of our times. *Second*, it identified five dimensions of development in its agenda for development: peace, the economy, environmental protection, social justice and democracy. *Third*, a new framework for international cooperation for development was recognized as a necessity. *Fourth*, within this new framework for development cooperation, the United Nations was to play a major role in both policy leadership and operations. 'An Agenda for Development' has brought to the fore three key objectives: to strengthen and revitalize international development cooperation generally to build a stronger, more effective and coherent multilateral system in support of development; and to enhance the effectiveness of the development work of the organization itself, its departments, regional commissions, funds and programmes, in partnership with the United Nations system as a whole.[65]

In spite of all these attempts, social inequalities between and within countries have remained and even widened; in terms of representation Asia, Africa and Latin America continue to be seen as falling short in relation to the developed countries. Slowly this has been changing, and their views and opinions are taken seriously in the environmental negotiations versus the developed countries in the later years.

Despite movements and writings about conservation which have influenced policies of several countries, there was not a whisper on environment, natural resources or ecology. In the report of the Commission on International Development entitled 'Partners in Development'[66] (1969), which was chaired by Lester Pearson, there is not even a remote reference to environmental issues.

Environment on the UN Agenda

The emergence of contemporary environmental issues on to the agenda of world politics has arisen as a response to environmental accidents and disasters, increases in scientific knowledge and activism, and lobbying by non-governmental organizations and grassroots movements. Environmental issues encompass such concerns as ozone depletion, climate change, deforestation, loss of biodiversity, desertification, management of toxic and hazardous wastes, pollution of the oceans and waters, acid rain, air pollution, land degradation and the depletion of resources.

The reason for concern is twofold: First, scientific evidence supports the proposition that ecological damage is increasing at a faster rate because of human activity. Second, environmental changes have the potential for severe and irreversible impact on the ecosystem and also on the economic and social development of peoples and states. Criteria for identifying environmental issues as global/multilateral are twofold. *First*, those issues which affect the global commons, and *second*, those parts of the ecosystem or the planet which are beyond national jurisdiction. Environmental degradation is also claimed as global when the common heritage of human kind is at stake. This principle invokes the idea of a shared global resources managed in the common interests of all. The forests are claimed as a common heritage, and a global environmental concern because the destruction of forests, the lungs of the world, affects the global climate system. Similarly, biodiversity loss is global because the world is being deprived of potentially valuable resources of pharmaceutical and agricultural genetic resources, and the impact of the loss of one component of the ecosystem has effects on the global ecosystem. These global environmental issues, which require collective action, fall within the realm of multilateral decision-making bodies such as the United Nations. The United Nations is credited as the only institution to strengthen global environmental governance and achieve environmental security. In this context, Maurice Strong (1991)[67] argues that 'the United Nations is the only multilateral organization that is universal in its membership and global in its mandate' is 'uniquely positioned' to help governments because of its multidisciplinary capacity and extensive experience. Within and under the auspices of the UN, global environmental issues have been addressed by the General Assembly, ECOSOC and the specialized agencies in accordance with their mandates, in a series of issue-specific two global conferences, at Stockholm (1972) and in Rio de Janeiro (1992),[68] 20 years later. Today, countries agree that sustainable development, development that promotes the environment, offers the best path forward for improving the lives of people everywhere. In this context, the significance and contributions made by Millennium Development Goals and Sustainable Development Goals are explained in the following section of this chapter.

Millennium Development Goals (2000–2015)

At the Millennium Summit in 2000, member states adopted the UN Millennium Declaration,[69] which was translated into a road map for achieving eight time-bound and measurable goals by 2015, known as the Millennium Development Goals (MDGs). The MDGs aimed to eradicate extreme poverty and hunger; achieve universal primary education; promote gender equality and the empowerment of women; reduce child mortality; improve maternal health; combat HIV/AIDS, malaria and other diseases; ensure environmental sustainability; and develop a global partnership for development.

In September 2015, world leaders adopted the 17 Sustainable Development Goals of the 2030 Agenda[70] for Sustainable Development. The 2030 Agenda officially came into force on 1 January 2016, marking a new course for the organization towards ending poverty, protecting the planet and ensuring prosperity for all by 2030. Three other accords adopted in 2015 play critical roles in the global development agenda, and they are highlighted as the Addis Ababa Action Agenda on financing for development, the Paris Agreement on climate change and the Sendai Framework on disaster risk reduction.

2030 Agenda

By 2015, significant progress had been made across all goals and millions of lives were improved due to concerted global, regional, national and local efforts. The data and analysis presented in the *Millennium Development Goals Report, 2015*,[71] proved that, with targeted interventions, sound strategies, adequate resources and political will, even the poorest countries can make dramatic and unprecedented progress. Building upon the achievements and lessons learned from the MDGs and the Earth Summit, at the UN Summit held on 25–27 September 2015 (the post-2015 development agenda), member states adopted General Assembly resolution 70/1, 'Transforming our World: the 2030 Agenda for Sustainable Development'. The 2030 Agenda seeks to address the unfinished task of the MDGs while committing to achieving sustainable development in its three dimensions – economic, social and environmental – in a balanced and integrated manner. It is the first agreement of its kind that integrates goals from the three pillars of the organization's work – peace and security, human rights and sustainable development – into a single agenda. It reflects a paradigm shift from previous approaches to development that focus on the economic or social structure separately. It is also stronger on environment compared to the MDGs. Bold and transformative, the 2030 Agenda seeks to end poverty by 2030 and pursue a sustainable future. It stands as a universal agenda, consisting of a *declaration*; 17 goals to be reached by 2030, the means of its implementation, including through a revitalized global partnership; and a framework for review and follow-up.

The world faces immense challenges ranging from widespread poverty, rising inequalities and disparities of wealth, opportunity and power, to environmental degradation and the risks posed by climate change. The 2030 Agenda provides a plan of action for ending poverty in all its dimensions, leaving no one behind and reaching the furthest behind first. It also builds on the outcomes adopted at three other conferences in 2015 on Climate Change, Disaster Risk Reduction and Financing for Development.

Sustainable Development Goals (2015–2030)

The 17 Sustainable Development Goals[72] are highlighted as: ending poverty and hunger in all its forms everywhere; achieving food security and improved nutrition and promoting sustainable agriculture; ensuring healthy lives and promoting well-being of all at all ages; ensuring inclusive and equitable quality education and promoting lifelong learning opportunities; achieving gender equality and empowering all women and girls; ensuring availability and sustainable management of water and sanitation; ensuring access to affordable, reliable, sustainable and modern energy; promoting sustained, inclusive and sustainable economic growth; building resilient infrastructure, promoting inclusive and sustainable industrialization and foster innovation; reducing inequality within and among countries; making cities and human settlements inclusive, safe, resilient and sustainable; ensuring sustainable consumption and production patterns; taking urgent action to combat climate change and its impact; conserving and sustainably using the oceans, seas and marine resources for sustainable development; protecting restoring and promoting sustainable use of terrestrial ecosystems, sustainably managing forests, combating desertification, and halting and reversing land degradation and biodiversity loss; promoting peaceful and inclusive societies for sustainable development and providing access to justice and building effective, accountable and inclusive institutions at all levels; and strengthening the means of implementation and revitalizing the global partnership for sustainable development.

These 17 time-bound and measurable goals, known as the Sustainable Development Goals, represent the core of the 2030 Agenda. Comprising 169 targets and 230 indicators, a blueprint for a sustainable future, these goals integrate the social, economic and environmental dimensions of sustainable development. They are not independent of each other but are implemented in an integrated manner.

The SDGs are also universal, for all countries, developed, developing and middle-income, while taking into account different levels of national development capacities. The SDGs acknowledge that ending poverty necessitates strategies that build economic growth and address a range of social needs, including education, health, social protection and job opportunities, while tackling climate change and environmental protection. The 2030 Agenda emphasizes national ownership of the agenda and its implementation, which is also inclusive. To be successful, action has to be taken on the global, national, regional and local levels, by all people and stakeholders. The SDGs unite everyone in the effort to make these positive changes.

Financing Sustainable Development

The holistic approach to financing sustainable development is rooted in the financing for development process, embodied in the (besides the 2002 Monterry Consensus, and the 2008 Doha Declaration) 2015 Addis Ababa Action Agenda on financing for development held in Addis Ababa in Ethiopia. The third International Conference on Financing for Development (Addis Ababa, 13–16 July 2015) resulted in the adoption of the Addis Ababa Action Agenda,[73] which provides a strong foundation to support implementation of the 2030 Agenda. It includes a new global framework for financing sustainable development by aligning all financial flows and policies with economic, social and environmental priorities. Member states and other stakeholders also agreed on a set of policy actions, with more than 100 concrete measures that draw upon all sources of finance, technology, innovation, trade, debt and data in order to support achievement of the SDGs. All these means of implementation of the 2030 Agenda are included in the Addis Ababa Action Agenda. The General Assembly on 27th July 2015 established an annual ECOSOC forum on financing for development, with universal, intergovernmental participation, as a follow-up process to the Addis Ababa Action Agenda. Further, the Global Environment Facility (GEF), established in 1991, helps developing countries fund projects that protect the global environment and promote sustainable livelihoods in local communities.

Sustainable Development Challenges (Post-2015)

In September 2000, global community adopted the United Nations Millennium Declaration, which provided the basis for the pursuit of the Millennium Development Goals. A global consensus was successfully forged around the importance of poverty reduction and human development. The world reached the poverty target five years ahead of the 2015 deadline. Still, results fall short of international expectations and of the global targets set to be reached. It remains imperative that the international community takes bold and collaborative actions to accelerate progress in achieving the Millennium Development Goals and Sustainable Development Goals. Continuation of current development strategies is not sufficient to achieve sustainable development beyond 2015. The risks and challenges involved are highlighted as follows:

- The impact of climate change threatens to escalate in the absence of adequate safeguards, and there is a need to promote the integrated and sustainable management of natural resources and ecosystems and take mitigation and adaptation action in keeping with the principle of common but differentiated responsibilities;
- Food and nutrition security continues to be an elusive goal for many;
- Income inequalities within and among countries have been rising and have reached an extremely high level, invoking the spectre of heightened tension and social conflict;
- Rapid urbanization, especially in developing countries, calls for major changes in the way in which urban development is designed and managed, as well as substantial increases of public and private investments in urban infrastructure and services;
- Energy needs are unmet for hundreds of millions of households, unless significant progress in ensuring access to modern energy services is achieved;
- Recurrence of financial crises needs to be prevented, and the financial system has to be redirected towards promoting access to long-term financing for investments required to achieve sustainable development.

Conclusion

Continuations with current development strategies are not going to help achieve sustainable development targets. Economic and social progress remains uneven, the global financial crisis has revealed the fragility of progress, and accelerating environmental degradation inflicts increasing costs on societies. There are a number of economic, social, technological, demographic and environmental trends that underlie these challenges – such as a deeper globalization, persistent inequalities, demographic diversity and environmental degradation – to which a sustainable development agenda has to respond effectively. These trends influence and reinforce each other in myriad ways and pose enormous challenges. Urbanization is proceeding rapidly in developing countries; globalization and financialization are perpetuating inequalities, while exposing countries to greater risks and crises; and food and nutrition as well as energy security are threatened by competing demands on land and water, as well as environmental degradation. Sustainable development requires transformative changes at the local, national and global levels. The SDGs' targets are aspirational, and the critical challenge is how 'each government sets its own national targets guided by the global level of ambition, taking into account national circumstances and how these aspirational and global targets are incorporated into national planning processes, policies and strategies'. The integrated SDGs and targets have taken into account different national realities, capacities and levels of development giving respect to national policies and priorities, but the major challenge remains transforming aspirations to implementation.

Effective, accountable and inclusive policies and actions are critical for realizing the 2030 agenda. Achieving the SDGs requires farsighted, holistic and participatory decision-making by governments. An unprecedented level of policy integration and institutional coordination and expertise is required so that progress is made on all the SDGs at the same time, building on the interrelations and synergies between them.

Actually, this aspect of sustainable development has been further explored in the context of South Asian countries where the integration of national conservation strategies into the development and planning process of these countries has been examined.

Summary

This chapter has analysed and explored some significant issues and they are: a brief review of the concepts related to economic development and environment in the evolution of the concept of sustainable development, and in the process Brundtland Report and the challenges presented by this report has been elaborately discussed. Some other important points dealt in this chapter are: Bretton Woods Institutions and United Nations and 'Environment on the UN Agenda'. The chapter has been updated by highlighting the significance of Millennium Development Goals and Sustainable Development Goals covering the 2030 agenda including the analysis of the challenges of post-2015 agenda.

Since the problems we face pose global challenges, to which no one single nation or group of nations alone can find solutions, they can only be tackled at the global level. So, the need of the hour is to secure an unprecedented high degree of international cooperation and mutual understanding between nations if the world community is to have a real chance of finding solutions that are not only environmentally sustainable but economically, politically and in terms of security as well.

Notes

1. United Nations, the United Nations Conference on Environment and Development: Preamble of Agenda 21, chapter 1, Rio de Janeiro, 3–14 June 1992.
2. In 1968, the first International Conference on Global Biosphere Protection, UNESCO's intergovernmental Conference on the Biosphere, took place in Paris. The Conference was a turning point in international environmental politics, as a result of which, in 1970, UNESCO launched its 'Man and Biosphere Programme' (MAB) to protect areas representing the central ecosystems of the planet as 'biosphere reserves'.
3. United Nations, The United Nations Conference on Human Environment, Declaration on the Human Environment, Stockholm, June 1972.
4. Ibid., Declaration and Principles on Human Environment, Stockholm, June 1972.
5. Ibid., Outcomes of the Conference on Human Environment, Stockholm, June 1972.
6. World Commission on Environment and Development, *Our Common Future* (New York: Oxford University Press, 1987).
7. Ibid., UN Conference on Environment and Development, Rio de Janeiro, 3–14 June 1992.
8. Ibid., Outcomes of the Conference, Rio de Janeiro, 3–14 June 1992.
9. Ibid., UN Conference on Environment and Development: Implementation of Agenda 21.
10. United Nations, The United Nations Conference on Human Environment, Declaration on the Human Environment, Stockholm, June 1972, p. 1.
11. United Nations Development Programme *(UNDP) Human Development Report, 1994* (Bombay: Oxford University Press, 1994), p. 2.
12. Gore, Al., *Earth in the Balance* (Bombay: Oxford University Press, 1992), pp. 295–302.
13. World Bank, *World Development Report, 1991* (Oxford: Oxford University Press, 1991), pp. 7–8.
14. United Nations, 'Report of the Secretary General, United Nations Development Decade: Proposals for Action', UNDOC, E/3613, pp. 2–3.
15. Aron, Raymond, 'An Ecological Approach to International Development, Lyndon K', in Taghi M. Farvar and John P. Milton (eds.), *Caldwell in the Careless Technology* (New York: Natural History Press, 1972).
16. Ul Haq, Mahbub, *The Third World and the International Economic Order* (Washington: Overseas Development Council, 1976).
17. Reddy, Amulya K.N., 'Alternative Energy Policies for Developing Countries: A Case Study of India', Discussion Paper, ASTRA, Bangalore 1999.
18. Kuper, Adam and Jessica Kuper (eds.), *The Social Sciences Encyclopaedia* (New York: Routledge, 1996).
19. De Monte, Darryl, 'Temples or Tombs', in *Industry vs Environment* (New Delhi: CSE, 1985).
20. Carson, Rachel, *Silent Spring* (Boston: Houghton Mifflin, 1962).
21. Meadows, Donella H., *Limits to Growth* (New York: Universe Books, 1972).
22. UNDP., op. cit., p. 4.
23. World Commission on Environment and Development, op. cit., p. 1.
24. Ibid., p. 9.
25. Ibid., p. 40.
26. This newly sophisticated approach to environment and development is well-reflected in the World Conservation Strategy, released in 1980 by IUCN.
27. Most of these conventions were regional agreements that included the Oslo Convention to control maritime dumping, the Helsinki Convention, the Paris convention, the Mediterranean Action Plan and the Basel Convention.
28. Gareth, Porter and Brown Janet, *Global Environmental Politics* (Boulder: Westview Press, 1991).
29. Carson, op. cit., p. 22.
30. Ehrlick, Paul, *The Population Bomb* (New York: Ballantine Books, 1990).

31 Meadows, Donella, Jorgen Randers, and William W. Behrens III., *Limits to Growth* (New York: Universe Books, 1972).
32 The communist countries withdrew from the preparatory process on the ground that '[p]ollution was the product of capitalism and consequently a problem from which they did not suffer' (Bretton, Tony, *The Greening of Machiavelli* (London: Earthscan Publications, 1994), p. 37.
33 Reed, David, *Structural Adjustment, Environment and Sustainable Development* (London: Earthscan Publications, 1996), p. 28.
34 International Union for Conservation of Nature and Natural Resources (IUCN) United Nations Environment Programme, *World Wide Fund for Nature – International World Conservation Strategy* (Geneva: IUCN, 1980).
35 Adams, W.M., *Green Development: Environment and Sustainability in the Third World* (London: Routledge, 1990), p. 47.
36 IUCN., op. cit., p. 1.
37 World Commission on Environment and Development, op. cit.
38 Ibid., p. 51.
39 Holmberg, John, Koy Thomson, and Lloyd Timberlake, *Facing the Future* (London: Earthscan Publications, 1993).
40 Reed, op. cit., p. 32.
41 Nicolas Georgescu, Reogen, *The Entropy Law and the Economic Process* (Cambridge: Harvard University Press, 1971); Kapp, W.K., *The Social Costs of Private Enterprise* (New York: Schocken Books, 1950); Kapp, W.K., *Towards a Science of Man: A Positive Approach to the Integration of Social Knowledge* (The Hague: Matinus Mijhoff, 1961); Ullman, John (ed.), *Social Costs, Economic Development and Environmental Disruption* (Lanham, MD: University Press of America, 1983); Boulding, Kenneth, *Beyond Economics: Essays on Society, Religion and Ethics* (Ann Arbor: University of Michigan Press, 1970); Daly, Herman and John Cobb, *For the Common Good* (Boston: Beacon Press, 1989).
42 International Union for Conservation of Nature and Natural Resources, United Nations Environment Programme and World-Wide Fund for Nature – International, *Caring for the Earth* (London: Earthscan Publications, 1991) p. 4.
43 Ibid., p. 10.
44 Kapp, op. cit., p. 141.
    Goodland, Robert and Herman Daly, 'Ten Reasons Why Northern Income Growth Is Not the Solution to Southern Poverty', in *Population, Technology and Lifestyle* (Washington, DC: Island Press, 1992).
45 Tolba, M.K., *The premises for building a sustainable society – Address to the World Commission on Environment and Development, October 1984* (Nairobi: United Nations Environment Programme, 1984a).
46 Jacobs, P., J. Gardner, and D. Munro, 'Sustainable and Equitable Development: An Emerging Paradigm', in P. Jacobs and D.A. Munro (eds.), *Conservation with Equity: Strategies for Sustainable Development* (Cambridge: International Union for Conservation of Nature and Natural Resources, 1987), pp. 17–29.
47 World Commission on Environment and Development, op. cit., p. 43.
48 Ibid., p. 49.
49 Ibid., p. 50.
50 World Commission on Environment and Development, op. cit., p. 311.
51 Ibid., p. 14.
52 Ibid., p. 95.
53 Ibid., p. 11.
54 Ibid., p. 9.
55 Kothari, Smitu, 'Where Are the People? The United Nations, Global Economic Institutions and Governance', in Albert J. Paolini, Anthony P. Jarvis, and Christian Reus Smit (eds.), *Between Sovereignty and Global Governance. The United Nations, the State and Civil Society* (Great Britain: Macmillan, 1998), p. 188.

56 Ibid., p. 189.
57 Lawyers Committee for Human Rights, The Bretton Woods Institutions and Human Rights (1994).
58 Adams, Nassau, 'The UN's Neglected Brief: The Advancement of all Peoples?' In Erskine Chiders (ed.), *Challenges to the UN* (London: Catholic Institute for International Relations, 1994), p. 29.
59 Jamal, Amir, 'The IMF and the World Bank: Managing the Planet's Money', in *Childers, Challenges to the United Nations* (New York: St Martin's Press, 1995), pp. 53–54.
60 Adams, 'The UN's Neglected Brief', p. 31.
61 Adams, op. cit., p. 31.
62 Ibid., p. 32.
63 Dodds, Felix, *Earth Summit 2002: A New Deal* (London: Earth Scan Publications, 2000), p. 300.
64 Accessed digital library on 30th June, 2020.digitallibrary.un.org., 'An Agenda for Development: Report of the Secretary General', 1994. Doc. A/48/935.
65 Ibid. Doc. A/48/935.
66 Pearson, Lester, *Partners in Development: Report of the Commission on International Development* (New York: Praeger Publications, 1969).
67 Strong, Maurice, 'ECO 92: Critical Challenges and Global Solutions', *Journal of International Affairs* 44(2) (1991), p. 297.
68 Williams, Maurice, 'Guidelines to Strengthening the Institutional Responses to Major Environmental Issues', *Development* 1992(2) (1992), p. 23.
69 UN Millennium Declaration: Doc A/RES/55/2, Accessed internet on 2 July 2020.
70 Transforming our World: The 2030 Agenda for Sustainable Development, 2015.
71 Millennium Development Goals Report, *Time for Global Action for People and Planet* (New York: United Nations, 2015).
72 Sdgs.Un.Org. (2015–2030) Accessed internet on 19 June 2020.
73 Montes, Manuel, *Five Points on the Addis Ababa Action Agenda*, Policy Brief, no.24 (New York: United Nations, 2015).

2
EVOLVING ROLE OF UNEP AT 50

Introduction

It took two centuries of environmental degradation before worldwide concern brought world leaders to Stockholm in 1972 for the first-ever summit dedicated to the environment. The meeting was a great boost to the still young environmental movement and saw the establishment of a new international agency – UNEP – to keep the environmental flame alive.[1] The web of international environmental treaties that have been negotiated over the past couple of decades owes a great deal to UNEP's role as advocate for greater environmental responsibility. Equally UNEP has quietly promoted environmental management capacity in developing countries and pushed to include environmental considerations in social and economic policies. The Earth Summit held in Rio in 1992 gave new impetus to this by placing the environment at the centre of the development debate. The great legacies of that Summit – Agenda 21 and the Conventions on Climate Change and Biological Diversity – helped forge linkages between natural ecosystems, people and the fight against poverty.[2] Greater international dialogue and cooperation is needed for a strong UNEP energetically engaged in assessing trends, harmonizing environmental standards and bringing the international community together. Technological advances, the demands of public opinion and a new sense of government and corporate responsibility for the health of the planet provide an opportunity for UNEP to champion new approaches to dealing with pollution and conserving natural resources. No other international organization has the track record or mandate in protecting the environment.

Leading Role of Environmental Movements and Environmentalists

The global environmental movement has brought three significant changes. *First*, it has prompted the rediscovery of one of the most fundamental realities of human

DOI: 10.4324/9781003370246-2

existence: that humanity is dependent on a healthy natural environment. The earliest environmentalists felt that the changes brought by the agricultural and industrial revolutions have exerted too great a cost on nature. With the rise of a new middle class, the expansion of education and the emergence of two-thirds of the world's population from the shadows of colonialism, questions of social, economic and political justice began to be addressed and the environment emerged as one of such questions.[3] The *second* change was manifested by a shift from an overwhelming emphasis on material values and physical security towards greater concern for the quality of life.[4] This change is characterized by Inglehart as 'a silent revolution'. Ophuls and others argued that many dominant social beliefs, formed in course of times, needed to be reassessed in the light of growing ecological scarcity.[5] In the words of Pirages, environmental problems in industrial societies had their roots in the Dominant Social Paradigm, a set of beliefs and values that included private property rights, faith in science and technology, individualism, economic growth, and the subjection of nature and exploitation of natural resources.[6] Pollution, energy shortages and even inflation, economic recession and unemployment posed challenges to the Dominant Social Paradigm.[7] Fears about the limits to growth and the implications of environmental mismanagement gave rise to a new world view more compatible with environmental issues. This view has been called the New Environmental Paradigm.[8] At its heart is a call for an entirely new kind of society based on carefully considered production and consumption, resource conservation, environmental protection and the protection of the basic values of compassion, justice and quality of life. Milbrath observes that environmentalists constitute a vanguard, using education, persuasion and politics to try to lead people to a new and more sustainable society.[9]

The *third* significant facet of environmentalism lies in the challenge it posed to orthodox models of economic growth, capitalist or socialist. In developed countries, it is a direct challenge to unregulated production and consumption patterns; in developing countries, it is a challenge to the assumption that the industrial model is the most effective route to rapid and equal development.

However, early environmentalism has forced a reconsideration of the priorities and principles of economic growth. From the late 1960s, the nature of the debate was fundamentally altered by the competing perspectives of 'more' and 'less' developed nations and communities within nations. Western environmentalists were concerned more with domestic issues, but at the global level, some of the most fundamental philosophical changes in environmentalism were derived from the influence of developing countries. According to Holdgate, Kassas and White, environmental problems are seen 'individually, simplistically and overwhelmingly, from a developed western country's standpoint',[10] where environmentalists warned of the evils of economic growth. For developing countries, faced with the immediate and visible problems of poverty and underdevelopment, environmental management was a distant concern.

While developing countries were convinced of the economic benefits of environmental management, their contribution to the international debate on the issue had the effect of injecting a note of realism into environmentalism. The simplistic initial view of some new environmentalists that all growth was wrong and that the aims of economic development and sound environmental management were incompatible was replaced by the more realistic view that the aims of the two had to be reconciled. The aims of development and environmental management were mutually dependent. The reactionary categorizations of environmental issues were gradually replaced by more conciliatory attitudes and attempts to achieve realistic compromises.[11] In this section, the role of environmental movements, leading environmentalists, and the impact of environmental publications and environmental disasters that led to the establishment/creation of United Nations Environment Programme in 1972 that marked the significant commitment of the world community to the environmental cause has been briefly analysed. UNEP has also acted as an important catalyst in bringing together different groups and in collaboration with other UN organizations to discuss environmental issues and develop plans of actions and promote international conventions.[12]

Role of Environmental Movements

The earliest environmental issues were local issues. Once the most immediate issues like costs of pollution or the loss of forests were raised, individuals formed groups and coalitions that became national movements and finally a multinational movement. According to John McCormick, the role of the 'environmental movement' in the second half of the twentieth century brought three developments of major importance. *First*, the scientific and nature protection components grew together, especially under the influence of professional ecologists. *Second*, appreciation of the environment grew in many countries outside Europe and North America. *Third*, and most important, the character of the approach changed. A much broader conception of the environment was emerging which encompassed all aspects of the natural environment – land, water, minerals, all living organisms and life processes, the atmosphere and climate, the polar ice-caps and remote ocean deeps, and even space – and emphasized the relationship between manmade and natural environments. This new environmental movement became concerned with a much wider range of environmental issues, which were strengthened during the 1950s and 1960s by a number of demonstrations. These widely publicized events caused many people in the developed countries to fear that these issues and events were going to jeopardize human nature. In this context, the origins and growth of British and American environmentalism were discussed.

British Environmentalism

The origins of British environmentalism lie in the age of scientific discovery. The *first* major influence on early British environmentalism was the study of natural history. The most notable was Gilbert White, whose seminal work *The Natural History of Selbourne*, published in 1788, epitomized the Arcadian view of nature, which advocated simplicity and humility in order to restore man to peaceful co-existence with nature.[13]

The understanding of the natural environment that emerged from the eighteenth and nineteenth centuries profoundly affected man's view of his place in nature. The Victorian age was one of great confidence and self-assurance, although the Victorian ideal of civilization almost always depended on the conquest of nature by science and technology.[14] Mastery over the environment was seen as essential for progress and for the survival of the human race. But a 'bio-centric conscience'[15] gradually emerged, supporting a sense of kinship between man and nature and acceptance of a moral responsibility to protect the earth from destruction.

For Lowe, the term 'balance of nature' in the eighteenth century had implied 'a robust pre-ordained system of checks and balances which ensured permanency and continuity in nature. By the end of nineteenth century, it conveyed the notion of a delicate and intimate equilibrium, easily disrupted and highly sensitive to human interference'.[16] By the 1880s, there were several hundred natural history societies and field clubs in the country, with a combined membership of about 100,000.[17]

Out of the humanitarian zeal that had generated the anti-slavery movement, came the *second* major influence on British environmentalism, the crusade against cruelty to animals. Although the Society for the Protection of Animals, founded in 1824 and given a royal charter in 1840, first campaigned against cruelty to domesticated animals, it soon turned its attention to wild animals; cruelty to animals was seen as an expression of the most ravage and primitive elements in human nature. Protectionists believed that in saving wildlife, they were helping preserve the very fabric of society.[18]

David Allen identifies a turning point in the 1860s when the protectionist crusade mustered its forces around the issue of the killing of birds, particularly gulls, to provide plumage for women's fashion.[19] The opposition to the killing of birds for plumage was led by women themselves, who made up most of the membership of the earliest bodies, that is, the Plumage League (1885), the Sel borne League (1885), the Royal Society for the Protection of Birds (SPB) and the Fur Fin and Feather Folk.[20] The SPB pledged its members not to wear plumage and set up a network of national and overseas branch, and this was behind one of the earliest pieces of legislation against international traffic in wildlife,[21] including the 1902 Indian government order banning the export of bird skins and feathers.[22]

Revulsion at the squalor of life in the industrial towns combined with the yearning for solace in open space and nature produced the *third* major thrust of early British environmentalism: the Amenity Movement. The world's first private environmental group – 'the Commons Open Spaces' and 'Footpaths Preservation Society' (1865) – campaigned successfully for the preservation of land for amenity, particularly the urban commons that were often the nearest 'countryside' available to urban workers. David Allen sees the concepts of recreation, preservation, sanctuary and wilderness combined in the 1860s and being overlaid with economic and scientific arguments to give the amenity movement its strength.[23]

The need for a body to acquire and hold land and property for the nation was met in 1893 by the creation of the National Trust, which aimed to protect the nation's cultural and natural heritage from the standardization caused by industrial development. The National Trust enjoyed some early success in acquiring land for preservation. By 1910 it counted 13 sites of natural interest among its acquisitions, but it was as much interested in sites of cultural and historic interest, and naturalists expressed concern at the almost random way in which potential nature reserves were acquired, with apparently little regard for the national significance of their plants and animals.[24] In 1912, the Society for the Promotion of Nature Reserves was created, not to own nature reserves but to stimulate the National Trust to give due regard to the creation of reserves. Nature reserves were regarded by most people as a subsidiary and expensive means of supplementing legislation.[25] Although the first had been created on the Norfolk Broads in 1888, it was not until after the Second World War that the idea of habitat protection won wider support in Britain.

Ironically, the National Trust was condemned by its own success in acquiring properties to expand more and more of its resources on land agency and management. This opened the way for the creation in 1926 of the Council for the Preservation of Rural England (CPRE), founded to coordinate the voluntary movement, promote legislation, give advice to land owners and 'make a single, simple and direct appeal to everyone concerned with the preservation of the countryside'.[26] The Addison Committee was set up and presented a report in 1931 that advocated the creation of national parks to protect flora, fauna and areas of exceptional natural interest and to improve public access.

American Environmentalism

From about 1620 to 1870, wood was the major source of energy in the United States and the primary building material, but there was little understanding of woodland management techniques. In the words of Huth, the axe was the symbol of early American attitudes towards nature. It was writer and poet Alexander Wilden, also known as John James Audubon, who published a nine-volume study of American birds which raised interest in ornithology. The publication (1827 and 1838) of Audubon's *The Birds of America*, the writings of Ralph Waldo Emerson and Henry David Thoreau further influenced the early American environmental philosophy of man and nature.

Two seminal events in American environmentalism occurred in 1864. The first was the publication of *Man and Nature* by George Perkins Marsh,[27] a book remarkable for espousing ideas on nature in which Marsh argued that 'wanton destruction and profligate waste were making the earth unfit for human habitation and ultimately threatened the extinction of man, who had forgotten that the earth was given to him not for consumption still less for recklessly wasting and destroying'.[28] Marsh's ideas had much influence in the later establishment of a National

Forestry Commission[29] and influenced French writers and Italian and Indian foresters. For Stewart Udall, the book represented the beginning of land wisdom in the United States.[30]

The second event was the 1864 'Act of Congress', which facilitated the transfer of the Yosemite Valley and the Mariposa Grove of Big Trees to the State of California on the conditions that 'the premises shall be held for public use, resort and recreation and shall be held inalienable at all times'.[31] Similar ideas were expressed by Henry David Thoreau in 1853 when he wrote of the desirability of 'national preserves'[32] and suggested that wilderness preservation was ultimately important for the preservation of civilization.[33] George Perkins Marsh took the argument further by suggesting that wilderness preservation had 'economical' as well as 'poetical' justifications; that is, it could be managed sustainably for the benefit of all.[34]

Muir Versus Pinchot

Initially American environmentalism was divided into two camps: the preservationists and the conservationists. The preservationists sought to preserve wilderness from all except recreational and educational use, and conservationists to exploit the continent's natural resources, rationally and sustainably.

A champion of wilderness preservation, the naturalist John Muir's[35] earliest campaigning was instrumental in the creation of Yosemite National Park in 1890, the first preserve consciously designed to protect wilderness.[36] Muir in 1892 found the Sierra Club, which worked to make the mountain regions of the Pacific Coast accessible to enjoy wilderness. The club became a rallying point for the preservationist cause. While Muir and the preservationists spoke of 'protecting' the environment, often implying that wilderness is totally excluded from anything but recreation, Gifford Pinchot spoke of conservation or sustainable use of resources like land, forests and water. One of the earliest conservation issues was the protection of forests, championed by Gifford Pinchot,[37] who had studied forestry and learned that forests could be both protected and managed sustainably and was the first one to champion the cause of a forest service of scientifically trained specialists. Muir believed that conservation should be based on *three main principles*: development (the use of existing resources for the present generation), the prevention of waste and the development of natural resources for all,[38] whereas Pinchot's contribution brought the tradition of progressive agriculture to the management of public lands, particularly forests.[39]

Early American Conservation Movement is often represented as a battle between morality (represented by the people) and immorality (represented by private interests set on exploiting the nation's natural resources for their own ends). For MacConnell,[40] conservation had become closely identified with the progressivism that swept the United States in the period 1900–1917, which was often depicted as a morality play.[41] Organized conservationists, according to Bates, were more concerned with economic justice and democracy in the handling of resources than with mere prevention of waste, conservation was regarded as a democratic crusade 'to stop the stealing and exploitation and to distribute more equitably the profits of the economy'.[42]

Unlike the amateur preservationists in contemporary Britain, American conservation leaders were professionals in fields such as forestry, hydrology and geology. They were influenced less by public opinion than by loyalty to their professional ideals in ensuring rational planning and efficient use of natural resources.

Pinchot's utilitarian philosophy was keenly supported by Vice President Theodore Roosevelt, who became president later (in September 1901), and Pinchot became Roosevelt's 'Secretary of State for Conservation', and resource management became a matter of public policy. Roosevelt also sought Muir's opinions, and the demands of the preservationists were met during the Roosevelt era by the addition of Yosemite Valley to the surrounding national park and the creation of 53 wildlife reserves, 16 national monuments and five new national parks.[43] On the whole, Roosevelt's term saw the promotion of professional conservation, based less on emotional principles of preservation and more on the rational management of natural resources. Like forestry, water became another inspiration of American conservation. Roosevelt, at the suggestion of Pinchot,[44] created the Inland Waterways Commission (IWC) (1907) based on a comprehensive plan for the improvement and control of American river systems. Further, on his advice, Roosevelt had not only created a National Conservation Commission, entrusted him with making the first survey of natural resources in the United States, but also held a Conference of Governors on Conservation. Pinchot, as head of the Executive Committee, remarkably succeeded in completing an inventory which became a basis for future legislation that was described by Roosevelt as 'one of the most fundamentally important documents ever laid before the American people'.[45]

He pursued and further held two conservation conferences, international in their scope: NACC and ICC. The first was the North American Conservation Congress (NACC), held in Washington DC (on 18 February 1909) under the chairmanship of Pinchot. Ten delegates (including Canada, Newfoundland, Mexico and the United States) discussed the principles of conservation and agreed that conservation was a problem broader than the boundaries of one's nation. The second, the International Conservation Conference, was convened by Theodore Roosevelt, the President of United States, in Washington DC, on the 'the subject of world resources and their inventory, conservation and wise utilization'. The Dutch government agreed to act as host and 58 nations were invited to meet in Hague in September 1909. These two conferences, North American Conservation Congress and International Conservation Conference, were the first attempts to discuss conservation at an inter-governmental level. By this time, the protection of nature and particularly of birds had been the subject of international discussion, and bird protection organizations had become politically active and had developed the argument for more universal protection of wild birds.[46]

The environmental movement in the West has grown and moved from strength to strength. The early days were full of writers like Rachel Carson, Barbara Ward and Barry Commoner and activists like Brice Lalonde. Over the years thousands of individuals, volunteers, and local groups have become a part of this powerful movement. But the centre stage today is occupied by a select group of mass membership-based organizations which bring issues and campaigns to the fore. The most well-known of these are Greenpeace, Friends of the Earth, the Sierra Club, the Environment Defense Fund and others. But equally effective and powerful are national NGOs like Germany's BUND and the Danish Society for Nature Conservation. Some national environmental NGOs in West claim 1–2 per cent of their country's adult population as their members. This broad base within society allows these organizations to make full use of their democratic rights to fight for change at all levels – from the national to the grassroots.

The roots of a broader 'movement' were first noted in the second half of the nineteenth century. The first protectionist groups were created in Britain in the 1860s. In the United States, a two-pronged movement of wilderness preservationists and resource conservationists began to emerge, and as people became more mobile and looked beyond their immediate surroundings, the movement grew and spread. With the emergence of a substantial global movement, in the United States alone, 17 million people describe themselves as 'environmentally active', and 55 per cent of the population claims to support the aims of the movement.[47] Three million Britons were members of environmental groups,[48] making the movement the biggest in British history.

Conservation was one of America's great contributions to world environmental movement and its ideas were eventually exported to other nations.[49] In his study of 'American environmentalism',[50] Joseph Petulla identifies three main traditions: the biocentrism (nature for and in itself), the ecologic (based on scientific understanding of interrelationships and interdependence among the parts of natural communities) and the economic (the optimal use of natural resources, described as the utilitarian approach to conservation). He lists different arguments and a variety of ethical bases, ranging from the puritan tradition to health and corporate ethics.[51] Riordan notes the divergent evolution of two ideological themes: the eco-centric (believing in a natural order and natural laws) and the techno-centric (believing that man is able to understand and control events to suit his purposes).[52]

The ultimate aim of the environmental movement is the maintenance of the quality of the human environment, and it must be seen not as a series of separate national movements but as part of a wider and more long-term change in human attitudes. The Second World War transformed values and attitudes towards internationalism, which, in turn, radically altered the agenda of environmentalism. Even before the war had ended plans were drawn up to promote reconstruction and economic assistance, particularly through the new United Nations and its specialized agencies and the emergence of two environmental initiatives: the convening of an international conference on the conservation of natural resources and the establishment of an international organization for the protection of nature were significant steps in this direction.

Addressing the American Scientific Congress (1940), Pinchot argued that every nation's 'fair access' to natural resources was 'indispensable condition of permanent peace',[53] and even Roosevelt held the same view when he wrote to his Secretary of State, Cordell Hull that 'Conservation is a basis of permanent peace'.[54] Pinchot presented Roosevelt with a detailed draft plan for setting up an international organization to promote the conservation of natural resources, providing fair access to necessary raw materials by all countries, facilitating information exchange and drawing an inventory of natural resources and a set of principles on their conservation.

Ironically, the United Nations had already begun thinking about an international conference on the conservation and use of resources, and in the meanwhile post-war attempts to set up an international nature protection organization were initially side tracked by a struggle between British and American protectionists on the one side and Swiss, Belgians and Dutch on the other. The latter were particularly keen to set up an organization independent of the new United Nations system; the former thought it ill-advised. The views of Europeans prevailed, thus denying environmentalists an effective input into UN affairs for nearly 30 years. And finally, in December 1962, the United Nations adopted a resolution supporting the argument that natural resources were vital to economic development and that economic development in developing countries could jeopardize natural resources if due attention was not given to their conservation and[55] restoration.[56]

Rachel Carson

A new book by Rachel Carson, *Silent Spring* (1962), which detailed the adverse effects of the misuse of synthetic chemical pesticides and insecticides, generated heightened public awareness of the implications of human activity on the environment and the cost in turn to human society and prompted the creation of a Presidential Advisory Panel on Pesticides.[57] This event signified the beginning of the environmental revolution that exposed some of the social economic and scientific infrastructure that had knowingly permitted ecological degradation to occur. *Silent Spring* was different from her other publications, like *The Sea Around Us* (1951) and *On the Edge of the Sea* (1955), which grew out of her observations of the threats posed to nature. Graham notes, 'She clearly saw that man was, more than ever before approaching the earth not with humility, but with arrogance.'[58] In his words, *Silent Spring* was essentially an ecological book; it was designed to mobilize people into action against the misuse of chemical pesticides.[59] *Silent Spring* stimulated changes in local and national government policy not only in the United States but also in several European countries like Britain, Sweden, Denmark and Hungary.[60]

In the year 1970, in the month of April, 3,00,000 Americans, perhaps more, took part in Earth Day, the largest environmental demonstration in history. *Time Magazine* reported environment as the most prominent issue of 1970s,[61] and for *Life Magazine*, it was a movement that promised to dominate the new decade.[62]

New Environmentalism

An environmental revolution that transformed the movement, known as new environmentalism, emerged, notably in the United States, which was more dynamic, broad-based, more responsive that won much wider public support, and with this a wave of new organizations emerged,[63] differing from their precursors in at least two major respects.

In the *first* place, if nature protection had been a moral crusade centred on the non-human environment and conservation, a utilitarian movement was centred on the rational management of natural resources and on humanity and its surroundings. For protectionists, the issue was wildlife and habitat; for the new environmentalists, human survival itself was at stake. Roderick Nash argues that fear underlay the growth of popular American concern for the environment, fear of running out of resources and losing its position in world politics, fear for the future of life and the vulnerability of man; man 'was rediscovered as being a part of nature'.[64] Americans not just were limiting their perception to 'the great outdoors' but also were applying their moral and aesthetic appreciation of nature to the total environment.[65]

Second, new environmentalism was activist and political. While the conservationists based their arguments on economics, the new environmentalists brought a more direct political impact, and their message was that environmental catastrophe could be avoided only by bringing fundamental changes in the values and institutions of industrial societies.[66] New environmentalism was seen as part of a wider social transformation taking place in Western society which emerged as a social and political movement, and the issues it addressed were universal. The *first* embodied in the writings of George Perkins Marsh and in Conservation Movement of Roosevelt and Pinchot, who saw the sustenance of a viable physical and biological environment as the priority, influencing policies by presenting a scientifically valid case. The *second* was found in new environmentalism, which was more concerned with 'humanism' and according to a professional ecologist, John Maddox, 'ecology is no longer a scientific discipline, it is an attitude of mind'.[67] In the words of Michael McCloskey, executive director of the Sierra Club, the components of this new movement in the United States include

> the consumer and the corporate reformers movement, the movement for scientific responsibility, a revitalized public health movement, birth control and population stabilization groups, pacifists and supporters of participatory democracy and direct action, and a movement in search of a new focus for politics.[68]

The environmental movement was a product of forces both internal and external to its immediate objectives. The elements of change were emerging before the 1960s; the result was a new force for social and political change. Several factors, in particular, have played a significant role in the change: the effects of affluence, the atomic testing, the book *Silent Spring*, a series of well-publicized works on environmental disasters, advances in scientific knowledge and the influence of other social movements. In the growing atmosphere of alarm about the welfare of the environment in the 1960s, there emerged in the United States environmental theorists and philosophers whose ideas and concerns have contributed immensely in galvanizing public awareness towards environmental issues and movements. Most of them were academicians, namely Paul Ehrlich of Stanford, Barry Commoner of Washington, LaMonte Cole of Cornell, Eugene Odium of Georgia, Kenneth Watt of the University of California at Davis, and Garret Hardin of UC Santa Barbara. The new environmentalism was part of a broad, cumulative process of social and political change which in later years culminated in the creation of UN Environment Programme.

By the mid-1970s, almost every society, whether rich or poor, industrial or agrarian, authoritarian or democratic, socialist or capitalist, felt compelled to reassess its attitudes towards resource management and the condition of the human environment. Environmental issues and challenges like marine pollution, whaling, fisheries, desertification, acid pollution, depletion of the ozone layer and carbon dioxide build-up could not be solved by individual governments acting alone; the obvious response was greater international cooperation, and Stockholm had provided that opportunity to internationalize these issues. The creation of the United Nations Environment Programme was a direct consequence of the United Nations Conference on the Human Environment, held at Stockholm in 1972.

Stockholm Conference and the Creation of UNEP

In 1968 and 1972, two international conferences were held to assess the problems of the global environment, to suggest corrective action. The first was the biosphere conference, held in Paris in September 1968, focusing on scientific aspects of the conservation of the biosphere, which was a result of the coordination of ecological research encouraged by the International Biological Programme.

The second was the United Nations Conference on the Human Environment, held in Stockholm in June 1972. Stockholm was the landmark event in the growth of international environmentalism. It was the first occasion in which the political, social and economic problems of the global environment were discussed at an inter-governmental forum with a view to actually taking corrective action. It aimed to 'create a basis for comprehensive action' within the United Nations of the problems of the 'human environment' and to 'focus the attention of governments and public opinion in various countries on the importance of the problem'.[69] It also marked a transition from the emotional and naive new environmentalism of the 1960s to the more rational, political and global perspective of the 1970s. It further brought the debate between developing and developed countries with their differing perceptions of environmental priorities into open forum and caused a fundamental shift in the direction of global environmentalism.

Pre-Conference Discussions

If there was any single issue that influenced Stockholm, it was acid pollution. Sverker Astrom, the Swedish ambassador to the United Nations, submitted a proposal for an international conference in a resolution put before ECOSOC in July 1968. Dowd suggests that the speed with which the resolution was adopted by the General Assembly reflected to some extent the impact of new environmentalism.[70] The resolution emphasized the environmental work already being undertaken by intergovernmental organizations (IGOs), non-governmental organizations (NGOs) and UN-specialized agencies. The conference provided a framework within which the United Nations could comprehensively assess the problems of the human environment and focused the attention on governments and public opinion.[71] The most significant outcome of the pre-conference discussions was the new role of developing countries in the environmental debate.[72] Pollution,[73] a developing countries problem, has been the spark for the conference, but developing countries used their General Assembly voting power to make sure their perspective be supported and appreciated from the beginning. Before discussing the official proposals put forward by the preparatory committee, one important proposal advocated by George F. Kennan[74] was discussed, which was based on the premise that the United Nations is not capable of handling the environmental issues and that alternative methods should be devised outside the United Nations system. Kennan wanted an agreement signed outside the UN system, and without any need for either an international government or a mechanism to use sanctions on countries violating the agreement. If any country violates the agreement, other nations would also violate the agreement, resulting in status quo ante where pollution levels are concerned and which all developed countries have found to be mutually disastrous.[75]

Two institutional proposals were put forward by chief executives of heavy industry and representatives of various international organizations from Western Europe, North America and Japan.[76] These proposals tried to establish, first, an International Environmental Institute in which representatives of industry and government could meet at a technical level to suggest jointly agreed standards and norms which would be socially and economically feasible and effective in creating improvement in environment; and second, they agreed for a mechanism either through General Agreement on Trade and Tariff or the Organization for Economic Cooperation and Development (OECD) machinery for intergovernmental negotiations on the general application of the costs of environmental controls and standards to processes and products in order to avoid distortion of trade and so on.[77] All these proposals were of the opinion that the United Nations, whose membership includes a substantial number of developing countries, are going to land in a jam of political impasse and will not be in a position to take the kind of quick decisions necessary.[78]

On the other hand, the developing countries were not too keen on suggesting any institutional structure either outside the UN system or within it. They were persuaded to participate in the conference with the hope that their developmental needs will be safeguarded and were keen that whatever institution entrusted with this responsibility would be supported by them.[79]

McCormick emphasizes the role of a single issue, acid-deposition or 'acid rain', stimulating the need for a global conference. He also cites the apprehension of the developing countries that an agenda led by pollution issues would marginalize them.[80] The preoccupation of developed countries to tackle the question of climate change was not the most pressing concern of the Third World. The developed countries' desire for a conference based on environmental agenda was linked by the developing countries to the agenda on development.[81] McCormick credits Maurice Strong for incorporating the developing countries' views to the agenda, to the extent that the agenda of the conference and the very concept of environment were broadened to include issues such as soil loss, desertification, tropical ecosystem management, water supply and human settlements.[82] The final declaration of the Stockholm Conference constituted an action plan and 109 recommendations and 26 principles.

Developed and Developing Countries' Views

The representatives of the developed industrialized countries participated in the Stockholm Conference with environmental pollution problem weighing heavily on their minds and with the need for a worldwide conservation programme to safeguard the planet's genetic and natural resources. Therefore, they agreed that curative measures had to be found urgently if disaster was to be avoided.[83] The Stockholm Conference was accordingly expected to lead a global campaign to curb pollution, conserve resources, and lay the foundation for a more careful management of these resources.

But the developing countries approached Stockholm with a different perspective. In these countries energy and resource consumption was not high and their industrial pollution problems were localized. But with them poverty was widespread, expectations of life were poor, infectious diseases took a terrible toll and human settlements commonly failed to provide the basic essentials of adequate shelter, clean drinking water and safe disposal of human body wastes. 'Debates on doomsday theories', 'limits to growth', 'the population explosion' and the 'conservation of nature and natural resources' were thought of as largely academic, of no great interest to those faced with the daily realities of poverty, hunger, disease and survival. Arguments were presented to show that environmental concerns could well retard development efforts in the developing countries.[84] India became one of the leading participants in preparations for the conference and a strong and effective proponent of the developing countries' position. Mrs. Gandhi attended the Stockholm Conference, and apart from Sweden's prime minister, Olaf Palme, she was the only head of government to do so. Her speech (poverty being the greatest polluter) was both at Stockholm and subsequently widely recognized as one of the most moving and incisive orations to be made on that occasion.[85]

An important event in the preparatory process of Stockholm Conference was a seminar on 'Development and Environment', held at Founex, Switzerland, in June 1971. At Founex and Stockholm the phrase 'the pollution of poverty' came into use to describe the world's environmental problems and it was recognized that the skills of all nations were needed to tackle it. 'Eco-development', a word coined to describe the process of ecologically sound development, a process of positive management of environment for human benefit, emerged as a theme from Stockholm. These redefined concepts made the Stockholm Conference more attractive to developing countries.

The concerns and efforts of developed countries were largely concentrated on the following two substantive themes in the negotiations before and during the Stockholm Conference: (1) 'the identification and control of pollutants of broad international significance' and (2) 'environmental aspects of natural resources management especially water resources'.[86] In addition developed countries also pushed hard for increased efforts and cooperation among existing international organizations to coordinate and stimulate new international action and advise and assist governments on environmental problems.

The report of the Founex[87] delineated clearly and cogently some of the principal environmental issues of special concern and interest to developing countries. This report articulated some of the concerns of developing countries and many of them took an active role in the preparatory meeting and in negotiations itself. In addition to increased financial assistance, developing countries also pushed for greater access to environmental technologies and technical assistance, the proposal for an International Referral System on sources of environmental information. These issues generally dominated the approach of developing countries at the Stockholm Conference, not because they were less concerned than the developed countries about many of the emerging international environmental problems but because national economic development was their over-riding priority and preoccupation. The Stockholm Action Plan and Declaration accurately reflect the major concerns and objectives of the developing countries.

Achievements of Stockholm Conference

The United Nations Conference on the Human Environment was held in Stockholm, Sweden, from 5 June to 16 June 1972. It was attended by the representatives of 113 countries, 19 intergovernmental agencies and 400 other intergovernmental and non-governmental organizations. The conference according to Maurice Strong launched 'a new liberation movement' to free humans from environmental perils of their own making.[88] The emphasis on this theme, that is, the concept of 'human environment', had distinguished Stockholm from previous international gatherings. Barbara Ward, a leading British environmentalist, observed, 'Before Stockholm people usually saw the environment as something totally isolated from humanity. Stockholm recorded a fundamental shift in the emphasis on our environmental thinking'.[89] There was an air of youthfulness about the conference, and for the first time, the environment was being discussed by the world's governments as a subject in its own right.[90]

The major breakthrough was the new perception of the position of developing countries and the Stockholm had encouraged developing countries to equate environment with pollution.[91] Because many saw pollution as external evidence of industrial development, efforts to control it were seen as efforts to constrain development. A theme running through many developing countries' speeches was that environmental factors should not be allowed to curb economic growth.[92] Aaronson noted that, following Stockholm, it would 'be difficult for western environmentalists ever again to view 'the environment' in a parochial way'.[93] The views of the developing countries dominated the discussions in almost every respect and forced developed countries and environmentalists to change their parochialism and begin to see environmental challenges and problems in a global perspective.

The Stockholm Conference produced a declaration, a list of principles and an action plan. It was remarkable that so many countries with differing political, economic and social systems, have been able to agree on such a broad-ranging and philosophical exercise and it was a success because it created tremendous public interest in the environment, and provided directions for international and national action.[94] The Stockholm Declaration on the Human Environment and the declaration of principles constituted a solid foundation for future work and the Stockholm Action Plan was embodied in the United Nations Environment Programme.[95] The institutional and financial arrangements set out in the Stockholm Conference provided the basis for the establishment of UNEP by the UN General Assembly. As originally conceived, the United Nations Environment Programme was just what its name implies: the complete agenda of the United Nations system for environmental matters.[96] The Secretariat guided by the Governing Council was to be a reference point for environmental matters throughout the UN system.[97] UNEP was not to be an executing agency and does not bear the prime responsibility in the UN system for executing environmental projects. UNEP has come to play a leading role in developing environmental policies and promoting their implementation.[98] Through the collection and dissemination of environmental information, the development of policy guidance and efforts to mobilize support for environmentally sound development projects, and through UNEP triangle of environmental assessment, environmental management and support measures, the programme has converted its limited authority and resources into a strong and highly visible environmental activism.[99]

The Action Plan

The action plan consisted of 109 separate recommendations, ranging from the specific to the general, falling into one of three broad groups: environmental assessment, environmental management and supporting measures. Almost half of the action plan dealt with the conservation of natural resources, while the rest covered issues relating to human settlements, pollution and marine pollution, development and the environment, education and information. Sandbrook has summarized the general intent of the action plan as launching

> a set of internationally coordinated activities aimed first at increasing knowledge of environmental trends and their effects on man and resources, and secondly, at protecting and improving the quality of the environment and the productivity of resources by integrated planning and management.[100]

The Stockholm Conference was the single most influential event in the evolution of the International Environmental Governance. John McCormick, in his book *Global Environmental Movements*, had divided the major achievements of the conference into four[101] major parts. *First*, the conference confirmed the trend towards a new emphasis on the human environment. The nature of environmentalism itself changed: from the popular, intuitive and parochial form which had emerged in developing countries in the late 1960s to a form that was more rational and global in outlook and emphasized working towards a full understanding of the problems and agreeing on effective legislative action. *Second*, Stockholm forced a compromise between the different perceptions of the environment held by developed and developing countries. During the early UN debates on the conference developing countries used their UN General Assembly voting power to compel developed countries to recognize the need to balance environmental management priorities with the aims of economic development. Before Stockholm, environmental priorities had been determined largely by developed countries, following Stockholm the needs of developing countries became a key factor in determining international policy. *Third*, the presence of a large number of NGOs at the conference and the part they played marked the beginning of a new and more insistent role for NGOs in the work of governments and intergovernmental organizations. *Finally*, the most significant outcome of Stockholm was the creation of the United Nations Environment Programme. It had limitations and deficiencies, but it was probably the best form of institution possible under the circumstances.[102]

The Creation of the United Nations Environment Programme

The Stockholm Declaration, Principles, and Action Plan would remain paper exercises until they had some practical result, and their true effectiveness would depend on the institutional arrangements made for turning principles into policies and active programmes. The institutional arrangement eventually confirmed by UN General Assembly Resolution 2997 of 15 December 1972[103] was not for the creation of a new specialized agency, but of a cross-cutting programme of policy coordination. This took form in the United Nations Environment Programme (UNEP), created in recognition that 'environmental issues of broad international significance' fell within the province of the UN network. The headquarters of every existing UN-specialized agency were in North America or Europe, so when it came to the question of a secretariat for the new body, there was a campaign to have it set up in a developing country.[104] The developing countries' option prevailed and UNEP was located in Nairobi, Kenya. Liaison offices were set up in New York, and Geneva, and regional offices in Bangkok, Beirut, Mexico City and Nairobi. Maurice Strong was appointed the first executive director of UNEP.

Institutional Framework for Environmental Governance

The new organization had four parts:[105] a Governing Council for Environmental Programmes (which was replaced by Environment Assembly of UNEP in 2012); a small Secretariat, which is 'focal point for environmental action and coordination within the UN system'; a voluntary Environment Fund, to which governments could contribute, and an Environmental Coordination Board made up of members of all relevant UN bodies. The Governing Council meets annually, promotes international environmental cooperation, provides 'general policy guidance for the direction and coordination of environmental programs within the United Nations system and ensures that governments gave emerging international environmental issues appropriate attention'. A significant recent development in this context was the creation of the Environment Assembly of the UNEP in June 2012, replacing the Governing Council (which was composed of 53 members), with a universal membership 193 member states, when world leaders called for UN Environment to be strengthened and upgraded during the United Nations Conference on Sustainable Development, also referred to as Rio+20.[106] The Environment Assembly embodies a new era in which the environment is at the centre of the international community's focus and is given the same level of prominence as issues such as peace, poverty, health and security. The establishment of the Environment Assembly was the culmination of decades of international efforts, initiated at the UN Conference on Human Environment in Stockholm in 1972 and aimed at creating a coherent system of International Environmental Governance.[107]

The blue print for UNEP was the Stockholm Action Plan. This is implemented through three functional components, namely environmental assessment, environmental management, and supporting measures.[108]

Global Environmental Assessment

This materialized in the form of Earth watch, a UN-sponsored network designed to research, monitor and evaluate environmental processes and trends, providing early warning of environmental hazards and determining the status of selected natural resources.[109] Earth watch has three[110] components. The first was the International Referral System (INFOTERRA), a decentralized 'switchboard' for the exchange of information. The second component, the Global Environmental Monitoring System (GEMS), puts environmental surveys into effect, collecting information from governments and building a picture of regional and global environmental trends.[111] The third component of Earth watch is the International Register of

Potentially Toxic Chemicals (IRPTC), which became operational in 1976. Based in Geneva it has built up a data bank on potentially toxic chemicals for the use of governments.

Environmental Management

As regards UNEP's coordinating role, the Stockholm Conference provided specifically for the establishment of the Environmental Co-ordination Board (ECB), chaired by the executive director and under the auspices of the United Nations Administrative Committee on Co-ordination (ACC).[112] The main function of this Board was to ensure cooperation and coordination among all bodies concerned with the implementation of environmental programmes. ECB was discontinued, and its task was entrusted to ACC, which consists of the heads of the specialized agencies and other bodies under the chairmanship of the Secretary-General. As regards the environment, ACC has established an Inter-Agency Committee on Sustainable Development[113] and another unit named Designated Officials for Environmental Matters. UNEP plays a significant role in both bodies and acts as the secretariat for the latter. A concrete step within the framework of UNEP's coordinating function is the adoption of a programme on Awareness and Preparedness for Emergencies at the Local Level (APELL) and the establishment in 1991 of a UN Centre for Urgent Environmental Assistance.[114] The purpose is to address assessment of and responses to manmade environmental emergencies, including industrial accidents.

Supporting Measures

These included education, manpower training, public information and financial assistance. In 1975, a Joint UNEP/UNESCO Environmental Education Programme was created, and in 1977 the UNEP/UNESCO Inter-governmental Conference on Environmental Education was held in Tbilisi, USSR.[115] Both events resulted in a series of training workshops and seminars, the creation of an information network and the training of environmental specialists. But education was ultimately a national issue, and the work of UN-specialized agencies was to be less productive in this area than the work of national NGOs. In 1975, it provided the funding to set up Earth scan, a news and information service within International Institute for Environment and Development (IIED), which went on to produce a substantial body of information for media and NGOs.

Role of UNEP: 1972–1992

The 1972 UN Conference on the Human Environment (Stockholm Conference) marked the beginning of the institutionalization of global environmental concerns within the UN system. The conference – whose theme was 'only one earth'[116] – was greeted as evidence of the commitment of governments to address these common concerns. Tensions between the richer industrialized countries and the developing countries surfaced early and shaped the outcomes of the conference. These differences have continued to be a defining theme of global environmental politics and the pursuit of environmental security.

The developed countries were reluctant to agree to a new institution that would require increased funding. Developing countries were cautious about supporting anybody who might hamper their development. Existing UN agencies were opposed to any new specialized agency that would undermine their responsibilities or competencies in the environmental field. UNEP's mandate was to monitor, coordinate and catalyse. It has no executive powers and no direct operational responsibilities. It has suffered from a paucity of funding, lack of political support from governments and from geographical isolation within the UN system.[117] In spite of these constraints, UNEP has been active and relatively successful in carrying out its mandate and contributing to the pursuit of environmental security. Through its Earth watch programme, UNEP has collected and monitored information about the environment. It has forged partnerships with other intergovernmental and non-governmental bodies, such as the International Union for the Conservation of Nature (IUCN) the Worldwide Fund for Nature (WWF) and the World Meteorological Organization (WMO). Responsibility for environmental concerns within the United Nations has not rested solely with UNEP. The General Assembly, ECOSOC and its various commissions have all contributed in various ways to the debates about environmental protection.

In the years since the Stockholm Conference, the UN, and especially UNEP, has pursued something of an activist agenda in the field of environmental governance and security.[118] In 1977, UNEP convened an ad-hoc committee of experts which produced the World Action Plan on the ozone layer. In the same year UNEP sponsored a conference on desertification, which resulted in the Plan of Action to Combat Desertification. WMO and UNEP convened the first World Climate Conference in 1979. In 1980, UNEP and the IUCN produced the World Conservation Strategy. Two years later the General Assembly adopted the World Charter for Nature. In 1983, the General Assembly established the World Commission on Environment and Development, an independent commission mandated to formulate a global agenda for change. The commission, chaired by Norwegian prime minister Gro Harlem Brundtland, completed its report, entitled 'Our Common Future' (echoing the Palme Commission's report 'Our Common Security'), in 1987. The Brundtland Commission put the concept of 'sustainable development' firmly into the lexicon of environmental politics and onto the UN agenda. The report is a mix of progressive reforms on debt, reduction of military spending, socio-economic justice and conventional solutions.

In 1985, the Vienna Convention for the Protection of the Ozone Layer was adopted, the culmination of negotiations conducted under UNEP's coordinating eye. The 1987 Montreal Protocol and subsequent negotiations also proceeded under UNEP's auspices. In the late 1980s climate change became a focus for the

UN. In conjunction with WMO, UNEP convened two workshops on climate change – one in Bellagio in Italy and the second in Villach in Austria – in 1987. In 1988, with the support of the Canadian government, UNEP and WMO sponsored the Toronto Conference on the changing atmosphere which brought together scientists, NGOs and policymakers to consider climate change issues. Among other things, the conference adopted the Toronto target, a non-binding commitment to reduce global CO_2 emissions by 20 per cent of 1988 levels by 2005.[119] In the same year, UNEP and WMO established the Inter-governmental Panel on Climate Change (IPCC) and in 1990 the General Assembly established an Intergovernmental Negotiating Committee (INC) for the negotiation of a Framework Convention on Climate Change (FCCC).[120] In both cases, the Rio Conference of 1992 provided a forum for completion of the negotiations.

Rio Conference

General Assembly resolution 44/228 adopted in December 1989 called for a second conference on the scale of Stockholm, which would specifically adopt the logic of the Brundtland Report and thus link the previously separate agendas of environment and development. The United Nations Conference on Environment and Development (UNCED) was scheduled for the first two weeks of June 1992, and the Rio-de Janeiro, Brazil, nominated as the host city. The UNCED agenda was developed over a period of two years by a Preparatory Committee (abbreviated as 'Prep com').

Four sessions of the Prep com were held: in August 1990 at Nairobi, March–April 1991 and August 1991 at Geneva, and finally at New York 2 March–3 April 1992. The sessions of Prep com therefore totalled 13 weeks of activity spread over 21 months. Prep com divided its work into three working groups, reflecting an attempt to divide the agenda on functional or sectoral lines. A provisional agenda was developed in collaboration with the Designated Officers on Environmental Matters (DOEM). This small secretariat, chaired by UNEP, brought together the programme officers in every specialized agency and UN programme with environmental responsibilities.[121]

Derived from UN General Assembly Resolution 44/228, the provisional agenda for UNCED reflected the attempts to forge the linkage between environment and development. The provisional agenda passed from detailed specificity on some issues to a very general level was itself the product of multilateral bargaining and compromise in the Prep com process. The Environment Programme met for over two years prior to the convening of the Plenary Conference, in which all the major legal instruments were drafted. Four sessions of the Preparatory Committees (Prep-coms) were held and in the last Prep-com held in New York in April 1992, key aspects of both the forestry and climate conventions which were not subject to prior agreement, provided the central disputes to be resolved at Rio Conference.

Outcome of the Conference: An Analysis of the Convention on Biological Diversity, Treaty on Climate Change (UNFCCC), Forestry Agreement and Reforming GEF

The United Nations Conference on Environment and Development, held in Rio de Janeiro in June 1992 was evidence of the commitment of the UN and the international community to the cause of environmental governance and security, and it was the largest UN Conference held to date on environment. Its purpose was to elaborate strategies and measures to halt and reverse the effects of environmental degradation in the context of increased national and international efforts to promote sustainable and environmentally sound development in all countries.[122] There were delegations from 178 countries; 1,400 NGOs had official observer status.[123] Over one-third of those NGOs were from developing countries, compared with one-tenth at Stockholm. Another 18,000 people from environment, conservation and development NGOs and grassroots movements attended the Global Forum, and this separate NGO forum became a hallmark of UN diplomacy.[124]

Three non-binding agreements emerged from this conference: the Rio Declaration, Agenda 21 and a statement on forest principles. The UN Framework Convention on Climate Change and the Biodiversity Convention were opened for signature at the Rio Summit. A new UN institution was also established at Rio: the Commission on Sustainable Development (CSD). The Commission on Sustainable Development was a compromise outcome. Views were divided between those who favoured strengthening UNEP rather than creating a new body, seeking to strengthen its coordinating activities through involving the authority of the Secretary-General, and those countries which wanted a new body to provide a specifically post-Rio focus.[125] The commission is mandated to enhance international cooperation and to examine the progress of implementation of Agenda 21.

According to UNCED secretary-general Maurice Strong, Rio Conference also produced an earth charter as a statement of ethics to guide environmental governance into the twenty-first century. The charter became a declaration of 27 principles. The Declaration on Environment and Development is referred to as the Rio Declaration. Some observers suggest that it was watered down because the earth charter was considered to be 'too environmental'[126] and it ignored the developmental concerns of the developing countries.[127] Besides Rio Declaration, Agenda 21, which is another landmark document, lengthy and non-binding programme of action, in its 40 chapters,[128] addresses a range of environmental issues, questions of participation and capacity and institution building.[129] The most contentious issues were regarded as those covering financial and technology transfers, climate change, institutional arrangements, high-seas, fisheries, biotechnology, poverty and consumption and forests.[130] The third document to arise from the UNCED preparatory process was the non-binding statement on forest principles.[131] Those countries which 'own' tropical forests were adamant that this was a sovereign resource issue and refused to countenance any kind of international binding treaty.

In this section, while dealing with the outcomes of the conference, developed and developing countries views were analysed in the process of the negotiations with regard to these conventions (Biodiversity and Climate Change), agreements (Forestry) and restructuring of GEF. It was observed how in many occasions UNEP was being dominated and manipulated by the developed countries, but at the same time UNEP emerged as that platform where developing countries were able to put forth their concerns powerfully and consistently. In the following section these treaties, conventions and agreements are analysed from the perspectives of developing countries.

The Convention on Biological Diversity

The Convention on Biological Diversity is an inter-governmental agreement which affirms that states are responsible for conserving and using their biological resources in a sustainable manner. The convention was a product of two years of hard negotiations between developing and developed countries on the methods of saving and sharing biological resources.

The convention attracted American hostility, as the argument was that it might affect adversely the biotechnology industry of the United States and thereby affect jobs and income, and it further objected to certain legal and financial implications particularly 'compensation for commercial exploitation of natural products in the pharmaceutical industry'. On the other hand, a number of other countries did not adhere to the biological diversity convention, including some leading tropical forestry countries, such as, Cameroon, Singapore, Vietnam, Brunei, Equatorial Guinea and Sierra Leone among others,[132] citing their displeasure with Western stigmatization over forestry questions. Ashish Kothari, an environmentalist and founder-member of Kalpavriksh, an Indian environmental NGO, while analysing the proposed convention, notes that it remains largely at a superficial level and fails to tackle adequately the root causes of global and national biodiversity destruction.[133] According to him, 'inequality in controlling biological resources', both between and within nations, resulted in cornering of a vast majority of these resources for the benefit of a small minority elite within poor nations, and the wasteful consumption patterns of the North was blamed for this.

The North-South divide was prominent on this very issue. By linking biodiversity to biotechnology, India, represented by Southern developing countries, put forth this argument that if the North wants access to the richer biological resources of the South, it must guarantee similar access to its relevant biotechnologies. Monopolizing such technologies by patenting them, or imposing other forms of intellectual property rights (IPRs) on them, is simply not acceptable for Southern developing countries. The demand for a stronger biodiversity convention that could become the South's most effective weapon against the increasingly monopolistic and North-dominated international trade regime was supported by using the biological assets as a bargaining lever in a world which is so heavily stacked against the developing countries of the South. The South has managed to incorporate some of its concerns vis-à-vis the North, including the demands for funding and technology transfer. Another significant aspect widely recognized by the governments of North and the South is that biodiversity, including wildlife conservation, will succeed with the involvement and protection of the interests of local communities.[134]

Another powerful argument put forth by the leading environmentalist representing the views of developing countries of the South was by Vandana Shiva. She argues that 'biodiversity' is another area in which control has shifted from the South to the North through its identification as a 'global problem'.

> As in the case of ozone depletion, biodiversity erosion has taken place because of habitat destruction in diversity-rich areas by building dams, mines and highways financed by the World Bank to help TNCs and substituting diversity based agricultural and forest systems with mono-culture of green revolution, where wheat and rice and eucalyptus plantations also supported and planted by World Bank, to create markets for seed and chemical industries.[135]

In her words, the most important step in biodiversity conservation is to control the World Bank-planned destruction. By treating biodiversity as a global resource, the World Bank emerges as its protector through GEF (domination and destruction carried out by Bretton Woods Institutions represented by World Bank). The allegation was that North demands free access to the South's resources through the biodiversity convention. However, biodiversity is a resource over which local communities and nations have sovereign rights. According to her, globalization has become a political means to ensure an erosion of these sovereign rights and shifting control over the access to biological resources from the gene-rich South to the North. 'Global environment' thus emerges as a principal weapon for the North to gain worldwide access to natural resources and raw materials and to force a worldwide sharing of the environmental costs it has generated while it retains monopoly on benefits reaped from the destruction.

Treaty on Climate Change

As Stockholm Conference had been galvanized by the question of acid rain, so climate change came to dominate the environmental aspect of the Rio/UNCED preparations. The climate issue was itself a deeply divisive one, both within the group of industrialized countries and between the industrialized and developing countries. As early as 1978, UN General Assembly Resolution 42/184 of 11th December had highlighted climate change and biological diversity as two priority issues. Two joint initiatives of UNEP and WMO were created in the late 1980s: the Intergovernmental Panel on Climate Change (IPCC) and the World Climate Conference (WCC). IPCC presented its first assessment report at Sundsvall, Sweden, in August 1990 and operated in three separate working groups, and they addressed different aspects of climate change and the outcome was the adoption of a Framework Convention on Climate Change. The United Nations Framework Convention on Climate Change (UNFCCC) established an international environmental treaty to combat 'dangerous human interference' with the climate system, by stabilizing greenhouse gas concentrations in the atmosphere. It was signed by 154 states at the Earth Summit, held in Rio de Janeiro from 3 to 14 June 1992. The treaty established different responsibilities for three categories of signatory states. These categories are developed countries, developed countries with special financial responsibilities and developing countries. The developed countries, also called Annex 1 countries, originally consisted of 38 states, 13 of which were Eastern European states in transition to democracy and market economies and the European Union. All of them belonged to the OECD countries. Annex 1 countries are called upon to adopt national policies and take corresponding measures on the mitigation of climate change by limiting their anthropogenic emissions of greenhouse gases as well as to report on steps adopted with the aim of returning individually or jointly to their 1990 emissions levels. The developed countries with special financial responsibilities are also called Annex II countries. They include the entire Annex I countries with the exception of those in transition to democracy and market economies. Annex II countries are called upon to provide new and additional financial resources to meet the costs incurred by developing countries in complying with their obligation to produce national inventories of their emissions by sources and their removals by sinks for all greenhouse gases not controlled by the Montreal Protocol. The developing countries are then required to submit their inventories to the UNFCCC Secretariat. Because key signatory states are not adhering to their individual commitments, the UNFCCC has been criticized as being unsuccessful in reducing the emission of carbon dioxide since its adoption. The Kyoto Protocol, which was signed in 1997 and ran from 2005 to 2020, was the first implementation of measures under the UNFCCC. The Kyoto Protocol was superseded by the Paris Agreement, which entered into force in 2016. As of 2020–2021, the UNFCCC has 197 signatory parties. Its supreme decision-making body, the Conference of the Parties (COP) meets annually to assess progress in dealing with climate change.

Restructuring Global Environment Facility (GEF)

At the heart of another conflict between the North and the South was the GEF, set up as a pilot project with the UNDP and the UNEP as co-sponsors, but controlled and run by the World Bank. Since the South has been a victim of World Bank policies for decades, the developing countries were reluctant to accept the GEF as the 'sole financial mechanism' for post-UNCED activity, including the implementation of the Convention on Climate Change and Biodiversity. But now they are ready to be financed through the GEF, on the condition that it is modified.

The choice of a financial mechanism to implement the decisions of the UNCED had been one of the most intractable issues. The conflict centres on the Global Environmental Facility (GEF), which was supposed to be managed jointly by the

World Bank, the UNDP and the UNEP but, in reality, is operated and controlled by the World Bank. Criticizing the bank's role in environmental destruction and poverty creation in developing countries, the leading Indian environmentalist Vandana Shiva compares the North's demand of the GEF to be made the sole financial mechanism for implementation of the UNCED agenda to 'recommending the wolf to protect the sheep'.[136] The South rightly fears the use of 'green conditionality' by the GEF.

The World Bank and other participant countries in the GEF agreed to a restructuring of the GEF to expand its lending programme to make its membership universal and the decision-making process a weighted one. With GEF controlling financial flows, there is a fear that environmental aid will be made conditional on a country's willingness to remould its economic and social policies to fit the World Bank's policies. Developing countries of the South (the G-77) included clauses in the finance chapter of Agenda 21 to democratize the structure of the World Bank-controlled GEF. Allegations and demands of the developing countries with regard to undemocratic functioning of the GEF were highlighted on these grounds: The *first* point of non-democratic functioning of GEF is its membership. UNCED is an inter-governmental process involving 175 countries. The implementation of Agenda 21 and the conventions demand that all member countries of the UN are involved in the follow-up in an active way and universal membership must be a precondition for any financial mechanism. Any mechanism that avoids this commitment is undemocratic. The *second* point of non-democratic functioning of the GEF is its decision-making and voting. The South wanted 'one-country one-vote' principle. The North has been insisting on weighted voting. The GEF preserves the 'weighted voting' by bringing 'double majority mechanism', whereby a 'participant assembly decision would require support of participant countries that represented both a simple majority of participants and a simple majority of financial contributions to the GEF'. The 'double majority mechanism' proposed as a step towards democratic decision-making is still flawed since both account for only one-third membership of the GEF. In the absence of universal membership, the decision-making structures will be weighted in favour of the developed countries.

And the *third*, most significant distortion that the GEF introduces into the UNCED process is that it transforms global responsibility for global problems into an exclusive burden on the South. However, by treating the GEF as the only mechanism for post-UNCED financing, the UNCED reduces itself to a catalogue of responsibilities for environmental action only for the South with no environmental actions and responsibility for the North. The developing countries of the South allege that through the GEF the North can regulate the South, but the South cannot regulate the North. This one-way control is unjustified in all contexts. The North wanted GEF to be accountable to its creditors, whereas the South wanted GEF to be accountable to its clients.

The basic question these issues pose is whether it will be the Bretton Woods Institutions represented by the World Bank, the International Monetary Fund and others or the UN system including UNEP and other agencies which are going to govern the world after Rio. The restructuring of the UN shows that the UN itself could be made more like the Bretton Woods system.[137]

The Forestry Agreement

The forestry agreement ran into difficulty because the industrialized nations look upon tropical forests as carbon sinks which can absorb the excess carbon dioxide emitted by them, and also, they are seen as a reservoir of biological diversity as two-thirds of global biological diversity exists in tropical rain forests. To the developing countries, however, forests provide the livelihood of a large number of people, including tribal families, and forests are not just carbon sinks but the very source of life. Hence, this polarization in perceptions, forests as pollution prevention and economic entities versus forests as the home of nature's creation and the source of livelihood dominated the discussions, and developing countries were strongly opposed to the globalization of forest resources and their prevention in earning foreign exchange through the sale of forest products.[138]

Desertification issues within the UNCED proceedings were raised by a group of Sahelian countries, who knew the integrated nature of the environmental catastrophe as climate change, land use patterns, deforestation, grazing and population pressures, civil war and refuge flows had each contributed to the devastation that they faced and demanded focus towards the unifying concept of sustainable development by linking the discussions on forestry issues with action on desertification. Hence, for the countries of the South the priorities were clear: trade and debt reform, action on desertification, water quality and reaffirmed rights to national sovereignty over natural resources, as well as demands for financial instruments and mechanisms.

Achievements and Criticism

Since the Rio process, both the Climate Change and Biodiversity Conventions have entered into force and conferences for parties for both conventions have been convened. Under resolution 47/188 1992 the General Assembly established the negotiating process for convention on desertification, which was opened for signature in Paris in October 1994.[139] The Commission for Sustainable Development has convened substantive sessions monitoring the implementation of Agenda 21 through sectoral process, based on reports submitted to it by governments.[140]

Maurice Strong claimed that the 1972 Stockholm Conference 'launched a new liberation movement', but it has not achieved the success he hoped for it. Two years after the 1992 Rio Conference, Strong admitted that 'the momentum of Rio has been lost at the government level and the fundamental changes needed to head off impending disasters are no closer to reality than they were at Rio – in fact things have gone backwards' (cited in Kirwan).[141] Maurice Williams suggests that 'environmental issues have continued to outpace the measures intended to deal with them' (1992, p. 22). According to Shanna Halpern (1992) 'the global environment is worse now than it was two decades ago, not one major environmental issue debated in Stockholm has been solved'. All these global conventions that have been adopted under the UN have been limited in their success. It has proved impossible to get any consensus on the need for forest convention. In the meantime, deforestation continues. While UNFCCC provides at least some kind of agreement on climate change, it contains no firm commitments. The UNCED process has not been able to address in a meaningful way the three issues that Maurice Strong argued would be the test of the conference: financial assistance, technology transfer and institutional reform.[142] The conference secretariat estimates that the costs to developing countries of implementing Agenda 21 were in the range of US$ 600 billion per annum, of which US$ 125 billion per annum could have come from the developed countries. No such commitments, in fact very little contributions in those cases were made, to reach the UN target of 0.7 per cent of Official Development Assistance (ODA).[143]

The Commission on Sustainable Development has not been favourably received. Rowlands points out that the commission has been 'so weakened that it is fated to be an ineffective body'.[144] Progress to date under the CSD has been primarily organizational rather than substantive, and little progress has been made on commitment to the cross-sectoral concerns of financial and technology resources, poverty alleviation or changes in patterns of production and consumption. The reason for this lack of success is that the United Nations has become unwieldy and unresponsive, the UN process is shaped by demarcation and duplication of responsibilities and it has been underfunded.[145] Hurrell and Kingsbury suggest that 'states have been singularly parsimonious in their contributions to UNEP'.[146] Maurice Williams argues that 'the organization has become too complex with proliferation of many bodies, overlapping agenda and fragmentation of efforts. As a result, authority and responsibility have become blurred and coordination an end in itself'.[147] Coordination on environmental issues has become increasingly confused since Rio.[148]

The challenge for the United Nations is perceived in institutional reform and managerial terms. There has been no scarcity of suggestions about how to reform and strengthen the United Nations, some of which could be achieved within existing structures and some of which require changes to the charter.[149] Some significant suggestions and proposals focus on strengthening UNEP, to provide it with more resources, to reform its mandate to give it more political authority, to strengthen the role of the executive director, possibly as a special commissioner for the global environment.[150] Other proposals have called for special Security Council sessions on environmental threats to peace and security, for the role of the Security Council to be enhanced to deal with environmental securities or for the establishment of an Environmental Security Council.

Mische argues[151] that we must recognize planetary sovereignty and the ethics of accountability and eco-responsibility in the process. Global environmental governance is that process of democratization where local rights and responsibilities should be strengthened, and this continued empowerment of civil society represents a 'return to the sovereignty of the people'.[152]

Post-Rio (Post-UNCED) Role of UNEP: 1992–2022

The 'international institutional arrangements' agreed in Rio in June 1992 have posed considerable difficulties to UNEP, both psychologically and practically. UNEP's 'co-coordinating and catalytic' mandate, as devised in Stockholm in 1972 and endorsed that same year by the United Nations General Assembly, was not clear, and the creation of the Commission on Sustainable Development – which was established by the UN General Assembly in December 1992 (Resolution A/Res/47/191) to ensure effective follow-up of UNCED, also known as the Earth Summit, to monitor and report on implementation of the agreements at the local, national, regional and international levels – had complicated things considerably.[153] Further, a five-year review of Earth Summit progress in 1997 by the UN General Assembly meeting, the implementation of Agenda 21 and the commitments to the Rio principles were strongly reaffirmed at the World Summit on Sustainable Development (WSSD), which was held in Johannesburg, South Africa, from 26 August to 4 September 2002.[154] However, the Commission on Sustainable Development was given the responsibility of reviewing progress in the implementation of Agenda 21 and the Rio Declaration on Environment and Development as well as providing policy guidance to the Johannesburg Plan of Implementation (JPOI) at the local, national, regional and international levels. The JPOI reaffirmed that the CSD is the high-level forum for sustainable development within the United Nations system.[155] Since its establishment in 1992, the commission has greatly advanced the sustainable development agenda within the international community, but a significant development and decision taken by member countries at the United Nations Conference on Sustainable Development (Rio+20), held in 2012 at Rio, led to the establishment of a high-level political forum that has subsequently replaced the Commission on Sustainable Development.[156] This is a major recent development which cannot be ignored with regard to the role of UNEP in the post-UNCED process.

The creation of Commission on Sustainable Development after the Rio Conference of 1992 reduced the political support for UNEP, which hit the organization with full force in the coming years. UNEP also suffered politically from the negative atmosphere resulting from falling ODA levels.[157] This increased uncertainties within the organization at a time when it was also troubled by management and governance issues, a precarious financial situation that worsened as confidence in UNEP and its overall role in the international system diminished.[158] During these years, UNEP also had difficulties showing concrete results, such as the earlier recognized achievements in international environmental negotiations. 'These factors, together with a lack of clarity in UNEP's relationship with CSD', made the task of the executive director (Elizabeth Dowdeswell) very difficult.[159] Under these circumstances, it was difficult for UNEP to live up to its post-UNCED role as the principal UN body in the field of environment. This problem was to last until the end of the 1990s. As a result, 'UNEP went through an inward-looking phase' that lasted until mid-1998.[160] Ultimately, progress was made and UNEP emerged with a new sense of purpose.

These difficult circumstances were further aggravated by the critical financial situation. For example, 'The overall budget was reduced from US$150 million for the 1992–1993 biennium to US$90 million for the 1996–1997 biennium'.[161] During 1996–1997, there also was a US$23 million shortfall between the contributions promised and those actually paid. For 1998/1999, the budget was further reduced toUS$75 million.[162] 'This brought UNEP's reliance on voluntary contributions and its dependence on donors into sharp focus'. This was a huge handicap for an organization with such an important global mandate. 'The seriousness of the challenge can be illustrated by the fact that, despite accomplishments after 1998, the financial contributions to the Environment Fund at the start of the new millennium stagnated at around US$40 million a year, of which some US$30 million were contributed by the ten top donors'.[163] This and also the resources available to the MEAs (Multilateral Environmental Agreements) did not stand in any realistic proportion to the problems they were mandated to solve. 'The top donors' also made the largest donations to earmarked trust funds that were established for particular projects. This part of UNEP financing was growing in relation to the stagnating contributions to the Environment Fund and the shrinking contributions from the regular budget of the UN.[164]

'In 2001, the top 11 donors in order of the size of their financial contributions were: the US, UK, Japan, Germany, Netherlands, Finland, Switzerland, Sweden, Denmark, Norway and Italy'.[165] Of these, the United States, Japan and Italy contributed well below the UN scale of assessment for obligatory contributions to the UN and most developing countries contributed less than their share of the assessed scale.[166] A gradual development occurred that signified a radical, and not very healthy, change. In the 1980s, the financing proportions were: regular budget at 10 per cent, Environment Fund at 75 per cent and earmarked contributions at 15 per cent. By 2002–2003, projected proportions were: regular budget at 3.4 per cent, Environment Fund at 56.2 per cent, and trust funds and counterpart contributions at 40.4 per cent.[167] Even if 'the trust funds' were welcome additions in the difficult resource situation, this big increase in relative proportions created a problem of legitimacy over time and also involved the risk of distorting the priorities of UNEP. When UNEP was created, its regular budget was intended to finance policy functions, the secretariat and coordination within the UN system. The Environment Fund was meant to finance projects and new environmental initiatives within the UN system. Trust funds and counterpart contributions were meant to supplement the work programme.[168] However, since Johannesburg Summit (2002) and continuing to the present, the Environment Fund has financed staff costs and trust funds finance programmes and projects. Apart from the legitimacy issue involved, this situation has created serious problems for long-term planning of UNEP's programme of work.[169] The resources of UNEP's Environment Fund had actually plummeted in real terms, and by 1992, in the Rio Earth Summit, the situation had improved, but by 1998–1999, another decline had set in. UNEP's donor base was a narrow one, which meant, as Maria Ivanova[170] has commented, 'fluctuations in government priorities and attention can be particularly impactful'. The Soviet

Union, the fifth-largest donor to the Environment Fund, was a case in point.[171] In the words of Ivanova: 'Soviet Union had been a significant contributor to UNEP in financial terms as well as through political, technical, and human resources, until it ceased to exist in 1991'. On average, the Soviet Union contributed approximately US$7.3 million a year to UNEP's Environment Fund from 1975 to 1991.[172] Soviet contributions accounted for 12.1 per cent of the Environment Fund during that period. (By comparison, the United States contributed 28.6 per cent of the Environment Fund during the same time, the United Kingdom 5.7 per cent, and France 4.0 per cent.)[173] From 1992 to 2009, the Soviet Union's successor, the Russian Federation, saw a decline in contributions to US$0.47 million a year (0.8 per cent of the Environment Fund). UNEP thus lost 'one of its most significant donor countries, and the downward trend in the Environment Fund can be explained partly by the disappearance of the Soviet Union.'[174]

So, post-UNCED voluntary contributions to UNEP (taking 'earmarked' contributions as well as contributions to the Environment Fund) fail to live up to expectations, and surely, part of the reason lies in the institutional fragmentation that had resulted, from UNEP and others' efforts in establishing multilateral environmental agreements (MEAs). These MEAs had their own mandates and governing bodies, and where UNEP had officially a role in providing the secretariat, the MEAs inevitably developed their own personalities and loyalties.[175] If 'implementation' of agreements was the priority, it wasn't clear what UNEP's role was. UNEP wasn't an implementing agency in the traditional UN sense (though it had that role in respect of GEF projects), and it did not have 'country' offices, as UNDP and the World Bank did, country programmes and projects.[176] On the other hand, The Global Environment Facility, which was established in October 1991 as a US$1 billion pilot programme in the World Bank to assist in the protection of the global environment and to promote environmentally sustainable development, as part of the 1994 restructuring, was officially confirmed as the financial mechanism for both the UN Convention on Biological Diversity and the UN Framework Convention on Climate Change. In partnership with the Montreal Protocol of the Vienna Convention on Ozone Layer Depleting Substances, the GEF started funding projects (which enabled many countries to phase out their use of ozone-destroying chemicals) and served as financial mechanism for two more significant international conventions: the Stockholm Convention on Persistent Organic Pollutants (2001)[177] and the United Nations Convention to Combat Desertification (2003).[178] After the 1994 restructuring, GEF increasingly evolved into a powerful independent agency, operating in its own right and reporting to its own governing body.[179] Despite the fact that the UNEP and the UN Development Programme (UNDP) were brought in to help the World Bank-implemented GEF projects, the Bank remained institutional parent and trustee of GEF funds and literally defined the terms under which global environmental aid was made available in the 1990s. Through its effective control of the GEF, the World Bank has been able to bring its economist vision of development into focus what used to be UN territory of global environmental protection.[180]

Addressing a special event held at UNEP's Nairobi office to celebrate UNEP's fortieth year, in 2012, ex-UNEP executive director Dowdeswell aptly notes that, besides the huge Agenda 21, the concept of sustainable development, convention on biodiversity and climate change are the real legacies of Rio, which she regards as a genuine opportunity for stakeholders and citizens to be heard as legitimate actors in the UN system. In the evolving story of this unique organization over the last 40–50 years, she gives credit by recognizing the contributions of the visionary executive directors (former and present)[181] and the dedicated and committed secretariat staff, who worked and played a significant role under challenging situations. In her words, 'actually, there was a huge gap between rhetoric and action, because of which ambiguity prevailed', on the one hand, a vibrant effective world class UNEP was the focus, but on the other hand, adequate tools were never provided to make it effective.[182] Further, it has been observed that everyone needs to be reminded of what UNEP has been able to achieve. In those five short years following Rio (1992–1997), in spite of not having answers to all of those questions, it was able to deliver a significant agenda.[183] What is equally important is that it was a time when the foundation (Stockholm and Rio) was built for, a very different future. It paved the way for actual delivery and realization of some of the promises of Rio, 20–30 years later. UNEP broke 'new ground, in dealing with environment and trade, environment and financial institutions' which were never part of UNEP's traditional agenda. It broadened considerably the chemicals agenda to persistent organic pollutants, and these are major achievements where major initiatives have matured. UNEP was also active in supporting the implementation of multilateral environmental agreements and assumed responsibility for the creation of the secretariat for the biodiversity convention. 'It promoted further consideration and inclusion of ozone-depleting substances in the Montreal Protocol and developed a world atlas on desertification and a global plan of action on the protection of the marine environment from land-based sources of pollution'. The GEF was created with UNEP as one of the three key founders and designed sustainable cities programmes, and the most important achievement was the UN Desertification Convention, adopted in 1994.[184]

Maurice Strong, under whose aegis UNEP's first efforts to tackle this massive subject had been undertaken, describes the disappointing progress during the early years.[185] This issue was spearheaded by special representatives appointed by Executive Director for Africa, who made important contributions to the process, as well as to the chapter on the subject in Agenda 21.[186] UNEP was not invited by the UN General Assembly to host the original negotiating process which led to the adoption of the UNCCD in 1994, as in the case of the Climate Change Convention, the General Assembly decided to keep such matters in its own hands. Nor does UNEP supply the Secretariat. But UNEP can be given the credit for work done in those early years.[187] UNEP assisted African countries in their preparations for the convention. Most of the African participants of the convention held national workshops and established desertification councils or committees, with the result that many were well-placed to begin programmes when funding became available.

Progress in the fight against desertification needs to be linked to progress on other fronts, for example, climate change and the protection of biological diversity.[188]

Another significant feather in the cap of UNEP was the publication of the first GEO report *Global Environment Outlook* (GEO-1)[189] in 1997, which marked a radical departure for UNEP because it provided a detailed region-by-region assessment of the state of the world environment. The report received worldwide publicity, not just for its conclusions but also for its assessment of the policy options required. The report was the beginning of a long-term continuous assessment process, with regional and local input from all over the world. A product of the collaboration between UNEP and internationally renowned scientists and other United Nations agencies and experts, GEO-1 was frank in its conclusions: that despite the efforts and commitments made at the Rio Earth Summit five years previously, the global environment had continued to deteriorate. 'Progress towards a global sustainable future is just too slow. A sense of urgency is lacking', said the report.[190] 'There is no longer a problem of lack of clarity of vision or institutional frameworks. . . . [W]hat is now needed is the will to act.' The report stressed four conclusions that pointed to priority areas for immediate action by the international community and they are highlighted as follows:

- Improving energy efficiency and using renewable sources of energy would result in major environmental improvements;
- Appropriate and environmentally sound technologies, applied worldwide, would greatly reduce natural resource use, waste and pollution;
- Global action on freshwater was needed to remove a major impediment to development in many regions;
- Improved benchmark data and integrated environmental assessments were essential for effective policy-making at all levels.[191]

GEO-1 broke new ground in establishing mechanisms for continuous expert environmental assessment and was commended worldwide, and it was acknowledged as an authoritative statement of the current environmental balance sheet. Governments also welcomed GEO-1's establishment of priorities for global and regional action. They gave broad acceptance to the need to act on the report's unequivocal conclusion that 'significant environmental problems' remain deeply embedded in the socio-economic fabric of nations in all regions.[192]

When Klaus Töpfer,[193] one of the most respected and influential leaders of the world environment movement who had been closely involved in UNEP's affairs, took over as executive director in February 1998, he was determined to confront the problems which have prevented UNEP from fulfilling its full potential. Apart from fund-raising, Töpfer was determined to deal with the problems caused by the proliferation of bodies in the international environmental field, an issue which – post-UNCED process – had become concerned with since the agencies involved in 'sustainable development' were not necessarily the same as those involved in the environment. One of the key issues was that different agencies had their own governing structures. When Kofi Annan became secretary-general of the United Nations in 1997, he had established a blue-ribbon UN Task Force on Environment and Human Settlements[194] and appointed Dr Töpfer as the chairman of the Task Force, and the report of the 'Töpfer task force' had addressed the proliferation of environmental organizations and treaties and the recommendations were very significant. Some of these recommendations are highlighted: (1) the secretary-general should establish an Environmental Management Group (EMG),[195] chaired by the executive director of UNEP, supported by a secretariat, and the core members that includes the main UN entities concerned with environment and human settlements, financial institutions, and organizations outside the UN system with experience and expertise and convention secretariats as well as a co-coordinating mechanism within UNEP in the chair. (2) UNEP should be able to exercise intellectual leadership in an ever-growing field. One of the ironies of successfully bringing an MEA or similar instrument was that the vital centre of competence within UNEP itself was weakened or hollowed out, while intellectual and administrative competence was built up elsewhere. Hence, to make UNEP's contribution useful and effective, it was recommended that UNEP's substantive support to global and regional conventions based on its capacities for information, monitoring and assessment must be strengthened substantially. (3) Further, 'UNEP was trying to build its capacity and its networks of support to ensure the scientific conventions, to respond to requests for specialized analysis and technological assessments, and to facilitate their implementation'. It was recommended to strengthen UNEP as a 'centre of excellence', which reflected Töpfer's point of view. (4) Another landmark proposal, recommended by the task force, was that the membership of the UNEP Governing Council be made 'universal'; in other words, instead of being a 58-member body elected by the General Assembly, the membership of UNEP's Governing Council would be co-terminus with that of the UN General Assembly itself. This landmark proposal, that the membership of UNEP's Governing Council should be 'universal', again resurfaced in the Rio+20 in the context of discussions on 'International Environmental Governance', also known as the 'institutional framework for sustainable development' (IFSD).[196] It actually took the United Nations General Assembly some time to absorb and react to the proposals of the Töpfer Task Force. When finally, it did so, in July 1999, these proposals on increased coordination and coherence with regard to the MEAs met with resistance.

On the one hand, the General Assembly 'supported the proposals for the facilitation and support for enhancing linkages and coordination within and among environmental and environment-related conventions, including the United Nations Environment Programme', but on the other hand it insisted that it has to be done 'with full respect for the status of the respective convention secretariats and the autonomous decision-making prerogatives of the conferences of the parties to the conventions concerned'.[197] But, at a time when the 'environmental' pillar was trying to regain coherence and credibility, the language and the intention behind it were not especially helpful. The General Assembly failed to endorse the Task Force's proposal that the membership of the UNEP Governing Council should be made universal. But this proposal had 'met major resistance from many quarters. In other words, the General Assembly was supportive. The UNGA, for example, emphasized the importance of strengthening the capacity of the United Nations Environment Programme and the United Nations Centre for Human Settlements (Habitat) in their Nairobi location and ensured support and 'stable, adequate and predictable financial resources necessary to both organizations for the fulfillment of their mandates'.[198] It also supported the establishment of an EMG for the purpose of enhancing inter-agency coordination and strengthening UNEP's capacity to fulfil its mandate. Another area where the General Assembly resolution was helpful was 'capacity building' in the developing countries.[199] The issue of implementation was increasingly becoming crucial where countries needed help to implement those treaties. Over the years since Stockholm, the pressure from the developing countries on UNEP to actually deliver on the 'technical assistance' part of its mandate had been steadily growing. Although UNEP was already implementing GEF projects in a number of countries, there was clearly a large demand for 'capacity building' at both regional and national level.[200]

Focusing on a revitalized UNEP, the Nairobi Declaration of 1997, adopted by UNEP's Governing Council, clearly specified that one of the 'core elements of the mandate of the revitalized United Nations Environment Programme has to provide policy and advisory services in key areas of institution-building to Governments and other relevant institutions'.[201] By 1999, two years later, the General Assembly reiterated 'capacity building' to be a priority and stressed

> the need to ensure that capacity-building and technical assistance, in particular with respect to institutional strengthening in developing countries, as well as research and scientific studies in the field of environment and human settlements, which remained important components of the work programs of both the United Nations Environment Programme and the United Nations Centre for Human Settlements (Habitat).

Finally, and importantly, the General Assembly endorsed the idea of an annual, ministerial-level, global environmental forum. Hence, the 1997 Nairobi Declaration had mainly concentrated on reaffirming UNEP's own mandate in the light of the post-UNCED process in confidence, and it would have been counterproductive for the Malmo Declaration (Sweden, 2000) to revisit that in detail. The first Global Ministerial Environment Forum,[202] though it called for UNEP's role to be 'strengthened and its financial base to be broadened and made more predictable', wisely avoided a detailed reassertion of UNEP's 'leading role', concentrating instead on the substantive challenges ahead.

The first meeting of the Global Ministerial Environment Forum (GMEF) was held in Malmo, Sweden, at the end of May 2000. Though the United Nations General Assembly had been concerned to 'maintain the role of the Commission on Sustainable Development as the main forum for high-level policy debate on sustainable development', the GMEF got off to a good start as a place for discussing and reviewing the 'environmental pillar' of sustainable development, where 100 environment ministers attended out of over 600 delegates. The forum adopted the 'Malmo Declaration'[203] and reiterated that '[i]ntegrating the principles of sustainable development into country policies and programs and reversing the loss of environmental resources' was basically what UNEP has been doing all these years. Over the previous almost five decades it had been helping countries develop environmental legislation or incorporate environmental considerations into national programmes and policies. Within the UN system, UNEP had been assigned special responsibilities for 'freshwater' which related directly to the designated target of 'halving, by 2015, the proportion of people without sustainable access to safe drinking water and basic sanitation issues'. Many other UNEP programmes, such as, the Global programme for the Prevention of Marine Pollution from Land-Based Sources (GPA), with its potentially beneficial impact on fisheries, were in the frontline as far as actions to achieve MDGs were concerned.

The relevance of the Millennium Declaration[204] and the Millennium Development Goals (2000–2015)[205] for UNEP's future programme went far beyond. The enduring importance of the Millennium Summit (2000) lay very much in the eight Millennium Development Goals (MDGs), with their associated 12 Targets. These were derived from the commitments in the Millennium Declaration and widely promoted in the years following the summit. For UNEP, MDG 7 'Ensuring Environmental Sustainability' was of obvious relevance. The adoption of that goal could be seen as an endorsement at the highest possible level of its work.

UNEP's potential contribution to the achievement of the MDGs had to be seen and assessed in the context of the work of the MEAs with which UNEP was associated, for example, the Desertification Convention or the Basel Convention on the Transboundary Movement of Hazardous Wastes. 'Capacity building', in the sense of helping countries actually implement the agreements they had signed up to, was an increasingly important aspect of not only UNEP's own work programme but of the ever-growing family of environmental entities. The challenge is to ensure that sound environmental management was not a constraint on development but the essential basis for it. 'Mainstreaming the environment' was a new phrase used as the new millennium dawned. The Brundtland Commission introduced the notion of 'inter-generational equity' as a guiding principle in the management of human affairs; the Millennium Declaration spoke of 'children of the world, to whom the future belongs'.[206] These were powerful and persuasive ideas which brought an extra moral dimension to the argument.

In the evolution of the implementation of sustainable development agenda and the role of international institutions and environmental governance, the Johannesburg World Summit on Sustainable Development (WSSD),[207] which took place in June 2002 on the tenth anniversary of Rio, was another important milestone conference. The United Nations General Assembly, on 20 December 2000, accepted South Africa's offer to host a conference in Johannesburg to review the progress achieved in implementing the decisions of the Rio Conference of June 1992. The conference took place 'at the summit level' as per the recommendation of the

General Assembly and was called the World Summit on Sustainable Development (WSSD), also referred to as Rio+10. The 'outputs' were very different from those at Rio in June 1992. Whereas Rio had been a scene-setting conference, with Agenda 21 as its massive centrepiece, and the two major environmental treaties as its ornaments, Johannesburg was all about implementation,[208] actually delivering on commitments. The Johannesburg Plan of Implementation[209] was hard-hitting and to the point and was only 60 pages in length, as compared with Agenda 21's 400 pages. The key priority of WSSD was the Plan of Implementation, which contained a section on 'the role of international institutions'. This was an issue which UNEP had taken very seriously in the Johannesburg Plan. Between April 2001 and February 2002, UNEP conducted a thorough-going review of 'International Environmental Governance',[210] the results of which were reflected in a consensus decision of the UNEP Governing Council when it met in its seventh Special Session in Cartagena in February 2002. The International Environmental Governance (IEG) process, as it became known, was complex and the issues considered included the role of the Global Ministerial Environmental Forum, the financing of UNEP, coordination and effectiveness of MEAs, UNEP's role in capacity building and training, and the relationship between the newly established Environmental Management Group and the GMEF. The very substantial investment of time and energy that UNEP, through the Governing Council process, had devoted to the issue of International Environmental Governance had paid off in a meaningful sense. But no progress was made in Johannesburg on some of the key issues of burning concern to UNEP, such as stable and predictable financing, or better coordination of convention secretariats.[211]

The Johannesburg Plan of Implementation (JPOI) of the WSSD in 2002 recognized Sustainable Consumption and Production (SCP) as an 'overarching objective of and an essential requirement for sustainable development' and for the 10-Year Framework of Programmes (10YFP).[212] The 'Marrakesh Process'[213] was launched in response to this challenge at the first international expert meeting on the ten-year framework held in Marrakesh, Morocco, 16–19 June 2003, organized by UN Division for Sustainable Development and UNEP. The Marrakesh Process included regular global and regional meetings, informal expert task forces and other activities to promote progress on the 10-year framework on sustainable consumption and production. UNEP and UN DESA's Division for Sustainable Development were identified as the leading agencies in promoting and developing the 10-Year Framework of Programmes at the global and regional levels. These actions were seen in the context of an even broader challenge: in promoting the 'green economy'[214] and its pathways to sustainable development and the eradication of poverty. Over the next decade, from 2002 to 2012, the effort to build the 'green economy became one of UNEP's major preoccupations'.

In placing the emphasis on the environment as the foundation for development, Klaus Töpfer had catalysed a new direction and a new understanding of the relevance of UNEP to the international community.[215] Whereas for Steiner[216] (executive director of UNEP who succeeded Töpfer), the initial focus was on implementation, to make UNEP a far more focused and responsive institution so that when governments request action, the system is already aligned to deliver'. To help improve its capacity to deliver, UNEP worked closely with the Nairobi-based UNEP Committee of Permanent Representatives or Ambassadors (CPR) to deliver a document known as the Medium Term Strategy (MTS).[217] Borrowing language from the 1997 Nairobi Declaration, the MTS recalls that the vision of UNEP for the medium-term future is to be

> [t]he leading global environmental authority that sets the global environmental agenda, and promotes the coherent implementation of the environmental dimension of sustainable development within the United Nations system and serve as an authoritative advocate for the global environment.[218]

UNEP's mandate, as defined in the MTS, continued to comprise five[219] overall, interrelated areas:

1 Keeping the world environmental situation under review;
2 Catalysing and promoting international cooperation and action;
3 Providing policy advice and early warning information, based upon sound science and assessments;
4 Facilitating the development, implementation and evolution of norms and standards and developing coherent interlinkages among international environmental conventions;
5 Strengthening technology support and capacity in line with country needs and priorities.

UNEP's Medium Term Strategy 2010–2013, as approved by the Governing Council in 2008, had six cross-cutting thematic priorities.

The selection of those thematic priorities was guided by scientific evidence, the areas in which UNEP has a comparative advantage, the UNEP mandate, priorities emerging from global and regional forums and an assessment of where UNEP could 'make a transformative difference'.

The six cross-cutting thematic priorities[220] are highlighted as follows:

1 Climate change;
2 Disasters and conflicts;
3 Ecosystem management;
4 Environmental governance;
5 Harmful substances and hazardous waste;
6 Resource efficiency – sustainable consumption and production.

Trying to re-orient UNEP's priorities, Steiner, executive director of UNEP, paid tribute to the practical usefulness of the Bali Strategic Plan[221] for Capacity Building and Technology Support. 'The Bali Strategic Plan' essentially gave UNEP a stronger mandate but also the added challenge, particularly from the G77, to be more responsive on the ground, nationally and regionally. This desire for UNEP to do more at the country level was clearly welcomed, but at the time, because of lack of resources, the Bali Strategic Plan was generating tensions and polarization between UNEP's 'normative role and a more down-to-earth role.' By making Bali Strategic Plan an integral part of delivering the overall programme of work, the executive director had come up with 'a solution-making equilibrium between these two functions' (normative and down-to-earth role).[222] The route Steiner chose was to enhance UNEP's relationships with the UN system and international community. And the partnership with United Nations Industrial and Development Organisation (UNIDO) in establishing cleaner production centres, of which there are over 50 today, is one good example. Another key partner for UNEP is UNDP, which gave the argument/logic that UNEP can engage at the country level in a more meaningful way.[223]

One of the most important contributions of Steiner was referred to as the 'strategic presence model',[224] 'which means UNEP is more relevant regionally and nationally'. The UNEP's regional offices have evolved from a representational role to a more substantive and programmatic role and are today an integral part of delivering the programme of work which has strengthened the regional offices with expertise despite restricted resources. 'The *second* part of this strategic presence model is the establishment of country offices in cities such as Brasilia, Beijing and Moscow that reflect these nations' regional and geopolitical importance. UNEP's presence in Addis Ababa and in Brussels allows UNEP to focus on the crucial processes and decisions being taken in the African and European Unions.

> And **thirdly,** the notion of the entire UN system 'delivering as one' includes the role of the UNDP-led resident coordinator system, which means more and more UNEP senior technical advisors embedded in UN country teams and who are engaged in 30 to 40 countries providing support and expertise and working with the UN teams. Hence, the depth of UNEP's support is far more meaningful and relevant for the countries concerned; but sadly, the resources have not been available to take this pillar of an enhanced strategic presence to a higher level.[225]

This also reflects the fact that UNEP's role within the United Nations system has inevitably changed over time. Hence, the strategy was not embedded in a notion of building UNEP as a standalone institution but rather as a service provider, building UNEP as the environment programme of the United Nations serving the mission and the entities of the UN family. One manifestation of this was establishing a SUN team, a Sustainable United Nations team,[226] which has been a service provider to the whole system, that is, climate neutrality, social environmental safeguards and supporting country teams, to assist in making their whole infrastructure more environmentally responsible. Rebuilding the Environment Management Group,[227] a UN system-wide body which UNEP hosts, is another example, where climate neutrality, biodiversity, desertification and other issues are mainstreamed within the UN family. This also reflects the way the world has changed. When UNEP was established, the focus was almost entirely on how to influence the UN system; to an extent this has happened and many agencies now have a larger environment-related portfolio than UNEP has, so influencing this reality is key to future environmental sustainability. A second aspect is that today the environmental issues are fewer issues of projects and more about environmental governance.[228] How does the UN provide a plan for countries to interact on governance issues that are by definition today much more global than they were in the 1970s, 1980s and 1990s? UNEP's role in terms of its relationship to the Secretary-General, to its sister agencies and programmes and funds remains crucial. But increasingly, now member states view UNEP as a platform for national governments' approaches to environmental issues.

'Over the past few years, a great deal of debate was going on about whether UNEP remains UNEP or whether it will take a new form, a UN Environment Organization, in order to achieve better delivery and improved governance. It was clear that 50 years on from Stockholm (1972–2022) and 30 years on from Rio 1992, the balance of emphasis between UNEP's 'normative' role and its other functions had been evolving. It was perhaps easier in the 1970s, 1980s and 1990s to achieve victories on one issue at the time, ozone being the biggest one, but also CITES, Ramsar, CMS and many others. But today, the real challenge is to integrate and reintegrate all these different strands in part because the ability to act on one challenge is now so clearly premised on a myriad of other factors and not just environmental ones. Biodiversity is predicated on action on climate change, and the atmosphere is keenly linked to the condition of the biosphere. And both are impacted by social and economic pathways. This is what has emerged as the greatest challenge.[229]

Another, very significant milestone in the evolution of the role of UNEP has been the launching of the Intergovernmental Platform on Biodiversity and Ecosystem Services (IPBES)[230] at the end of April 2012, which added to the momentum for the UN Conference on Sustainable Development (Rio+20), which was held in June 2012, in Rio de Janeiro, Brazil, which is compared to UNEP's role in launching, together with WMO, the IPCC back in 1988. It was hoped that IPBES would become the key focal point for all agencies and organizations involved in the conservation and sustainable use of biodiversity, long-term human well-being and sustainable development. This Platform would raise the issue on the political agenda in the same way that the IPCC raised the climate change issue. Even after 50 years from now, the establishment of IPBES is seen as another significant 'milestone' for UNEP.

Another turning point was work towards achieving a green economy and sustainable consumption and production[231] which are mutually supportive, and both are currently high on the international agenda. The 10-Year Framework of Programmes on Sustainable Consumption and Production is one of the key themes of the Commission on Sustainable Development's agenda, developed as a consequence of the World Summit on Sustainable Development (2002). Constructing a green economy in the context of sustainable development and poverty eradication is one of the two central themes of the UN Conference on Sustainable Development (Rio+20), which took place in 2012.[232] It was recognized at CSD that the 10 YFP was an important input to the UNCSD, serving as a key building block for the transition to a green economy. This can be achieved through the deployment of cleaner energy technologies and improved access to energy services, improved resource efficiency through investments in cleaner production approaches, increased food security through the use of more sustainable agricultural methods and access to emerging new markets for their green goods and services in developing countries.[233]

In this context, another significant event highlighted was the Belgrade-Helsinki process[234] (UNEP GC in February 2009 in Belgrade, and in November 2010 in Helsinki, held a plenary session on IEG and set up an ad-hoc ministerial process which became known as the Belgrade-Helsinki process), which was running as the preparatory sessions for the Rio+20 Conference (June 2012), where the IEG issue was taken up in a broader perspective regarding an appropriate institutional framework for sustainable development. The authors of the Zero Draft[235] were serious about the proposals regarding the future status of UNEP, and 15 years of discussions on IEG had led to a clear consensus about the future role of UNEP. The proposal for a World Environment Organization (WEO) was supported by the European Union and many African countries. The alternative option of 'strengthening UNEP' also had its advocates. In the absence of a consensus, the authors of the Zero Draft decided to table both options: the creation of a WEO, on the one hand, and 'strengthening UNEP', on the other.[236]

On 21 February 2012 (after the Zero Draft had been tabled), UNEP welcomed the 26th of the Governing Council/Global Ministerial Environmental Forum in its Nairobi headquarters which coincided with UNEP's 40th-anniversary celebrations.[237] One highlight of these anniversary celebrations was undoubtedly the Governing Council's session of 21 February, at which three former executive directors (Tolba, Dowdeswell and Töpfer) joined Steiner, to reflect on UNEP's past achievements and future challenges. Various proposals were debated, and some of them were (1) consensus on global reform for sustainable development where about 120 countries have endorsed establishing UNEP as a specialized agency, emphasizing that its focus should be on helping member states meet their environmental commitments, and affirming that governance reform must nurture a robust, green economy (Malaysia),[238] (2) there was a call to transform UNEP into a specialized agency, noting that the current system of governance includes many binding agreements, but without the systems to monitor and enforce implementation (a powerful proposal from Peru),[239] (3) a proposal that called for a specialized agency on environment that would provide financial, technical and scientific support to developing countries and 'this kind of architecture' would best coordinate all MEAs, stating that UNEP's current mandate is not broad enough to fulfil this function (Republic of Congo).[240] (4) Another proposal noted that a combination of assessed contributions, voluntary contributions and private sector funding is imperative to the running of a new 'anchor institution' that would enhance oversight and coordination of MEAs(Switzerland).[241]

> During discussions, some delegates favoured establishing UNEP as a specialized agency, calling for: an institution with a strong mandate, political visibility and universal membership; effective use of resources; strengthening the scientific basis for decision making; and improving the science/policy interface. Others dissented; calling instead for 'strengthening' UNEP, with some emphasizing that UNEP should remain the 'voice of the environment' and not broaden its scope into sustainable development as a whole.[242]

As the 26th session of the UNEP Governing Council ended, at least on two key issues of the Rio+20 agenda, namely the Green Economy and IFSD, no consensus emerged.

Another turning point was (before Rio+20 Conference was due to begin) UNEP published its fifth edition of the Global Environmental Outlook (GEO-5).[243] GEO-5 assessed 90 of the most important environmental goals and objectives and found that significant progress had only been made in four, namely eliminating the production and use of substances that deplete the ozone layer, removal of lead from fuel, increasing access to improved water supplies and boosting research to reduce pollution of the marine environment.[244] GEO's goal, historically, has been to produce scientifically credible and policy-relevant assessments of the state of the global environment and to enhance the capacity of a wide range of actors to perform integrated environmental assessments[245] to establish a UN-specialized agency for the environment with universal membership of its Governing Council, based on UNEP, with a revised and strengthened mandate, supported by stable, adequate and predictable financial contributions and operating on an equal footing with other UN-specialized agencies.

As the negotiations continued in the UNCSD Preparatory Committee, the Zero Draft turned into another turning point with regard to the future of UNEP. The Institutional Framework for Sustainable Development or IFSD in the Zero Draft of 10 January 2012 – had contained two options as regards UNEP's possible future, namely 'strengthening UNEP' or 'creating a World Environment Organization (WEO)'.[246] Member countries agreed to establish a UN-specialized agency for the environment with universal membership of its Governing Council, based on UNEP, with a revised and strengthened mandate, supported by stable, adequate and predictable financial contributions and operating on an equal footing with other UN-specialized agencies. This agency, based in Nairobi, would cooperate closely with other specialized agencies, whereas the text of 'The Future We Want', dated 22 May 2012,[247] retains the two basic options of strengthening UNEP: on the one hand, the setting up of a World Environment Organization, on the other, strengthening UNEP. As compared with the 10 January 2012 'Zero Draft', the proposals for strengthening UNEP are worked out in greater detail, while the central thrust of achieving 'universal' membership and 'increased' financial resources is maintained. In the final outcome document, the WEO option has disappeared. Only one option, that is, the 'strengthening' of UNEP, emerged. The relevant text of 'The Future We Want', as it emerged from the Brazil-led negotiating process before the conference began, was formally endorsed in its entirety when the conference ended on 22 June (The Future We Want).

The United Nations Conference on Sustainable Development, which is also known as Rio+20, took place in Rio de Janeiro, Brazil, from 20–22 June 2012, which resulted in a political outcome document that contains clear and practical measures for implementing sustainable development. Member states decided to launch a process to develop a set of Sustainable Development Goals, which were building upon the Millennium Development Goals (MDGs) and converged with the post-2015 development agenda.[248] The conference also adopted groundbreaking guidelines on green economy. *Green economy*[249] in the context of sustainable development and poverty eradication contributes to meeting key goals, in particular the priorities of poverty eradication, food security, sound water management, universal access to modern energy services, sustainable cities, management of oceans and improving resilience and disaster preparedness, as well as public health, human resource development and sustained, inclusive and equitable growth that generates employment, including for youth.[250] It is based on the Rio principles, in particular, the principle of common but differentiated responsibilities,[251] and is people-centred and inclusive, providing opportunities and benefits for all citizens and all countries. Further, the green economy is viewed as a means to achieve sustainable development, which is the overarching goal, and a green economy in the context of sustainable development and poverty eradication also protects and enhances the natural resource base, increases resource efficiency, promotes sustainable consumption and production patterns, and moves the world towards low-carbon development.[252] Governments also decided to establish an intergovernmental process under the General Assembly to prepare options on a strategy for sustainable development financing which strengthens the UNEP on several fronts and agreed to establish a high-level political forum for sustainable development. Governments also adopted the 10-Year Framework of Programmes on Sustainable Consumption and Production patterns, as contained in document A/CONF.216/5, and invited the General Assembly, at its 67th session, to designate a member state to take necessary steps to fully operationalize the framework.[253] The Rio+20 Conference also galvanized the attention of thousands of representatives of the UN system and major groups. In this context, the United Nations Summit for the adoption of the post-2015 development agenda,[254] which is a significant development in synthesizing the inputs on sustainable development, was held from 25 to 27 September 2015, in New York, which was convened as a high-level plenary meeting of the General Assembly. The process of arriving at the post-2015 development agenda was member state-led with broad participation from major groups and other civil society stakeholders. Numerous inputs to the agenda have emerged, notably a set of Sustainable Development Goals (SDGs) proposed by an open working group of the General Assembly, the report of an intergovernmental committee of experts on sustainable development financing, GA dialogues on technology facilitation and many others. A synthesis report before the end of 2014 as a contribution to the intergovernmental negotiations in the lead-up to the Summit, emerged where the United Nations and UNEP have played a facilitating role in the global conversation on the post-2015 development agenda and supported broad consultations. In this process of developments, on the post-2015 UN development agenda; it brought together more than 60 UN agencies and international organizations.

A very significant recent development has been the universal membership of the Environment Assembly of UNEP. Earlier, UNEP was administered by a 58-member Governing Council, but at the Rio+20 UN Conference on Sustainable Development in 2012, 'universal membership' was established for UNEP, to include the full 193 member states of the UN. This is a significant development in the evolution of environmental governance, which replaced UNEP's Governing Council with the UN Environmental Assembly (UNEA)[255] of UNEP, composed of the full 193 UN members. UNEA acts as a parliament of the environment and sets UNEP's agenda. The first meeting of UNEA (UNEA-1) was held in June 2014. UNEA-2 was held in May 2016, UNEA-3 in December 2017 and UNEA-4 in March 2019. UNEA-5 is scheduled for February and was held in February 2021.[256]

The most recent UNEA session was held during the raging pandemic. The overall theme of UNEA-5 is 'Strengthening Actions for Nature to Achieve the Sustainable Development Goals'.[257] The theme calls for strengthened action to protect and restore nature and the nature-based solutions to achieve the SDGs in its three complementary dimensions (social, economic and environmental). UNEA-5 provides member states and stakeholders with a platform for sharing and implementing successful approaches that contribute to the achievement of the environmental dimension of the 2030 Agenda and the SDGs, including the goals related to the eradication of poverty and sustainable patterns of consumption and production. UNEA-5 also provides an opportunity to take ambitious steps towards building back better and greener world by ensuring that investments in economic recovery after the COVID-19 pandemic contribute to sustainable development. In light of the restrictions related to the COVID-19 pandemic, and based on broad consultations with the UNEA Bureau, member states and stakeholders, the fifth session of the UN Environment Assembly (UNEA-5) has decided that UNEA-5 will take place in a two-step approach.[258] The first session (UNEA-5.1) was held online on 22–23 February 2021 with a revised and streamlined agenda that focused on urgent and procedural decisions. Substantive matters that require in-depth negotiations are deferred to a resumed in-person session of UNEA-5 on 28 February– 2 March 2022 (UNEA-5.2). UNEA-5.2 is followed by a Special Session of the UN Environment Assembly (UNEA-SS), held on 3–4 March 2022, which was devoted to commemorate the 50th anniversary of the creation of UNEP in 1972.[259] UNEA-5 is an opportunity for member states to share best practices for sustainability. It creates a momentum for governments to build back better after the COVID-19 pandemic through green and sustainable recovery plans.

Summary

This chapter has raised various important issues while dealing with the evolution, growth and the role of UNEP covering the period from 1972 to 2022. Role of environmental movements, that is, British and American environmental movements, influence of Rachel Carson, significance of new environmentalism and its impact on the creation and evolution of UNEP, has been highlighted. Developed and developing countries' views on various conventions and agreements, achievements and outcomes of both Stockholm and Rio Conferences, which are foundational to future environmental negotiations and governance, have been analysed. Further, the chapter has explored the role of UNEP, highlighting the major achievements and shortcomings/challenges covering the period from 1972 to 2022. In conclusion, it has been observed that strong political commitment is required as support for a strengthened UNEP, which has demonstrated skill in the pursuit of environmental governance and security, in spite of its limited mandate and lack of funding. Greater financial commitments are required, both for the UN environmental security network and for overcoming the causes and symptoms of environmental insecurity. For the environment, this is a time for dynamic change where policy tools are continuously developed and adapted to new issues and challenges. As the United Nations organization is mandated to play a central role on environmental matters, UNEP must also adapt and respond to new priorities and solutions.

Since UNEP's beginnings, there has been a substantive shift in environmental priorities and the policies used to address them. Although much remains to be done in the face of the traditional pollution-related problems, the policies needed to prevent and cure these are increasingly being refined and implemented. In recent years, a more complex set of environmental challenges have come to the top of the agenda, including climate change, biodiversity and the sustainable management of forests, oceans, freshwater and land resources.

The Stockholm Declaration/Rio Declaration set out a clear focus for UNEP's work, and the practical implementation of this will be a major mark of UNEP's progress in becoming a powerful global player in the twenty-first century.

Notes

1 Johnson, Ian, 'Keeping the Flame', *Our Planet* 10(2) (1999), p. 32.
2 Ibid., p. 31.
3 Inglehart, Ronald, *The Silent Revolution: Changing Values Among Western Publics* (Princeton: Princeton University Press, 1977).
4 Ibid.
5 Ophuls, William, *Ecology and the Politics of Scarcity* (San Francisco: W.H. Freeman, 1977); Pirages, Davis C., *The Sustainable Society* (New York: Praeger, 1977); Rifkin, Jeremy Entropy, *A New World View* (New York: Viking Press, 1980); Robertson, J., *The Same Alternative: A Choice of Futures* (St. Paul, MN: River Basin, 1979).
6 Pirages, op. cit., p. 26.

7 Yankelovich, Daniel and Bernard Lefkowitz, 'The Public Debate on Growth: Preparing for Resolution', *Technological Forecasting and Social Change* 17(2) (June 1980), pp. 95–140.
8 Dunlap, Riley E. and Kent D. Van Liere, 'The New Environmental Paradigm': A Proposed Measuring Instrument and Preliminary Results', *Journal of Environmental Education* 9 (1978), pp. 10-19; Cot grove, Stephen, *Catastrophe or Cornucopia: The Environment, Politics and the Future* (Chichester: John Wiley and Sons, 1982); Milbrath, Lester W., *General Report: US Components of a Comparative Study of Environmental Beliefs and Values* (Buffalo: State University of New York Environmental Studies Centre, 1981).
9 Milbrath, Lester W., *Environmental Beliefs and Values in Political Psychology,* (ed.) Margaret G. Hermann (San Francisco: Jossey-Bass, 1986).
10 Holdgate, Martin, Gilbert White and Mohammed Kassas, *The World Environment, 1972-82* (Dublin: Tycooly, 1982), p. 626.
11 Sandbrook, Richard, 'The UK's Overseas Environmental Policy', in *The Conservation and Development Programme for the UK: A Response to the World Conservation Strategy* (London: Kogan Page, 1983), p. 322.
12 Ibid.
13 McCormick, John, *The International Environmental Movement: Reclaiming Paradise* (London: Belhaven, 1989), p. 92.
14 Worster, Donald, *Nature's Economy* (San Francisco: Sierra Club Books, 1977), pp. 170–171.
15 Ibid., p. 179.
16 Lowe, Philip, 'Values and Institutions in the History of British Nature Conservation', in Andrew Warren and F.B. Goldsmith (eds.), *Conservation in Perspective* (London: Wiley, 1983), p. 337.
17 Ibid., p. 333.
18 Sheail, John, *Nature in Trust* (London: Blackie, 1976), p. 9.
19 Allen, David Elliston, *The Naturalist in Britain* (Harmondsworth: Penguin, 1978), pp. 197–198.
20 Lowe, op. cit., p. 331.
21 Sheail, op. cit., p. 12.
22 Ibid., p. 13.
23 Allen, op. cit., pp. 199–200.
24 Sheail, op. cit., p. 60.
25 Ibid., p. 63.
26 Ibid., p. 70.
27 George Perkins Marsh (1801–1882) was born in Vermont and taught briefly before entering law, then politics. As a wing member of Congress (1843–1859), he helped found and shape the Smithsonian Institution. He was United States Minister to Turkey (1848–1854) and to Italy (1861–1882).
28 Marsh, George Perkins, *Man, and Nature* (Reprint; Cambridge: Harvard University Press, 1965), p. 36.
29 Lowenthal, David. In Introduction, Ibid., xxii.
30 Udall, Stewart L., *The Quiet Crisis* (New York: Holt, Rinehart and Winston, 1963), p. 94.
31 Huth, Hans, *Nature and the American: Three Centuries of Changing Attitudes* (Berkeley: University of California Press, 1957), p. 148.
32 Thoreau, Henry David, *The Maine Woods* (Reprint; New York: W.W. Norton, 1950), p. 321.
33 Nash, Roderick, *Wilderness and the American Mind* (New Haven: Yale University Press), 1973, p. 102.
34 Marsh, op. cit., p. 203.
35 John Muir (1838-1914) was born in Dunbar, Scotland, and emigrated with his family to Wisconsin in 1849. In mid-life he took to writing and was instrumental in the

protection not only of Yosemite but of petrified forest and Grand Canyon National Parks in Arizona.
36 Nash, op. cit., p. 132.
37 Gifford Pinchot (1865–1946) was born in Connecticut and brought up in Paris and Pennsylvania. After education at Yale, he studied forestry in Germany and France before setting up a forestry consultant in New York. He was appointed to the board of the new National Forestry Commission in 1896. Pinchot first met John Muir in 1863; the two were to become bitter opponents.
38 Pinchot, Gifford, *The Fight for Conservation* (New York: Doubleday Page and Co., 1910), pp. 40–52.
39 Worster, Donald, *Nature's Economy* (San Francisco: Sierra Club Books, 1977), pp. 267–268.
40 McConnell, Grant, 'The Conservation Movement: Past and Present', *Western Political Quarterly* 7(3) (Sept. 1954), pp. 463–478.
41 Jones, Maldwyn A., *The Limits of Liberty* (New York: Oxford University Press, 1983), p. 369.
42 Bates, James L., 'Fulfilling American Democracy: The Conservation Movement, 1907-1921', *Mississippi Valley Historical Review* 44 (June 1957), pp. 29–57.
43 Fox, Stephen, *John Muir and His Legacy: The American Conservation Movement* (Boston: Little Brown and Co., 1981), p. 128.
44 Theodore, Roosevelt, *An Autobiography* (New York: Macmillan, 1913), p. 402.
45 Report of the National Conservation Commission, Senate Doc. 676, 60th Cong., 2nd Sess., 1 quoted in McGeary, Martin N., *Gifford Pinchot: Forester Politician* (Princeton, NJ: Princeton University Press, 1960), p. 100.
46 Boardman, Robert, *International Organization and the Conservation of Nature* (Bloomington: Indiana University Press, 1981), p. 27.
47 Editors, *Executive Office of the President: Eleventh Annual Report of the Council on Environmental Quality, 1980* (Washington, DC: US Government Printing Office, 1980), pp. 418–419.
48 Lowe, Philip and Jane Goyder, *Environmental Group in Politics* (London: George Allen and Unwin, 1983), p. 37.
49 Worster, op. cit., p. 261.
50 Petulla, Joseph M., *American Environmentalism: Values, Tactics, Priorities* (College Station: Texas A and M University Press, 1980).
51 Ibid.
52 Riordon, Timothy O', *Environmentalism* (London: Pion Limited, 1981), pp. 1-19, 375-377.
53 McGeary, Martin N., *Gifford Pinchot: Forester-Politician* (Princeton: Princeton University Press, 1960), p. 426.
54 Nixon, Edgar B. (ed.), *Franklin D. Roosevelt and Conservation 1911-1945*, Vol.2 (New York: Franklin D. Roosevelt Library, 1957), p. 599.
55 UNSCCUR, UNSCCUR Memorandum from Chairman to Members of NRC Co. on UNESCO, 8–9 quoted in Boardman, Robert, *International Organization and the Conservation of Nature* (Bloomington: Indiana University Press, 1981), p. 39.
56 IUCN Bulletin, No.6 (Jan./Mar. 1963), p. 7.
57 Carson, op. cit., p. 2.
58 Graham, Frank, *Since Silent Spring* (Boston: Houghton Mifflin, 1970), p. 17.
59 Ibid., p. 308.
60 Carson, op. cit., p. 8.
61 Time, Jan. 4, 1971, pp. 21–22.
62 Life, Jan. 30, 1970, p. 23.
63 Zinger, Clem L., Richard Dalsemer, and Helene MA gargle, *Environmental Volunteers in America* (Washington, DC: EPA, 1973), p. 5.
64 Nash, op. cit., pp. 251–253.
65 Ibid., p. 252.
66 Lotgrove, Stephen, *Catastrophe or Cornucopia* (London: Wiley, 1982), p. 5.

67 Maddox, John, *The Doomsday Syndrome* (London: Macmillan, 1972), p. 135.
68 Pursell, Carrol (ed.), *From Conservation to Ecology: The Development of Environmental Concern* (New York: Thomas Y. Crowell Co., 1973), p. 4.
69 United Nations Economic and Social Council, Annexes, Agenda Item 12 (Doc. E/4466/Add.1) at 2 (New York: ECOSOC, 1968).
70 Doud, Alden L., 'International Environmental Development: Perceptions of Developing and Developed Countries', *Natural Resources Journal* 12 (October 1972), pp. 520–529.
71 Stone, Peter, *Did We Save The Earth at Stockholm?* (London: Earth Island, 1973), p. 18.
72 United Nations, Resolution 2398 (XXIII) of the General Assembly, Dec. 3, 1968.
73 United Nations, *Yearbook of the United Nations 1970* (New York: Office of Public Information, 1970), p. 449.
74 Kennan, George F., 'To Prevent a World Waste Land: A Proposal', *Foreign Affairs* 48 (1970), pp. 401–413.
75 Contini, Paolo and Peter H. Sand, 'Methods to Expedite Environmental Protection', *American Journal of International Law* 66 (1972), pp. 37–59.
76 Brian, Johnson, 'The United Nations Institutional Response to Stockholm: A case study in the International Politics of International Change', *International Organization* 26 (1972), pp. 255–301.
77 Ibid., p. 300.
78 United Nations, Resolution, A/CONF. 48/PC. 6, of General Assembly New York, 20 March 1970.
79 United Nations, Resolution, A/CONF. 48/PC. 9, of General Assembly Geneva, 19 February 1971.
80 McCormick, John, *Reclaiming Paradise: The Global Environmental Movement.* (Bloomington: Indiana University Press, 1989), p. 92.
81 United Nations, General Assembly Resolution, A/CONF. 48/PC. 17, New York, 17 March 1972.
82 McCormick, op. cit., p. 94.
83 Meadows, D.H., *The Limits to Growth* (New York: University, 1972), p. 32.
84 UNEP, *Compendium of Legislative Authority' United Nations Environment Programme* (Nairobi, 1978), p. 17.
   Review of the Areas of Environment and Development and Environmental Management, UNEP Report No.3, Nairobi, 1978, p. 1.
85 See Stanely Johnson's The First Forty years of UNEP: A Narative, op. cit. (Chapter-2: Finding a Home), p. 33. (Extracts from Indian Prime Minister Indira Gandhi's speech to the Plenary session of Stockholm Conference on june 8, 1972 are printed in Environment, Stockholm published by the Centre for Economic and Social Information, UN Geneva,1972).
86 Trivedi, Priya Ranjan, *International Environmental Laws* (New Delhi: Aph. Pub. In Association with Indian Institute of Ecology and Environment, 1996), p. 29.
87 Almeida, Miguel Ozorio de, *Environment and Development: The Founex Report on Development and Environment*, Issue 586 of International Conciliation (New York: Carnegie Endowment for International Peace, 1972).
88 United Nations, *Year Book of the United Nations 1972* (New York: Office of Public Information, United Nations, 1972), p. 319.
89 Eckholm, Erik, *Down to Earth* (New York: W.W. Norton, 1982), xii.
90 Talbot, Lee M., 'A Remarkable Melding of Contrasts and Conflicts', *Uniterra* 1 (1982), p. 2.
91 Holdgate, Martin, 'Beyond the Ideals and the Vision', *Uniterra* 1 (1982), p. 3.
92 Keith, Johnson, 'A Second Copernican Revolution', *Uniterra* 1, (1982), pp. 4–5.
93 Aaronson, Terri, 'World Priorities', *Environment* 14(6) (July/Aug. 1972), pp. 4–13.
94 Landsberg, Hans H., *Reflections on the Stockholm Conference* (Unpubl. Paper, Washington, DC, Aug. 1972), p. 2.
95 Aaronson, op. cit., p. 14.
96 Ibid., p. 20.
97 Ibid., p. 100.

98 Haley, Mary Jean (ed.), *Open Options: A Guide to Stockholm's Alternative Environmental Conferences* (Stockholm: 29 May 1972), p. 3.
99 Landsberg, op. cit., p. 4.
100 Talbot, op. cit., p. 3.
101 Holdgate, op. cit., p. 2.
102 Barbara Ward, Interview with John Tinker, 1987 (Unpublish.), p. 5.
103 Ibid., p. 11.
104 Eric, op. cit., pp. xi–xii.
105 Talbot, op. cit., p. 1.
106 Holdgate, op. cit., p. 7.
107 Trivedi, Priya Ranjan 1996, 'Global Change and our Future', op. cit., p. 27.
108 Ibid., p. 28.
109 Sandbrook, op. cit., p. 390.
110 United Nations, Resolution 2997 (XXVII) of the General Assembly, Dec. 15, 1972.
111 United Nations, Resolution 3004 (XXVII) of the General Assembly, Dec. 1972.
112 Sandbrook, op. cit., p. 390.
113 Ibid., p. 391.
114 Ibid., p. 392.
115 United Nations Environment programme, *Earth watch: An In-depth Review*, UNEP Report No.1 (Nairobi: UNEP, 1981).
116 This was the title of a report subtitled 'Caring for a Small Planet' written by Barbara Ward and Rene Dubos, commissioned by the secretary-general of the Stockholm-Conference, Maurice Strong.
117 UNEP has its headquarters in Nairobi, in response to developing countries concerns that yet another UN body would be located in New York, Geneva or Vienna.
118 Birnie, Patricia, 'International Environmental Law: Its Adequacy for Present and Future Needs', in Andrew Hurrell and Benedict Kingsbury (eds.), *The International Politics of the Environment* (Oxford: Clarendon Press, 1992), and Imber, Mark, 'Too Many Cooks? The Post Rio Reform of the United Nations', *International Affairs* 69(1) (Jan. 1993), pp. 55–70.
119 The Toronto Conference proposed that there should be a World Atmosphere Fund financed through a levy on the fossil fuel consumption of the industrialized countries.
120 Preliminary discussions about a climate change convention began under IPCC, until the General Assembly took this over through its formal establishment of the Intergovernmental Negotiating Committee.
121 'Environment and Development', The Founex Report, International Conciliation (New York), no.586, Jan. 1972.
122 Grubb, M., M. Koch, A. Munson, F. Sullivan and K. Thompson, *The Earth Summit Agreements: A Guide and An Assessment* (London: Earthscan Publications, 1993), p. 15.
123 Paragraph 1 and Paragraph 3 of Resolution 44/228.
124 The issue of NGO participation had been a contentious one in the early Prep-coms. (Preparatory Committees). Agreement was finally reached to extend the narrow ECOSOC accreditation categories and enable any NGO which could demonstrate a relevance to the issues under discussion to apply for accreditation.
125 'The Global Forum' was held at Flamengo Park, 40 km from the main conference at Rio Centro Convention Centre, which was on the outskirts of the city.
126 Hurrell, Andrew and Benedict Kingsbury (eds.), *The International Politics of the Environment* (Oxford: Clarendon Press, 1992), p. 34, pp. 217–218.
127 Halpern, Shannan L., *The United Nations Conference on Environment and Development: Process and Documentation* (Providence, RI: Academic Council for the UN System, Report no.2 1992).
128 Fletcher, Susan, *Earth Summit Summary: United Nations Conference on Environment and Development, Brazil* (Washington, DC: Congressional Research Service, The Library of Congress, Report no.92–347 ENR, 1992), p. 2.

129 In print some 800 pages.
130 Tickell calls it a 'rag-bag of points judged to be important in the next century'. Tickell, Crispin, 'The Inevitability of Environmental Insecurity', in Gwyn Prins (ed.), *Threats without Enemies: Facing Environmental Insecurity* (London: Earthscan Publications, 1993), p. 23.
131 Its full title is 'Tickell, the "Non legally Binding Authoritative Statement of Principles for a Global Consensus on the Management, Conservation and Sustainable Development of all Types of Forests"', p. 23.
132 There were some creative diplomatic compromises in the wording of Agenda 21. The Rio declaration contains references to the need to protect the environment of 'People under Occupation', a phrase initially included in Agenda 21 as well. To forestall the United States and Israel's opening debate on the Declaration, the words were removed from Agenda 21 which simply says that Agenda 21 should be carried out in respect of all the principles in the Declaration. Similarly, because of Saudi Arabia and Kuwait's complaint the words 'environmentally sound energy systems' were removed and the words 'environmentally sound energy systems' were added.
133 Ibid.
134 Kothari, Ashish, 'A Mixed Bag: The Biodiversity Convention', *Frontline*, June 5, 1992, p. 113.
135 Ibid., p. 114.
136 Shiva, Vandana, 'Earth Summit: Agenda against 'Green Imperialism', *Frontline*, 5 June 1992, p. 110.
137 Ibid., p. 111.
138 Shiva, Vandana, 'The Road from Rio: Greenwash at the Earth Summit', *Frontline*, 3 July 1992, p. 106.
139 The Convention's full title is the 'UN Convention to Combat Desertification in those countries experiencing serious drought and desertification, particularly Africa'.
140 Environmental concerns have been raised in other UN Conferences, including the Population Conference in Cairo, the Copenhagen Social Summit and the fourth World Conference on Women in Beijing.
141 UN Information Centre, 'Earth Summit Summary 1992', 15 July 1992. Strong has argued that the disparity between rich and poor, unsustainable patterns of production and consumption and population growth had not really changed as a result of Rio.
142 Strong, Maurice F., *Text of Opening Statement, Secretary-General of the United Nations Conference on the Human Environment* (Stockholm, Sweden: First Plenary Meeting, 5 June 1972). (Association for Progressive Communications, UNCED-92, Global Electronic Network: UNCED documents).
    Tickell, 'The Inevitability of Environmental Insecurity', p. 23. Tickell, among others, is critical of the UNCED outcomes. He argues that 'final documents were badly drafted stuffed with jargon, politically in correct on the canons of UN-speech, shot through with ambiguities and lacking measurable commitment'.
143 Not all developed countries have affirmed this target of the United States – and therefore 'reaffirmation means little in those cases'. The target year for achieving the 0.7 per cent figure is 2000. Concerns have also been raised at the confirmation of the Global Environment Facility as the key funding mechanism for the UNCED conventions.
144 Rowlands, Ian H., 'The International Politics of Environment and Sustainable Development: The Post-UNCED Agenda', *Millennium* 21(2) (Summer 1992), p. 223.
145 Childers, Erskine, 'The Future of the United Nations: The Challenges of the 1990s', *Bulletin of Peace Proposals* 21(2) (1990), p. 147.
146 Hurrel and Kingsbury (eds.), op. cit., p. 31.
147 Williams, Maurice, 'Guidelines to Strengthening the Institutional Response to Major Environmental Issues', *Development* 1992(2) (1992), p. 24.

148 Imber, Mark, 'Too many Cooks? The Post-Rio Reform of the United Nations' *International Affairs* 69(1) (Jan. 1993), p. 115. Brenton, The Greening of Machiavelli, p. 50.
149 MacNeil, Jim, 'The Greening of International Relations', *International Journal* XLV(1) (Winter 1989–1990), pp. 1–35, PRIO/UNEP, *Environmental Security: A Report Contributing to the Concept of Comprehensive International Security* (Oslo: Peace Research Institute Oslo/United Nations Environment Programme, 1989).
150 Hurrell and Kingsbury (eds.)., op. cit., p. 22.
151 Mische, Patricia, 'Ecological Security' and the Need to Reconceptualize Sovereignty', *Alternatives* XIV (1989), pp. 389–427.
152 Tinker, Catherine J., 'NGOs and Environmental Policy: Who Represents Global Society?' Paper presented to the Annual meeting of the International Studies Association, Mar. 1993, p. 16.
153 UNGA Resolution A/Res/47/191, Dec. 1992, UNEP: Finance Initiative, Chapter II in Stanley Johnson in UNEP: The First 40 Years: A Narrative 2012, p. 143.
154 UNGA 5 Year Review of Earth Summit in 1997 WSSD, Johannesburg, South Africa August 26–4 September 2002.
155 See Johannesburg Plan of Implementation, paras 127–132.
156 United Nations Conference on Sustainable Development Rio+20 2012 (High Level Political Forum replaced CSD).
157 Johnson, op. cit., the United Nations Conference on Environment and Development (Chapter 10), pp. 139–141.
158 Ibid., Chapter 11, p. 142.
159 Ibid., Chapter 11, p. 143.
160 Lars-Goran Engfeldt 'From Stockholm to Johannesburg and Beyond' published in 2009 by the Swedish Ministry of Foreign Affairs, pp. 217.
161 See Stanley Johnson, Desertification POPs, GEO-1, UNEP Finance Initiative (Chapter 11), pp. 143–157.
162 Ibid., p. 144.
163 Ibid., para 8, p. 144.
164 Ibid., para 9, p. 144.
165 Ibid., para 10, p. 144.
166 Ibid., para 11, p. 144.
167 Ibid., p. 145.
168 Ibid., para 2, p. 145.
169 Ibid., para 3, p. 145.
170 'Financing International Environmental Governance: Lessons from the United Nations Environment Programme' by Maria Ivanova, assistant professor of global governance and co-director, Centre for Governance and Sustainability, John W. McCormack, Graduate School of Policy and Global Studies, University of Massachusetts, Boston, USA.
171 Ivanova, op. cit.
172 Ibid.
173 Stanley, (Chapter 11), op. cit., para 6.
174 Ibid. (Chapter 11), para 7.
175 Ibid., para 8.
176 Stanley, op. cit., See Chapter 11.
177 Ibid., p. 76.
178 Ibid., p. 75.
179 The GEF Council is the main governing body of the GEF. It functions as independent board of directors, with primary responsibility for developing, adopting and evaluating GEF programmes. Council members representing 32 constituencies (16 from developing countries, 14 from developed countries and 2 from countries with transitional economies) meet twice each year for three days and also conduct business by mail. All decisions are taken by consensus.

180 Young, Zoe, *A New Green Order? The World Bank and the Politics of the Global Environmental Facility* (London: Pluto Press, 2002), pp. 7–8.
181 In addition to Elizabeth Dowdeswell, former executive directors Tolba and Töpfer were present at the UNEP 40th Anniversary Celebrations as well as Mr. Achim Steiner. Mr. Maurice Strong was unable to attend but sent a message which was delivered on his behalf to the assembled ministers and delegates.
182 Stanley, op. cit., p. 147.
183 Ibid., pp. 147–148.
184 Ibid., pp. 148–150.
185 The World Environment, 1972–1992, Two Decades of Challenge, edited by Mostafa Tolba, Osama A. El-Kholy et al., Chapman and Hall, London, 1992, p. 271.
186 The World Environment, 1972–1992, p. 134.
187 See Stanley, op. cit., pp. 148–149.
188 Ibid., pp. 148–150.
189 UN Environment (1997) Global Environment Outlook: For Life on Earth (https://wedocs.unep.org).
190 Accessed internet on 25 May 2022, www.unenvnrionment.org/Globalenvironmentoutlook.
191 Ibid., Conclusions.
192 Accessed internet on 15 May 2022, www.unenvironment.org/Globalenvironment outlook.
193 Klaus Töpfer took over from Elizabeth Dowdeswell as executive director of UNEP in February 1998. From 1987 to 1994 he had been federal minister for the Environment, Nature Conservation and Nuclear Safety in the German government and in that capacity had been closely involved with UNEP affairs. He played a major role during 1992 Rio Conference in formulating forest principles and also had chaired in 1994 the second session of CSD.
194 United Nations General Assembly Resolution A/RES/53/242 of 28 July 1999. Report of the Secretary-General on Environment and Human Settlements.
195 The Environment Management Group was approved by the United Nations General Assembly in paragraph 5 of its resolution 53/242 of 1999 on the basis of proposals by secretary-general and the report of a UN Task Force headed by executive director of UNEP, Mr. Klaus Töpfer. UNEP is to provide the Secretariat for the Environment Management Group.
196 Some progress was made in this direction during the Rio+20 process, which recommended a broadening of GEF's mandate to include financing national actions designed to implement international environment commitments. See 'The Future We Want' A. Conf/216. L.1 of June 19, 2012.
197 Stanley Johnson's 40 years of UNEP, op. cit., (Chapter 12), p. 163.
198 Ibid., pp. 162.
199 Ibid., pp. 160–171.
200 See 'The Future We Want', A/Conf/216 L.1 of 19 June 2012.
201 Nairobi Declaration on the Role and Mandate of the United Nations Environment Programme, UNEP, Governing Council (19th Session: 1997: Nairobi).
202 The First Global Ministerial Environment meeting was held in Malmo, Sweden, from 29 to 31 May 2000, in pursuance of United Nations General Assembly Resolution 53/242 of 28 July 1999 to review important and emerging environmental issues and to chart the course for the future.
203 Malmo Ministerial Declaration of 31 May 2000 was adopted by the Global Ministerial Environment Frozen, Sixth Special Session of the Governing Council of the United Nations Environment Programme, Fifth Plenary Meeting, 31 May 2000.
204 UN General Assembly Resolution a/Res/55/2 of 18 September 2000: United Nations Millennium Declaration.

205 United Nations (2004), Millennium Development Goals: Progress Report, UN Dept. of Economic and Social Affairs.
206 UN General Assembly Resolution A/Res/55/2 of 18 September 2000: UN Millennium Declaration.
207 Report of the World Summit on Sustainable Development, Johannesburg, South Africa, 24 August–4 September 2002 (United Nations Publication, Resolution 1, Annex 1).
208 UNEP/GC/SS VII/1 of 15 February 2002.
209 See Johannesburg Plan of Implementation pp. 127–132.
210 Engfeldt, op. cit., pp. 245–247.
211 Ibid., p. 271.
212 The First Meeting developed to developing IOYFP took place in Marrakech, Morocco, in June 2003.
213 The Marrakesh process was launched in 2003 and is a multi-stakeholder process to support implementation of Sustainable Consumption and Production (SCP) and for the elaboration of a 10-Year Framework of Programmes in SCP. UNEP and UNDESA are serving as the Secretariat to co-ordinate their global process.
214 Towards a Green Economy: Pathway to Sustainable Development, Poverty Eradication, UNEP 2011.
215 On taking over as UNEP's executive director Topper had decided upon a substantial reorganization of UNEP shifting the emphasis from theme-based (e.g., air, water, waste management) programme activity centres (PACs) towards process goals, e.g., early warning and assessment, implementation and public information.
216 Achim Steiner took up his position as UNEP's fifth executive director in June 2006. A graduate of Oxford University Steiner had an outstanding record in the field of conservation and sustainable development. Before coming to UNEP, he had served for five years as the director-general of IUCN with which UNEP had a long and productive relationship. (See Johnson Stanley Chapter 16.)
217 UNEP SS X/3, Medium Term Strategy for the Period 2010–2013.
218 Stanley, op. cit., pp. 155–156.
219 Medium Term Strategy UNEP's Mandate.
220 Ibid.
221 Stanley, op. cit., p. 208.
222 Ibid., p. 209.
223 Ibid., p. 210.
224 Ibid., pp. 210–230.
225 Ibid., pp. 210–212.
226 Ibid., p. 211.
227 Ibid., p. 212.
228 Ibid., pp. 211–215.
229 Ibid., p. 213.
230 www.ipbes.net/See pp. 215–217, Accessed internet on 28 May 2022.
231 www.unep.org/greeneconomy/AboutGEI/ See pp. 218–228, Accessed internet on 29 May 2022.
232 Report of the WSSD 2012 (United Nations Publication).
233 www.unep.org/greeneconomy, Accessed internet on 28 May 2022.
234 Belgrade-Helsinki Process.
235 United Nations, 12 January 2012 'The Future We Want', Rio+20.
236 John Scanlon in 'Enhancing Environmental Governance for Sustainable Development: Function Oriented Option' Governance and Sustainability Brief Series, Mar. 2010, John. W. McCormack School of Policy University of Massachusetts, Boston, USA.
237 Life After Rio: A Commentary by Mark Halle, Published by International Institute for Sustainable Development on June 23 2012. An assessment of Rio conference of 1992 as a whole and the Rio 2012 conference in particular by the author is most readable.

238 (UNEP/GCSS.XII/13/Add.2).
239 See 40 years of UNEP, op. cit., p. 236.
240 Ibid., para 3, p. 236.
241 Ibid., Chapter 17, para 4, p. 236.
242 Ibid., pp. 236–237.
243 For a full list of goals and status of implementation, visit www.unep.org/geo/pdfs/geo5/progress towards goals.
244 The full report is downloaded here: www.unep.org/geo/GEDI Report.
245 UNEP performs little direct monitoring and surveillance. Rather it collects, collates, analyses and integrates data from UN agencies, other organizations, national statistical offices to form broader environmental assessments. In the case of Geo-5 some 600 experts were involved over a three-year period.
246 http:/: www.guardian.co.uk/environment/2012/June/21/rio21-world environment organization.
247 United Nations, 12 January 2012: 'The Future We Want', Rio+20.
248 Post 2016 development agenda was a process initiated from 2012 to 2015 led by UN to define the future global development framework that would succeed the MDGs. The new framework, starting from 2016, is called Sustainable Development Goals.
249 In the UN Conference on Sustainable Development (Rio+20) in 2012, green economy in the context of sustainable development and eradication of poverty was recognized as a tool to achieve sustainable social, economic and environmental development. In its simplest expression a green economy can be considered as one that is low in carbon, resource efficient and socially inclusive.
250 Stanley, op. cit., pp. 218–237 (Chapters 17 and 16).
251 CBDR, Principle of International Law establishing that all states are responsible for addressing global environmental destruction yet not equally responsible.
252 Stanley, op. cit., see Chapters 16 and 17.
253 Ibid., pp. 218–237.
254 Accessed internet on 5 June 2022, http://sustainabledevelopment.un.org.
255 Accessed internet on 8 June 2022, https://www.unep.org/environmentassembly/ Assembly of the UNEP (UNEA).
256 Accessed internet 25 May 2022.
257 Accessed internet 10 June 2022.
258 Accessed internet 15 June 2022.
259 Ibid.

3
INDIA AND UNEP

Introduction

Environment in the Indian Cultural Tradition

Cultural traditions and thought have provided the ideological underpinning and legitimacy to the present environmental movement in India. The cosmological view of the Vedic, Upanishad and Puranic traditions and literary imagination was enriched by fascinating symbols and idioms of the relationship of people with nature, which have provided the main mode of communication. Vandana Shiva, the renowned ecologist, observed: 'For the cultures of Asia, the forest has always been a teacher, and the message of the forest has been the message of interconnectedness and diversity, renewability and sustainability, integrity and pluralism.'[1]

Even, poet Makarand Dave selected some of the Sanskrit *shlokas* from the ancient scriptures like *Varaha Purana* and *Agni Purana* to bring out how nature was perceived with reverence and understanding for its benevolence towards mankind.[2] The high level of eco-tradition in our cultural heritage and the Puranic advice for protection of trees and environmental protection are illustrated:

- The man who plants shady trees yielding fruits and flowers, for the gratification of all living creatures, attains the highest bliss;
- Gods, demons, Gandharvas, Kinnaras, Nagas, Yakshas as well as animals, birds and human beings always take shelter under trees;
- The man who grows trees with fragrant flowers and fragrant fruits is blessed with a comfortable dwelling in a prosperous country;
- The trees by providing shade and shelter to passers-by, nests to birds', medicines to living beings through their leaves, roots and bark benefit them all. This is called the fivefold sacrifice, *panchayagna*, of the tree;

DOI: 10.4324/9781003370246-3

- India is a riparian civilization which has flourished along the banks from Ganga to Cauvery and from Narmada to Brahmaputra. The story of the river Ganga is, metaphorically and in actual life history, the best description of the vital eco-processes of the water cycle.

Further, Rabindranath Tagore, India's cultural seer, observed:

> Contemporary western civilization is built of bricks, iron and wood and is rooted in the city. But Indian civilization has been distinctive in locating the source of regeneration, material and intellectual in the forest, not in the city. India's best ideas have come where man was in communion with Nature, the trees, rivers and lakes, away from crowds. The unifying principle of life in diversity, of democratic pluralism, thus became the principle of Indian civilization.

> Rivers have been perceived and used in the total integration of their relationship with rainfall, mountains, forests, land and sea. Natural forests in catchment areas have been viewed as the best mechanism for water control and flood control in Indian thought.[3]

In the Indian philosophical and mythological traditions every living being is constituted by five basic elements – *panchamahabhuta*,[4] that is, sky, air, fire, water and soil. The sentiment created out of these five elements of nature is the philosophical statement of ecology. The entire living world – from the grass to the forests, rivers, lakes, oceans, hills and the birds and animals living in or around them – comprises the natural resources or the constituents of the *panchamahabhuta*. In Indian philosophy, a living being with *atman* is born out of the *panchamahabhuta*. The whole living world, the grand cosmos, the resources comprising the five elements, is the common resource of the global family. This concept is crucial to the ecological balance and well-being on this earth. It suggests the organic relationship, interdependence and continuing exchange between the five elements of cosmology. These elements are also related through a cyclical phenomenon of transformation of each of these five elements into one another.

The world of human beings, animal world, the flora and the fauna, according to this tradition, is possible because of the existence of these five elements. They are 'the life-supporting systems' and are the genuine treasure of every country. The *panchamahabhuta* has thus universal appeal, and 'ecology implies interdependence' – global, human and physical cosmology irrespective of race, ethnicity, caste, community, religion or gender. In the Indian tradition through constantly playing up of symbols and idioms, the man's mind was cultivated to treat nature respectfully and with a sense of humility. Like all the major cultures and civilizations, the earth is respectfully referred to as the Mother Goddess. Prithvisutra[5] is one of the greatest hymns dedicated to the Mother Earth. In the Atharva Veda, she represents the ecological balance.[6]

The nourishing earth (*prithvi*), the life-giving water (*jeevan* in Sanskrit), the bountiful *vriksha (trees)* and *vana (the forest)*, the emerging *vayu (air/the atmosphere)* and *surya (the sun)* are all expressed through the arts and aesthetics of India's rich culture. The seasons as the changing moods of nature, like the ragas in Indian classical music, are all in consonance with nature, and therefore artistic expressions of ecology.

According to Kapila Vatsyayan, in physical (architectural) term, the metaphor of the *meru*, the mythical centre, is expressed as *shikhara* and *garbhagriha* (the hypothetical centre or navel), thus the ascending continuum from the centre to the summit, the core of the ecological process.[7]

Indians from the ancient times were taught to respect various forms of nature and knew the interrelated, interdependent and interwoven nature of all wild and aquatic life and vegetative world. In the Vedas and Upanishads, the seers observed that at dawn and dusk, the changing seasons and various moods and constituents of nature are meditated to fathom the underlying meaning of such a complex phenomenon. What emerges is that India had the most complete holistic perspective of the universe. The cosmology, the science and philosophy, the total worldview, has been sustained by this civilization.[8]

The lotus is another dominant symbol of an ecological manifestation of various processes of nature and permeates in the literary and other cultural forms, like the Hindu, Jain and Buddhist art forms, which are connected with the mythical centre of the earth through its stem. To the Indian creative psyche conquest over nature is always regarded as a self-defeating goal and it is harmony and peace which are the cherished goals.[9]

The Environmental Movement in India

The environmental movement in India had the weaker sections of the society as its primary constituency of supporters. The tribal dwellers in forests and on hills, peasants, fisher folk and women were and are the carriers of environmental movements. It was from the lower strata of the society that it trickled up to the sensitive parts of the middle class and informed urban citizens, who conceptualized 'environmentalism' and provided the intellectual and communicational infrastructure to the movement through their newly formed non-governmental organizations. It was the traditional cultural seeds of values of reverence to nature and man's harmonious relationship with her that became the voice of the environmental discourse and movement in India.

Survival vis-à-vis Quality of Life

In her celebrated keynote address to the United Nations Conference on the Human Environment (UNCHE) held in Stockholm in 1972, Mrs. Indira Gandhi eloquently outlined the hopes and fears of the developing countries in response to the prevailing understanding of environmental priorities. She presented a sharply defined perception of environmental concerns and their relationship to the imperatives of economic growth and development. Issues that were thought to stem largely from industrial development were put into a more realistic framework so that the basic causes of environmental degradation in the poor and underdeveloped countries could be identified and tackled. According to her:

> The environmental problems of developing countries are not the side effects of excessive industrialization but reflects the inadequacy of development. The rich countries may look upon development as a cause of environmental destruction, but to us it is one of the primary means of improving the environment, for living or providing food, water, sanitation and shelter, of making the desert green and the mountains habitable.[10]

Thus, the degradation of environment is not merely a question of pollution generated by industrial activities; rather it probes the whole concept of the quality of human life. The destruction of forests, erosion of soils, the depletion of wildlife, the accumulation of wastes and the plight of urban areas are some of the illustrations of underdevelopment. On the other hand, diseases, squalor, hunger and malnutrition are the result of it. The concept of environment in developing countries, like India, brings a vision of society, where the human settlement will be healthy, drinking water would be easily accessible, sanitary conditions will be on acceptable level and the society will be able to provide basic opportunities to its members to live in dignity. Therefore, the first components of priority are in preserving and sustaining life than in improving the quality of life. Mrs. Gandhi, therefore, speaking before a gathering of distinguished Indian scientists on 3 January 1981 proclaimed:

> Development with conservation means that growth priorities do not sacrifice the needs of tomorrow for immediate compulsions. . . . [P]overty and economic backwardness are themselves the biggest constraints to growth.[11]

The environmental movement in India began in 1972, the year of the Stockholm Conference, and spread over 500 voluntary organizations and non-governmental organizations. They have an impressive record of confrontation and cooperation in promoting environmental education, creating environmental awareness among the people by interaction and consultation with government officials and people from all walks of life, such as consumers, contractors, industrialists, MNCs, engineers and even academicians. The contribution of these NGOs in developing and enhancing people's attitudes, scientific knowledge, technical skill, judgement and orientation regarding the environment has been, and is continuing to be, significant. They have also helped considerably in mobilizing public opinion for the promotion of environment in some trouble-spots of the country. Some of them are Conservation and Environment Group-India International Centre, Environmental Service

Group, Centre for Science and Environment, Association of Voluntary Agencies for Rural Development (AWARD) and Association of Scientific Workers in India, with headquarters in Delhi and liaison centres at different regions of the country.

The role of ecological considerations in rural social movements is certainly ambiguous and ambitious. In India, those who constitute the movement are engaged in a livelihood struggle, and they recognized that this livelihood struggle can be successful only if the environment is managed in a sustainable way. In this context, some environmental movements, like, Chipko in Garhwal Himalayas, Silent Valley in Kerala, Narmada and Tehri Dam movement in Madhya Pradesh, are briefly analysed.

The Chipko Movement

The state has been continuously encroaching upon the rights and privileges of the people to forest resources. The resistance to this encroachment has taken traditional Gandhian form, by following a peaceful noncooperation or Satyagraha. In the forests of the Gharwal Himalayas, the kind of protest that had originally been directed at the British was revived and used against the state government, when 5050 ash trees were allotted to sports goods company of Allahabad in Mandal Forest. When a single tree was refused on silvicultural grounds to the villagers for making yokes, people questioned this decision and declared their determination to save the trees by hugging the tree. This was repeated in other valleys and thus a non-violent technique of resistance emerged. The word *Chipko* means hugging the trees and was popularized through the folk songs of Chipko activist Ghanshyam Sailani and the civil disobedience practised by Chipko adherents has taken on an increasingly ecological character. 'Although it had its roots in a movement based on the politics of the distribution of the benefits of resources, it soon became an ecological movement rooted in the politics of distribution of ecological costs.'[12]

Chipko is post-independent India's most powerful ecological movement. Chipko's first battle took place (1973) in Chamoli district when the villagers of Mandal, led by Chandi Prasad Bhatt and the Dasholi Gram Swarajya Mandal (DGSM), prevented the Allahabad-based sports goods company Symonds from felling 14 ash trees. In the same year, the villagers again stopped Symonds agents from felling trees in the Phata-Rampur Forest, about 60 km from Gopeshwar. Another turning point for Chipko was that for the first time women took an active part in it.[13]

In Tehri Garhwal, Chipko activists led by Sunder Lal Bahuguna organized villagers (May 1977) to oppose tree-felling in the Henwal valley. They resorted to direct action to protect the forests, and 23 volunteers, including women, were arrested for opposing a forest auction. The struggle in Henwal marked the transformation of Chipko from an economic struggle to a fight for conservation.[14] Behind Chipko lies a century of peasant protests against commercial forestry in the Himalayas. It was Chandy Prasad Bhatt who formed the DGSM in 1964 in Gopeshwar inspired by the Sarvodaya movement and worked to promote Vinobha Bhave's concepts of gramdan and a non-violent, self-reliant society comprising villages based on rural industries. At both national and global levels, Chipko helped to consolidate the awareness that deforestation, soil erosion and floods are linked to the fragility of mountain ecosystems. Chipko was also interpreted as a critique of the overall development process and as an assertion of the rights of villagers whose labour and resources are subordinated to the demands of the urban industrial sector.[15]

Chipko's potential has been severely undermined by serious differences within the movement. The differences were caused partly by a clash of personalities between Chandy Prasad Bhatt and Sunder Lal Bahuguna. Bhatt coordinated its early protests and then involved village women with great success in the reclamation and afforestation of degraded land. Bahuguna organized important Chipko protests in Tehri Garhwal between 1976 and 1980 but has since concentrated largely on global issues. The third Chipko faction is the Uttarakhand Sangharsh Vahini, which provided a radical thrust to the movement in the late 1970s, but its dedicated core of student activists has gone their individual ways.[16] In interacting with the rest of the world, Chipko assumed a deep conservationist bearing and its developmental stance steadily eroded. International ecologists saw Chipko as a cultural response of the people's love for their environment.

The Swedish Parliament presented an award to Chipko activists on 9 December 1987. The citation reads: 'The Chipko Movement of India, whose members "hug trees" to prevent their felling, and have revived traditional agro-forestry, is honoured for its dedication to conservation, restoration and ecologically responsible use of India's natural resources.' Chipko Movement emphasizes two themes: (1) ecology is permanent economy, and (2) the most important forest products are not timber but soil, water and air. The message of the Stockholm Conference was taken to the hill villages in September–October 1972 by Swami Chidanandji, a spiritual leader, who had travelled widely all over the world and had seen the ecological crisis in the so-called developed world. In Uttarakhand, the Chipko Movement has gone a step further from protecting and planting trees to challenging mining and dam building. There has been a people's movement against mining in Garhwal and Kumaon. The main demand of Chipko in 1973 was an end to the contract system of forest exploitation and allotment of raw materials for local, forest-based industrial units, on concessional rates; it has developed into an ecological movement of permanent economy from a movement of short-term, exploitative economy.

Today, the Chipko Movement is interpreted as the first true ecological movement of the people. Undoubtedly it has the following successes:

- It has slowed down, if not totally stopped, the process of deforestation in Uttarakhand;
- It has exposed the vested interests involved in the mass-destruction of forests;
- It has given the people of the region a rallying point to collectively oppose the exploitation of their national resources;
- It has created awareness among the people regarding the implications of ecological destruction;
- It has demonstrated the workability of a powerful model of people's participation, a model that can be emulated and replicated.[17]

The Silent Valley Movement

The Silent Valley Movement stands out as one of the most important milestones in the shaping of public opinion as well as the formulation of official policy. The Silent Valley hydroelectric project in Kerala is the only scheme which has been halted on environmental grounds. The Silent Valley project, which would have dammed the Kunti river in Palghat district, was seen as a perfect site from the engineering and hydrological point of view. The power produced could have helped to start industries and create employment in Kerala. It is difficult to convince anyone that a dam, by creating a reservoir and in the process flooding a forest tract, may be causing ecological damage.

The people of Kerala felt that the construction of the dam and hydroelectric power station in Silent Valley has to be shelved and pursued under umbrella organizations like Kerala Sahitya Parishad,[18] the Society for the Protection of Silent Valley and 'Da Vinci' environmentalists. The people petitioned the government to educate the common people and fought hard to get a proper assessment of the project done.

Silent Valley occupies an area of only 8,950 hectares which is surrounded by the Nilgiris and Nilambur forests and the Atapattu forests that together comprises 40,000 hectares of pristine forest. The Kuntipuza river originates at a height of almost 2,400 m on the outer run of the Nilgiris, descends on the northern edge of the plateau and pursues a gentle southwardly course for 15 km before cascading down to the Mannarghat plains through a gorge at an elevation of 1,000 m. Although the British identified Silent Valley as an ideal site for the generation of hydel power as early as in 1929, technical investigations were carried out in 1958. The project was sanctioned by the Planning Commission in 1973. In 1977, a report titled 'Report of the Task Force for the Ecological Planning of the Western Ghats'[19] referred to what was envisaged in Silent Valley. This report pointed out, quite rightly, that project costs have to be revised taking into account capital losses like the forests submerged which had not been taken into account, and the disturbance to the forest because of the construction of the dam was also cited. The report concluded by highlighting the fact that, 'if the project is not shelved, the last vestige of natural climate vegetation of the region and one of the last remaining in the country will be lost to posterity and various adverse ecological consequences will follow'. The task force feels very strongly that the project should be shelved and the area declared a biosphere reserve. It seems unjustifiable to sacrifice this unique environment for the sake of generating '120 MW of power'.

Save the Silent Valley Day was observed throughout the state. Noted journalists wrote articles to create public awareness, public meetings were held, the Gandhi Peace Foundation joined the campaign, and organizations were formed by academics, scientists, environmentalists and public-spirited men throughout the country to serve as a focal point of public action against the project. An attempt was also made by the Friends of Trees and the Society for the Protection of Silent Valley at the High Court of Kerala to halt the execution of the project. A stay order was obtained from the court and this gave a temporary respite.

The second group was formed by people outside the state of Kerala who were concerned about the issue. They were vocal through organizations like the Bombay Natural History Society and the Save the Silent Valley Committee. The third group comprised the scientists and experts, who were agitated over the project and offered their expertise and views and served in various governmental and non-governmental committees. Points and counterpoints were made, pamphlets and books were published by each side, and counter-allegations made. However, the one remarkable result of this was that every possible aspect of the issue was examined in the greatest possible detail and the scientific debates were made public. The movement offered a unique foresight of people's participation, calling for right to information as an integral part of right to liberty.

Narmada Bachao Andolan

Much controversy has been created in the last few years about the environmental and social impacts of the upcoming Narmada Valley Development Project, which plans to build 30 big, 135 medium and 3,000 minor dams on the Narmada River and its tributaries, though the people of Narmada Valley spread over the states of Madhya Pradesh, Gujarat and Maharashtra oppose it bitterly. Their struggle, led by Baba Amte, Medha Patkar and others, is focused on two major dams under construction: the Sardar Sarovar Project in the Bharuch District of Gujarat and the Indira Sagar Project in Khandwa district of Madhya Pradesh. These projects will submerge almost as much area as it is meant to irrigate.

The major deficiencies were found in terms of the environmental assessment of the two dams.[20] Large-scale deforestation, loss of the flora and fauna in the submergence zone, water logging and salinity are some of the irreparable ecological damages the dams have caused. Construction of the dams which is going on in full swing and the continuing environmental assessments show that the region and the country cannot afford such ecological damages, as it is now politically impossible to stop the dams.[21] The Narmada Bachao Andolan intensifies its resistance on central government by peaceful means to withdraw this project as it is not going to help the people but it is going to destroy, deteriorate and displace their source of living.

The construction of the Sardar Sarovar and the Indira Sagar Project means the eviction of over three lakh people from the Narmada Valley, and according to official figures, these projects will submerge about 1,30,482 hectares, of which 55,681 hectares are prime agricultural land and 56,0666 hectares forests. The path of mutual reconciliation and dialogue has still not been open to the participants of this movement. The sense of conflict still prevails in the minds of people. The government should review its decision, which is going to affect the lives of millions of people.[22]

The Tehri Project

The 260.5-metre-high Tehri dam on the Bhagirathi in the Garhwal Himalayas has generated similar controversy since its inception. The cost of the project has accelerated to Rs. 4,142 crores in 1990 from the original Rs. 128.8 crores in 1967. The

Chipko leader Sundar Lal Bahuguna has been successfully leading it since its very inception and prevented the further development of the project.

The environmental movement emphasizes the importance of nature as a resource for survival and subsistence as well as the imperatives of economic redistribution. While the Gandhians, appropriate technologists and ecological Marxists represent the most forceful stance in the environmental-development debate, two points about their nature should be considered. One looks to protect the environment, while excluding development from its horizon and the optimists who view development in isolation from the environment in the belief that there are no physical limits to economic growth and that rapid industrialization on the Western model can be brought about in a matter of decades. The Gandhian Marxist and the appropriate technologists as well as conservationists argue for the future development of the movement. While Gandhians totally reject modernity, Marxists strongly believe in mobilizing action. Perhaps, the most promising group is the appropriate technologists whose perspective is most consistent with reality whereas their political orientation is a balanced mix of activism and reconstruction. Nevertheless, conservationists have a crucial role in widening the horizons of the movement.[23]

The movement has further influenced the judiciary (Bhopal disaster case), government and administration considerably. The need of the hour is a positive approach to environmental issues based on controlled consumerism and an emphasis on sustainable development. Communication between scientists and society and people's participation are urgent requirements for designing strategies for development. The challenges of environmental management are so colossal in nature, as compared with the resources available for meeting them, that there is no choice but to be highly selective and practical on approaching them. An immediate stoppage of all illegal felling of the trees, protection of all non-agricultural lands against the ravages of uncontrolled grazing, and the conservation of soil, water and other natural resources on a nation-wide scale should be given top priority.

Moreover, regarding conservation and prevention of pollution, the national environmental doctrine should draw inspiration from the cherished values that society has continued to progress over centuries as a code of conduct for environmental ethics. Ancient Indian scriptures have stressed that man and nature need to live in close harmony and plants and animals should be the object of unlimited kindness since they make no demand for their sustenance.

India has had a great tradition of environmental protection and conservation, as is evident from its ancient scriptures and Hindu traditions. The Vedic hymns deal with natural phenomenon, animals and birds, flora and fauna, pollution and ethical-moral responsibility of state and citizens to protect it. The Upanishads, Puranas, Brahmanas, commentaries of law givers, such as, Manu, all have profoundly emphasized from the Vedic times that *Prakrit* (Nature) and *Purush* (Humankind) form an inseparable part of the life support system. The Hindu view of reverence for nature and all forms of life, represent a powerful tradition which needs to be re-nurtured and re-applied in our contemporary conservation programmes and policies, with Vedic lesson.

> The earth is our mother, and we are all her
> Children and let us protect our mother.[24]
> The Assisi declaration speaks of ecological and environmental values in the passage 'O Wise One Uphold me in grace and splendour'.[25]

The Constitution of Democratic Republic of India further clarifies these goals with a special mandate to states to work for the conservation and protection of human environment. It also provides a fundamental duty to every citizen to protect and preserve the natural environment.

The UNEP and Management of India's Environmental Issues

According to Schumpeter 'man's relation with nature is not a cost-equation, but a life equation',[26] and the 'balance of nature' is related to the balance of 'human nature'[27] and our developmental activities have to be reoriented so that they will be in harmony with the natural environmental functioning.

India is a vast country with unparalleled diversity in eco-geography and biological resources. Obviously, our problems are equally diverse and gigantic, as it appeared in the Country Report of Pitamber Pant in the Stockholm UNCHE Report 1972.[28] The growing population, under-development and the limited option to choose 'appropriate technology' friendly to the environment, the curse of natural disasters, are few instances which contribute to the problems faced in the field of environment in India. Another side of environmental problems faced is the unintended side effects of development. These are mismanagement of natural resources, large-scale deforestation, unplanned discharge of wastes, handling of toxic chemicals, indiscriminate construction, expansion of settlement activities and so on. It is to this type of problems that tools and methodologies of environmental planning are primarily addressed through the training, strategies and guidelines of UNEP. The UNEP is guiding Indian initiatives and programmes.

The main contribution of the Stockholm Plan of Action is understanding the complexity of ecosystem and changes in attitudes towards environmental conservation. Wise management of natural resources has been an integral part of UNEP's conservation strategy. Thus, the alarming trends have forced a change in the governmental attitude to transform their action constructively for environmental protection. By integrating the inseparable economic and environmental systems and by creating environmental awareness among the masses, an ecologically compatible path to development was chosen.[29]

There is a growing feeling among the nations that environmental problems can best be solved through regional and sub-regional cooperative efforts. This feeling

86 India and UNEP

is reflected in the establishment of South Asian Co-operative Environment Programme (SACEP) in 1982, to share the conviction that environment is one area where nations will have to come together and resolve all their differences to save the common heritage. To quote Mrs. Gandhi –

> The inherent conflict is not between conservation and development, but between the environment and the reckless exploitation of man and earth in the name of efficiency.[30]

So, an integrated cooperative approach is needed to manage the increasing relationship between resources, eco-development, distribution and use of friendly technology on environment, and minimal need for sustaining decent standards of human life and protecting the natural systems on which life depends.[31]

According to Gopeshnath Khanna, environmental problems in India arise from a number of causes and they are as follows:

- The growing human and animal population is making increasing demands on natural resources in an unsustainable manner;
- The general indifference of industries on different aspects of environmental safety and protection, leading to the spread of avoidable air, water and soil pollution;
- Economic growth has necessitated a corresponding expansion in energy availability for educational, agricultural and domestic purposes;
- The level of environmental literacy is low and there is gross under-valuation of the economic and ecological aspects of biological diversity;
- The policies of the governments have not incorporated environmental accounting principles, with the result that many development projects have been short-term gains without considering their long-term ecological and social impacts.

Thus, 75 years after independence, while the country has achieved great gains in industry and agriculture, it has not achieved much on the environmental protection front. In this section, efforts of the UNEP in the management of various environmental issues in a variety of areas are analysed, which included desertification, deforestation, aquatic ecosystem, multipurpose valley projects, biological diversity, climate change, natural resource management, urbanization, human health and welfare, energy, industry and transportation.

Desertification

Desertification is defined as a process, which reduces the productivity of land and increases social distress. It also includes problems of deforestation, aridity and semi-aridity, land management, water logging and salinity and degradation of soil due to erosion, chemical action and possible factors.[32] Thus, it is both a major environmental hazard and a serious obstacle to development. More than 20 per cent of the earth, home to 80 million people, is directly threatened by it, and some 100 countries are affected, whereas manmade deserts stand at 9.1 million sq. km in area.[33] The arid zone of India occupies an area of 3.2 lakh sq. km of hot desert mostly in Rajasthan, Haryana, Gujarat and Karnataka, and 0.7 lakh sq. km of cold desert in Ladakh. Thus, about 200 million hectares are degraded and deteriorated.

In 1977, the UN Secretariat convened a UN Conference on Desertification[34] which adopted a plan of action to initiate a set of studies on the feasibility of transnational projects to combat desertification. These were:

1 The development and management of livestock in the Sudano-Sahelian Region;
2 The establishment of a greenbelt across the five countries of the north of Sahara;
3 A similar greenbelt in Africa, South of the Sahara extending from the Atlantic to the Red Sea.

In Thar Desert of India, UNEP initiated an interdisciplinary research project and water resource development, which had significant results. Regional networks and working groups, specialists, technical and field personnel were trained annually.[35]

The general objective of UNEP is to manage arid and semi-arid ecosystems for sustainable productivity to prevent desertification and to reclaim decertified land for productive use. The specific objective is to ensure the establishment of sustainable institutional arrangements and the adoption of national plans to combat desertification, integrated with overall development plans.[36] In India, the scientific and technical basis for dry land agriculture, horticulture, grassland management, livestock farming, afforestation and water harvesting have been explored with a fair degree of success; nevertheless, the available scientific information on the rational use of ecosystem resources is yet to be converted into action programmes.[37]

As the co-ordinator of the Plan of Action to Combat Desertification, UNEP has mobilized the support of all international organizations concerned and has utilized appropriate mechanism to reinforce co-ordination of bilateral and multilateral assistance programmes. Through GEMS it has evolved better systems of early warning against severe drought and other dry land disasters. It has paid particular attention to the demonstration of environmentally sound and economically profitable methods of farming, livestock raising and agro-forestry projects designed to promote afforestation, agro-forestry, water and soil conservation provided by Cairo Programme for African Co-operation, which responds to local needs for food, fodder and fuel, while enhancing the productivity of arid ecosystems. UNEP further emphasizes the significance of public participation at the grassroots and farm community level to control the problem of desertification.

These programmes are primarily the responsibility of governments and their success or failure is largely determined by the government priorities, institutional efficiency and the extent of public commitment and participation. Assistance has been provided to India by mobilizing resources and arranging for coordination of action by preparing national plans and priority projects through joint missions, training programmes and maintenance of technical cooperation networks.[38]

In pursuance of the guidelines of UNEP, the Government of India established, in May 1985, the National Wasteland Development Board (NWDB) primarily to formulate, coordinate and catalyse programmes for the management and development of wastelands in the country. Computer-based monitoring cells were set up in 15 states to monitor and evaluate various projects on afforestation, social forestry and so on.[39]

Compared to the magnitude of the problem, the achievements of the Plan of Action to Combat Desertification (PACD) have been marginal. Looking into this grim situation, the UNEP evolved a System-Wide Medium Term Environmental Programme for the period 1990–1995.[40] The main thrust of the strategy is summarized as follows:

1. Encouraging the application of existing knowledge and experiences in desertification control;
2. Co-ordinating and mobilizing support of the UN system and the rest of the international community to reinforce bilateral and multilateral assistance programmes;
3. Concentrating on the most seriously threatened areas;
4. Supporting the government in developing national action plans and sub-regional programmes for combating desertification. The implications of this strategy were highlighted by the following UN activities:

    i. Mapping of sand dunes and degraded arid and semi-arid regions and compilation of World Atlas, thematic maps on desertification control, development of computerized geographical information systems for use in assessing and monitoring of desertification through FAO/UNESCO;
    ii. Development of methodological and hydrological networks and improvement of climatological database, research and assessment of climate aspects of desertification, including droughts through WMO;
    iii. In the promotion of integrated development, the UN emphasizes the role of women to solve problems of food production and development of water resources in drought-affected areas and related training through FAO, ILO and ESCAP;
    iv. ILO provides training for the development of vocational and technical skills and appropriate technologies to prevent erosion;
    v. Production of audio-visual material, practical guidelines and hand books, promotion of exchanges of experience in drought and desertification control through demonstration projects, and research on soil management;
    vi. Promotion of the world soil policy at regional levels. Looking into the spectres of UNEP activities in this field, it can be assessed that given the will and the appropriate action, the desert can bloom.

Techniques for soil and water conservation and sand dune stabilization are available today, and large-scale planting of shelter belts to help minimize wind erosion is done. An Integrated Plan of Action backed with training and assistance on the national level by UNEP is successfully implemented. In India, a National Technology Mission on Wastelands Development was formed (on 5th October 1989), which contains the same objectives as contained in the IInd and IIIrd System-Wide Medium Term Environment Programme of UNEP. Under this mandate mission the following activities are highlighted:[41] (1) checking land degradation, (2) putting wasteland to sustainable use, (3) regenerating degraded forests, (4) promoting green public lands, (5) promoting farm forestry, (6) increasing biomass availability especially fuel, food and fodder and (7) restoring ecological balance. The key elements of this new approach are the development of low-cost technologies and the involvement of the people at the grassroot level.

Deforestation

Deforestation symbolizes the situation of overexploitation of natural resources. The forests of the world are losing about 15 million hectares of tropical, moist forest annually. The total forest area in the world was 7,000 million hectares; in 1980, it declined to 4,200 million hectares or 30 per cent of the world area in 1990. About 11 million hectares of forests are cleared for agriculture, destroyed and degraded for fuel-wood gathering, shifting cultivation, overgrazing and burning each year.[42]

Fuel-wood harvesting, irrigation and hydroelectric power development, mining, the expansion of urban and industrial infrastructure and railways, and natural disasters like earthquakes and cyclones have also contributed significantly to deforestation. Forest fires recently in many parts of the world, including India, have also had a substantial impact.[43] Thousands of square kilometres of dense forest cover are lost every year. The planet is losing the best of its natural treasures. Commenting on India's environmental crisis, Valmik Thapar[44] says, 'we have failed globally to act in time, our international organizations are not strong enough to deal with the crisis that engulf us'. Further, he notes that India's wilderness is heading towards disaster and a strong global and national political will combined with effective international cooperation can only save us from this.[45]

Today we are faced with soil erosion due to inadequate tree cover, drought due to loss of groundwater storage and landslides due to the denudation of hills. Waterlogging, salinity and the extinction of species are the major challenges which contributed substantially to the qualitative decline of the terrestrial ecosystem of the region containing some of the most devastated lands on the earth.[46] Thus the forests in the region are now depleted at the rate between 0.2 per cent and 2.0 per cent per year with the result that forest amounts for less than 20 per cent of the land area today.[47] It is estimated that about 6,000 million tonnes of top soil is wasted (soil erosion) every year in India. Ecologically, forests influence natural phenomena like rainfall, atmosphere quality, floods, landslides and natural disasters. Even, India has the worst fuel-wood problem, and only 39 million tonnes of fuel-wood is available against the total requirement of 133 million tonnes. Then the gap between consumption and production was filled by unrecorded removals from the forests resulting in heavy deforestation.[48] India has the highest amount of forest degradation by grazing; it has 15 per cent of the world cattle, 46 per cent of buffaloes, 17 per cent of goats and 4 per cent of sheep, hence most of the Indian forests are overgrazed and greatly degraded.[49] Moreover, forest degradation by fire is very common.

The UNEP has warned[50] that unless management of the tropical forest and woodland ecosystem is significantly improved, more than half the population of the developing world will be short of fuel wood, many indigenous people will be displaced with lost cultures, some 10–20 per cent of the world's plant and animal species will become extinct and many more watersheds will be severely degraded.

- India is one of the few countries having a forest policy since 1894. It was revised in 1952 and again in 1988, and the main objective of this new policy is reservation, conservation and development of the forests, whose aims are: maintenance of environmental stability through preservation and restoration of ecological balance; conservation of natural heritage; check on soil erosion; denudation in catchment areas of rivers, lakes and reservoirs; substantial increase in the forest/tree cover through massive afforestation and social forestry programmes; encouragement of efficient utilization of forest produce; and creation of a massive people's movement, with the involvement of women to achieve objectives and minimize pressure on existing forests.

Thus, the principal aim of the 1988 Forest Policy was to ensure environmental stability and maintenance of ecological balance, including atmospheric equilibrium.

The Forest Conservation Act, 1980, enacted primarily to check indiscriminate deforestation for non-forestry purposes, was amended in 1988, to make it more stringent by prescribing punishment for its violation.[51] A National Forest Fund has been set up to achieve this goal. Moreover, a project aimed at the prevention, detention and suppression of forest fire has been launched with UNDP assistance. The 1988 policy statement concedes that the customary rights of and concession to the people in forest areas must be protected. Despite recent commendable initiatives aimed at combating the decline of tropical forests, including the Silva Conference (1986), the Tropical Forestry Action Plan (1985) and the International Tropical Timber Organization (1986),[52] the situation remains alarming throughout the world. All these initiatives contain the basic strategy required and plans necessary for corrective action, but practical implementation remains slow.

The Tropical Forestry Action Plan has provided the much-needed international co-ordination mechanism for active collaborative action at the international level. Baseline data are sufficient to activate national plans for increased afforestation. Criticism had been launched regarding low involvement of NGOs and the grassroots people.

The general objective of UNEP is to achieve sustainable development of the tropical forest and woodland ecosystem, while ensuring their continuing functions of regulating water and climate, safeguarding biological diversity and providing goods and services to the local people. The specific objectives of UNEP include the following:

1. To develop and apply suitable methods for monitoring tropical forest and woodland resources and to assess their global status and trends;
2. To develop and dismount practical knowledge on the sustainable use of tropical forest through integrated approaches incorporating ecological, social and economic factors;
3. To achieve a better balance between deforestation and afforestation, while rehabilitating degraded areas.

System-Wide Strategy of UNEP

It reflects the principle that sustainable maintenance of forests is of major interest not only to the countries concerned but also to the community of nations as a whole. It concentrates on innovative action between forestry and agriculture with a view to conserving the resource base, integrating forestry in agriculture systems, ensuring rational and sustainable land use and giving due consideration to the needs of rural people. It emphasizes forest-based industrial development and the stabilization of fuel-wood and energy supply. It puts particular emphasis on conservation of tropical forests, genetic resources and ecosystems, simultaneously seeking to bring degraded areas back into productive use. It supports the development of increased capacity in national institutions to deal with sustainable development of forest land in close collaboration with FAO, UNESCO, ITTO and other IGOs and NGOs. The strategy further envisaged the development of the International Tropical Timber Organization (ITTO) and implementation of timber agreement to assist in afforestation, representation and production of industrial tropic timbers. These strategies have been incorporated to a limited extent in the 1988 Forest Policy. Besides, Forest Research Institute in Dehradun receives the training curricula of UNEP and selects personnel for the training organized by the UNEP and associate organizations.

Multipurpose Valley Projects

In a developing country like India, where three-fourths of the total population is dependent on agriculture, the execution of river valley projects is regarded as an important element of national development. Except the northern states, all other parts of the country are dependent on monsoon rains available only from July to September. Unless it is stored in an artificially created reservoir, it cannot be available for the whole year. In view of this situation, the implementation of river valley projects occupies an added significance.

India has the dubious distinction of having the largest number of river valley projects in the world. The 157 major dams have been built by 1985 at a cost of

Rs. 15,026 crores.[53] All these dams cover nearly 1 per cent of India's total land area. These projects have become a symbol of national development as they are located in backward and tribal areas. The dams have the potential of solving most of the famines, unemployment, urban water shortages and power shortages. On the one hand, they have created several benefits like providing additional irrigation facilities, increased crop production, power generation, increased availability of power supply for domestic as well as industrial purposes, checking floods, generating pisciculture, providing additional employment opportunities, infrastructure development and so on.[54] On the other hand, wide displacement, nonjudicious use of finances, submergence of vast tracts of arable and forest land, widespread water logging, soil erosion and siltation are some of the worst damages caused by these projects. All these have great economic, social and environmental impacts on the society. The magnitude of displacement caused by these river valley projects over the last 50–75 years can be as high as 20 million. For example, the Hirakud Dam, one of the largest in the country executed in the 1950s, has displaced more than 20,000 people residing in 249 villages.[55] The Srisailam Project completed in 1983 has displaced over one lakh people.[56] Another major irrigation project, Tungabhadra, executed in Karnataka, has affected 90 villages, displacing about 15,000 people, belonging to 12,000 families.[57] In Maharashtra alone, about 2,220 villages have been affected due to execution of 233 minor, medium and major projects, displacing more than 1.25 lakh families.[58] These projects mostly affect the poorest sections of the population consisting of scheduled castes, tribes and landless labourers. Out of 70,000 people displaced by Sardar Sarovar Project, nearly 69,000 are Scheduled Tribes.[59] The Report of the Commissioner for Scheduled Castes and Tribes says that about 119 large irrigation and hydro-electric projects implemented in the various parts of the country had displaced about 16.94 lakh people, and out of these about half (8.14 lakh) are tribals.

The rehabilitation measures adopted under various projects are very inappropriate and inadequate.[60] The displaced villagers consist of cultivators who have deep attachment to their lands and home. Thus, river valley projects have become the most controversial environmental issue, having serious implications in the long term. It includes depletion of forest resources, soil erosion, siltation/sedimentation, water logging, pollution, wildlife destruction and so on.[61] Between 1951 and 1971, river valley projects have denuded 4.01 lakh hectare of forest areas. The Narmada Sagar Project alone had submerged 3.5 lakh hectares of India's best deciduous forest, comprising rich teak and bamboo forests.[62] Construction of dams under these projects has taken a heavy toll on wild animals.[63] Moreover, the Tehri Dam on the Bhagirathi River in Gharwal Himalaya has submerged 1,000 ha. of forests, which has affected 427 species of plants, according to a survey carried out by the Botanical Survey of India.[64] The seismicity is another danger to dams. According to the National Geographical Research Institute, Hyderabad, dams may trigger off seismic activity. Sedimentation or siltation is another major danger caused by the river valley projects. It has also been estimated that the process of siltation has reduced the water-holding capacity of dams by 1.5–2.0 per cent per annum.[65] According to a remote sensing survey, Hirakud, one of the biggest river valley projects constructed in the mid-1950s in Orissa, is now threatening with fast siltation and sedimentation leading to a premature end of its life span. The survey also revealed that there has been a loss of 1.3 per cent of storage capacity per year along with 0.5 per cent loss of live storage.[66] The river valley projects also cause waterlogging, which leads to salinity and in turn reduces the natural productivity of the land. According to a study conducted by the Indian Institute of Science, Bangalore, almost 40 per cent of the total command area, one lakh ha. of black cotton soil, under Narmada Project would suffer from chronic water logging, which increases the salinity.[67]

Preventive and control measures to minimize environmental damages are essential for both optimization of natural resources and achieving sustainable development. The UNEP, in this direction, has the specific objective to improve and apply environmentally and economically sound methods to minimize waterloss, salinity, waterlogging and other environmental hazards in existing or planned irrigation projects. It intends to promote integrated water basin development wherever possible and endeavours to find an environmentally appropriate balance between large-scale and small-scale projects.[68] It has some publications on global assessment of environmental impact of large dams and other surface water schemes and decisions to strengthen national monitoring capabilities.[69] It provides assistance to specialized water resources training institutions to train experts for integrated water development projects, strengthening research in this field.

Aquatic Ecosystem: Water Resources and Freshwater Ecosystem

The aquatic environment is divided into island freshwater, estuarine and marine water. Certain major impacts are brought to bear in the aquatic ecosystems by the process of development. In India, 96 per cent of the water is used for agriculture, 3 per cent for municipal water supplies and remaining 1 per cent for the industry.[70] The increase in population has caused severe stress on aquatic resources.

The main lines of activities related to water resources undertaken by UNEP have been based on the action plan adopted by the United Nations Water Conference in 1977.[71] Considerable progress has been made with regard to some aspects of the action plan, particularly for the drinking water supply and sanitation. Phase III of the International Hydrological Programme has promoted an integrated scientific approach to the management of water resources, training and transfer of technology.

Although Indian government responded to the UN water conference by developing a National Water Policy[72] in 1986, most planning processes still fail to provide the necessary linkages between water management and national development programmes. UNEP recommended for setting up a water commission and improving and applying environmentally and economically sound methods to minimize waterloss, salinity, waterlogging and other environmental hazards in existing or planned irrigation projects. The strategy of the UNEP aims at ensuring adequate management, sustained use and effective control of surface and groundwater resources in all countries and, in particular, at alleviating potential water shortages, restoring or maintaining water quality and protecting freshwater ecosystems. The Operational

Hydrology Programme (OHP)[73] placed special emphasis on training of policymakers, managers, technicians and other specialized personnel; on the creation and strengthening of appropriate governmental and non-governmental institutions; and on addressing the issues by the residents of the communities who are affected by the water development projects. Four plans for international water systems were developed and proposed for adoption by 1995.[74] It aimed at providing assistance to government in the formulation and implementation of water plans for sewage disposal, irrigation and hydropower, and assistance to river basin development authorities.

Efforts to control water pollution by legislation are much old in India. One of the oldest laws against discharging industrial wastes into rivers was promulgated in 1898. The most comprehensive act was drafted in 1970 by Maharashtra, known as Water Pollution Control Act.

The Central Pollution Control Board (CPCB) is the national apex body for assessment, monitoring and control of water pollution. It has the executive responsibility for the enforcement of acts meant to prevent pollution. The states have also their own pollution control boards. The CPCB in collaboration with the Department of Ocean Development has identified 173 monitoring stations along the Indian Coast for water quality measurement with involvement of State Pollution Control Boards. The Board is also engaged in the study of existing surface water quality under the programme Monitoring of Indian National Aquatic Resources (MINAR), which includes UNEP's Global Environmental Monitoring Systems. There are altogether 400 water quality monitoring stations.[75]

Traditional water harvesting systems exist all over India, but after serving the nation for several millennia, they are slowly dying. India, after having gone through an extended 70-year phase of constructing big dams and canals, is once again forced to look at its traditional, small-scale water harvesting and management systems, especially among grassroots organizations working with the people to develop water management systems that people can themselves manage. It is time to revive people's wisdom.[76]

Ganga Action Plan

In February 1985, a Central Ganga Authority (CGA) was set up under the chairmanship of the then prime minister Shri Rajiv Gandhi, to oversee the implementation of the action plan drawn up for cleaning polluted stretches of the Ganga. The monitoring and Steering Committee identified different schemes to be taken up by Uttar Pradesh, Bihar and West Bengal through which the river flows.[77] The action plan has secured sufficient public participation on its implementation. The main objectives of this plan are as follows:

- Promoting awareness of the problems of pollution of rivers in general and Ganga in particular;
- Promoting right attitudes towards use of rivers;
- Promoting awareness of related issues – such as the siltation of rivers, soil erosion, afforestation, industrial pollution and conversion of water resources.

The Ganga Action Plan was the first major initiative taken by the Government of India which involves public organizations and NGOs as well as guidelines/directives of UNEP and other IGOs in this regard. It is a fact that a large part of India suffers from lack of safe drinking water and sanitation facilities. And one of the objectives of UNEP was to disseminate information, especially on the assurance of water quality requirements, low-cost technologies for water supply and environmentally appropriate sanitation services and their operation and maintenance. During the period 1990–1995, the emphasis of the system-wide strategy of UNEP was sectoral cooperation, involving environment, public health, rural development, housing agencies and water resource development. Linking drinking water supply and sanitation with primary health care at the community level continues to be a challenge for national agencies and national health authorities.

Air Pollution

Air pollution is caused by higher concentration of sulphur-dioxide, nitrogen oxides, carbon monoxide and dust. Although air pollution originates in the atmospheric ecosystem, its effect on the terrestrial and aquatic ecosystems through transport of sediments and pollutants, and acid rain on the occupational and living environment through exposure to fumes and toxic gases, can be dangerous. Among the factors responsible for such a situation are increased industrial activity, increase in the number of motor vehicles and burning of fossil fuels and fuel woods.

Among a host of environmental crimes, the dramatic rise in air pollution in most Indian metropolises over the last two decades is a direct result of an inefficient state, in terms of balancing both responsibilities and precautionary actions.[78] Delhi and Calcutta are already among the worst polluted cities in the world. Others like Bangalore, once a 'garden city', are rapidly deteriorating.

According to Mr. Anil Aggarwal, ex-director of Center for Science and Environment (CSE), in many cities, especially in Delhi, vehicular air pollution is the key culprit which has led to the formulation of a report called 'Slow Murder: The Deadly Story of Vehicular Pollution in India'.[79] A survey done by the Center for Science and Environment that assessed the importance of implementing the Supreme Court orders of 28 July 1998 warned that more deaths will occur over the next coming years from toxic particles in Delhi if the Supreme Court orders on moving the entire public transport to compressed natural gas (CNG) is not implemented as scheduled. The order states that among other vehicles, three-wheelers, taxis and all buses must run on CNG by 31 March 2001. CNG moved ahead to assess the impact of implementation of the court order. The study shows that implementation of the court order to move all buses to CNG is critical to get anywhere near achieving clean air targets by 2010.[80] This model has been developed by CSE with the help of a large number of experts from India and abroad. This model[81] helps in predicting the future vehicular pollution load in two possible scenarios. Impacts of interventions such as those ordered by the Supreme Court are measured against (1) a no-change scenario where the vehicles continue to emit as of now. This baseline scenario, however, includes implementation of Euro I and Euro II standards for cars and a ban on commercial vehicles over 15 years old. (2) Impact of other interventions like improvements in vehicular standards, fuel quality, changes in transport modes, improvements in emissions and emergency measures like restricting traffic on highly polluted days. This is the first policy tool of its kind developed in India for air quality management.

Air pollution in cities and industrial areas has been studied by the UNEP since its very inception in 1972. Assessments in the 1980s show that levels of sulphur-dioxide in many cities in the developed world were decreasing in response to the various control measures, but it is increasing in many metropolises of developing countries, including India. UNEP co-ordinates the activities needed for the assessment of stratospheric ozone depletion in close cooperation with WMO, WHO, FAO, IACO and other NGOs. These assessments provide the scientific basis for action to protect the ozone layer, which is represented by the Vienna Convention for the Protection of the Ozone Layer and the Montreal Protocol on substances that deplete the ozone layer. As a part of the world climate impact studies programme (WCIP), UNEP is coordinating national climate impact studies programmes in an international network, encouraging better understanding of climate and its effective utilization of resources. In its catalytic role, UNEP is assisting selected regions, which lack the necessary scientific or financial resources to undertake the policy exercise in preparation for climate change. At the same time, efforts to improve knowledge of air pollution, acid precipitation and other atmospheric issues to identify solutions to these problems are being undertaken by UNEP. It gives high priority to the transfer of technology and training for dealing with air pollution in developing countries.

The future strategy of UNEP includes continued monitoring of ozone and other gases in the atmosphere, reduction of inaccuracies in measurements both from space and with ground-based instrumentation, development of long-term series of global ozone data for analysis and providing relevant data and information for regular assessment of ozone layer.

Climate Change

Climate change is one of the most significant environmental threats of the present time, affecting global ecosystems, agriculture, water resources, sea-level rise and so on. It is predicted that global temperature will rise between 1.5°C and 4.5°C and consequently the sea level between 20 cm and 140 cm as early as the 2030s.

The general objective of UNEP is to encourage widespread application of climate knowledge and data to human activities in an effort to assist governments to adopt policies that would mitigate adverse impacts on climate change. Further, the aim is to improve the capacity to monitor and predict the environmental and socio-economic impacts of climate variability and develop policies to minimize variability. The GEMS is monitoring the world's glaciers and icecaps as related to climate development during the last 25 years. Emphasis is laid down on combating the effects of drought by agro-meteorological monitoring and improving the application of climate information to provide early warning of drought and climatic aspects of desertification. India is actively involved and associated with all these programmes of UNEP and its perception and attitude regarding treaties and conferences are very positive.

Biodiversity

The conservation of natural ecosystems and the maintenance of biological diversity constitute one of the main issues of our time. The objectives of UNEP in this field are to protect the natural heritage of mankind through conservation of ecosystems, to improve scientific standards and methods within the gene and data bank networks, to improve the conservation of the genetic resources of domesticated animals and plants, to improve methods and train personnel for the management of protected areas, and to promote and support the formulation and implementation of natural conservation strategies.[82]

The UNEP strategy, based on the World Conservation Strategy (1980), the World Charter for Nature (1982) and the action plan for biosphere reserve (1984), consists of the following:

- Formulating and effectively implementing National Conservation Strategies and international convention for the conservation of the world's biological diversity and genetic resources;
- Conserving ecosystems and wild animals and plants in selected areas of the world's biographic provinces;
- Promoting conservation of plants and animal genetic resources and development of national, regional, global information system;
- Expanding and improving related professional and institutional capability through appropriate training programmes in the conservation of biological diversity.[83]

India has a geographical area of 329 million hectares. The 45,000 species of plants constitute roughly 12 per cent of the global plant wealth. There are about 15,000 species of flowering plants and 33 per cent of this are endemic, located in 26 endemic centres.[84] The Indian region is placed seventh with 167 species.[85] From biospheres' point of view, the region falls into 2 realms and 12 biographical provinces – Ladakh, Himalayan highlands, Malabar rain forests, Bengal rain forest, Indus–Ganga monsoon forest, Assam-Burma monsoon forest, Mahanadi, Coromandel, Deccan forest, the Thar desert, Lakshadweep Islands, Andaman and Nicobar Islands.

There are 148 national parks, 503 sanctuaries with a total area of 1,51,342 sq. km, which includes 13 biospheres.[86] The plants and animal species in a given ecosystem coexist and interact with each other to maintain equilibrium. The single most important factor in maintaining this is the optimum level of population of each species. The rise and fall of any can lead to ecological imbalances with disastrous effects on all components.[87] India is well known for large-scale environment degradation and the consequent impact on the species; it is also known for its efforts to rehabilitate some important species. The populations of the tiger, elephant, the Gir lions, hongals, crocodiles and so on have gone up. The success stories of the tiger project, elephant project and crocodile farming show that with proper implementation of conservation measures, the loss of flora and fauna could be minimum. The attempts and endeavours taken by the Ministry of Environment, Forest and Wildlife in this direction can be summarized as follows:

- India is a signatory to World Conservation Strategy of 1980 and implements its guidelines through UNEP training, research and guidelines;
- The National Forest Policy of 1988 aimed at forest coverage of one-third of the total land areas, thereby protecting the vast genetic resources as the better management strategy[88];
- The Wildlife Directorate has its strategy for the establishment of an effective representative network of protected areas and the implementation of the National Wildlife Action Plan (1983)[89];
- The enforcement of the Wildlife Protection Act, 1972, through the regional offices, has strengthened the state governments, the police and the custom authorities to prevent the export or destruction of protected species. India is signatory to the Convention on International Trade in Endangered Species of wild flora and fauna.

The Indian Board for Wildlife is an advisory body established in 1982 with the guidelines for control of visitors to wildlife reserves, national policy or management of zoos, the posting of competent, motivated and well-trained personnel, while at the same time taking measures to reduce animal conflicts and implementing a stricter control of poaching and illegal trade in wildlife. Criticizing the Wildlife Board, a prominent scientist, T.N. Khushoo, comments:

> It is time the Indian Wildlife Board has to be scrapped and replaced by interdisciplinary National Biodiversity Conservation Board not merely in name but in substance and actual functioning. This board should be the guardian of our biodiversity and deal with the whole gamut of issues connected with biodiversity.[90]

India has been recognized as one of the world's top 12 mega-diversity nations. Conserving its biodiversity has become an important local concern with worldwide implications. The National Natural Resource Management system has been established for an accurate and updated inventory of bio-resources such as land, water, forests, mineral resources, oceans and so on. The UNEP gives support for conservation of plants and tree species in selected biosphere reserves, promoting the conservation and development of protected areas through the formulation and implementation of management plans and improving the legislation, in the framework of the World Conservation Strategy, the action plan for biosphere reserve and the Tropical Forestry Action Plan It also provides regional training programmes for the management of wildlife and protected areas in developing countries.

Urbanization

The National Commission on Urbanization, which submitted its report in August 1988, has estimated that by the turn of the century the urban population will be about 340 million, constituting about a third of the total population of India.[91] The overall population trends suggest that the population living in urban areas will increase in South Asia from 233 million in 1980 to 500 million in the year 2000, an increase of 125 per cent.[92] The urban areas are suffering from squatter settlements, lack of sanitation and water supply, overcrowding, congestion and population, and hence the cities in India face environmental problems such as lack of sanitation, shortage of services, polluted air and water, lack of open and recreational areas, traffic congestion and so on.

The governments have realized the need for looking at the urban problems from an overall national perspective. The UN Centre for Human Settlement (Habitat) had declared 1987 as the International Year of Shelter for the Homeless. A well-managed process of urbanization can be a moving force behind sustainable national development. The objective of UNEP is to be incorporated in order to develop land and to adopt innovative and environmentally appropriate techniques for housing, water supply and waste management facilities. In collaboration with other UN bodies, notably the United Nations Centre for the Human Settlement and World Health Organization, UNEP supports the efforts of governments to establish policy concepts and set up institutions to tackle the environmentally sound planning and management of human settlement, including the development of appropriate infrastructure for drinking water supply and sanitation.[93] The catalytic and coordinating role of UNEP and Habitat must be strengthened in the demand for national planning for Housing and Settlement.

Human Health and Welfare

In India, where poverty and malnutrition predominate, the problem of human health is a major concern. Environmental and genetic factors have a serious impact on human health. Influenza and pneumonia are more prevalent in colder regions than in warmer regions. Seasonal temperature, humidity and rain often contribute to the increase in contagious diseases. Tuberculosis is encountered in crowded urban conditions and where malnutrition prevails. Typhoid and Cholera are both endemic in large regions of Asia. India, which is largely rural with traditional societies, is characterized by contaminated water and food, malnutrition and infections, and high birth and death rates. And the country still faces a two-pronged problem of the environmental health of both unplanned and imbalanced development as well as underdevelopment.

The research studies conducted by UNEP show that environmental quality and human health and welfare are intimately related. Initially, the hazards of pollution were associated with urban air and inland waters but the scope of the concern has widened. Hazardous wastes have become an important concern in developing countries like India; besides chemical and nuclear accidents, noise pollution in the urban environment is disturbing and damaging. In this regard, International Programme on Chemical Safety, a joint programme of three cooperating organizations, WHO/ILO/UNEP, coordinate with each other in implementing activities related to chemical safety and in this regard, UNEP-envisages support, for strengthening the capacity of national institutions for assessment, prevention and control of pollution hazards at the country level.[94]

Union Ministry of Health and Family Welfare plays a vital role in the national efforts to help citizens lead healthy and happy lives. The ministry is responsible for implementation of programmes of national importance like family welfare, primary health care services, prevention and control of disease and so on, which form the main plank of the development programme. National commitment to attain the goal of 'health for all by 2030 A.D.' is another significant goal to be achieved.

The government has the National Malaria Control Programme, National Immunization Programme, National Filaria Control Programme, National Leprosy Control and Control of Blindness Programme, and these programmes are restructured in the light of UNEP guidelines so far as they are concerned with the environment.

Summary

This chapter has explored the role of environmental movements in the Indian cultural tradition, and in this context the role of Chipko, the Silent Valley Movement, Narmada Bachao Andolan and Tehri Project and so on are analysed in the first section. In the second section of this chapter, the role of UNEP in managing various environmental issues in India – such as desertification, deforestation, aquatic ecosystem, multipurpose valley projects, biological diversity, climate change, urbanization, human health and welfare – has been raised.

National governments have many obligations on the environmental front. While they may work in close collaboration with others, they have the ultimate responsibility for national policy development and implementation, enforcing national environmental legislation, ensuring national-level compliance with international agreements, public education and awareness building and so on. A climate of stability is required before national decision-makers turn their attention to the environment and make real headway on environmental problems. Likewise good governance and security are prerequisites for sustainable development, and it is up to citizens, governments and multilateral organizations to bring these about, both within and across national boundaries.

One of the measures that could assist governments with their environmental mandates includes setting up a dedicated system for mediation on environment-related disputes, by establishing the increasing role of the civil society for achieving environmental protection and sound management. Establishing clear responsibilities, eliminating overlap and duplication and improving information exchange remain major challenges. Increased support to International Environmental Governance organizations will enable national governments to improve international cooperation at regional and global levels, strengthen conflict-resolution mechanisms and implement environmental programmes and projects more effectively.

In the end, it has been observed that environmental movements played a crucial role in creating environmental consciousness in the minds of the people and helped in changing people's and governmental attitudes in transforming their action constructively for environmental protection and sustainable development. Under the guidance of UNEP, by integrating the inseparable economic and environmental systems and by creating environmental awareness among the masses, India has chosen ecologically compatible path to development. However, greater commitments must be shown by the governments with regard to implementation.

Notes

1 Vandana, Shiva, 'The Green Movements in Asia', in Matthias Finger (ed.), *Research in Social Movements. Conflicts and Change* (London: Jai Press Supplement 2, 1992), p. 195.
2 Dave, Makrand, *Vriksha Mahima* (Ahmedabad: Sadvichar Parivar, 1990).
3 Shiva, op. cit., p. 207.
4 Sheth, Pravin, *Environmentalism: Politics, Ecology and Development* (New Delhi: Rawat Pub, 1997), p. 18.
5 Ibid., p. 22.
6 Ibid., p. 23.
7 Vatsayana, Kapila, 'Ecology and Indian Myth', in Geeti Sen (ed.), *Indigenous Vision: People of India, Attitudes to the Environment* (New Delhi: Sage. 1992) p. 176.
8 Ibid., p. 15.
9 Mishra, Vidya Niwas, 'Man, Nature and Poet', in Vidya Niwas Mishra (ed.), *Creativity and Environment* (New Delhi: Sahitya Academy, 1992), p. 63.
10 Gandhi, Indira. 1972 Address to the Planetary Session of UNCHE, 'Man and His Environment' Sweden, Brochure, Ministry of Health and Family Planning, Govt of India, New Delhi.
11 The Times of India, New Delhi, 4 January 1981, p. 1.
12 Bandyopadhyay and Shiva, 'Development, Poverty and the Growth of the Green Movements in India', *The Ecologist* 19(3) (May/June 1989), pp. 111–117.
13 Mitra, Amit, 'Popular Movements: Chipko', in *Down to Earth*, (special issue) (New Delhi: CSE, 2001), pp. 26–29.
14 Ibid., p. 27.
15 Guha, Ramachandra, *The Unquiet Woods: A History of Social Movements in Uttarakhand* (New Delhi: Oxford University Press, 1991).
16 Ibid., p. 20.
17 Khanna, Gopesh Nath, *Global Environmental Crisis and Management* (New Delhi: Ashish, 1990), p. 230.
18 The Kerala Shastra Sahitya Parishad (KSSP), a committed group which fought the silent valley project, consists of school and college teachers, scientists and progressive-minded citizens.
19 Report of the Task Force for the Ecological Planning for the Western Ghats, NCEPC headed by Zafar Futehally, 1977.

20 Kothari, Ashish, 'Environmental Aspects of the Narmada Valley Project', *Indian Journal of Public Administration* (1989), pp. 480-484.
21 Amite, Baba, 'At What Price the Big Dams?' *The Hindu Survey of Environment* (1991), pp. 25–27.
22 Ibid., p. 27.
23 Gale, Richard P., 'Social Movements and the State: The Environmental Movement and Government Agencies', *Sociological Perspective* 29(2) (Apr. 1986), pp. 202–240.
24 Mundaka Upanishad, 2:13:7, Gita Press, Gorakhpur (1969), p. 31.
25 Abstract from the discussion in 'The Assisi Declaration on Religion and Nature', at Assisi, Italy WWF (London) 29 Sept. 1986 and WWF International 1196, Gland Switzerland, 1986, p. 17.

 Note: The Assisi Declaration was the first in history where five of the major ethical systems of the world have categorically stated that their belief leads them to conservation and protection of human environment. The declaration speaks of values and ethics which challenge many of the assumptions – such as the anthropocentric nature of conservation.
26 Schumpeter, F.F., 'Modern Pressure and the Environment', *MANAS*, Feb. 28, 1975, p. 14.
27 Marx, Leo, 'Nature of Environment', *Science*, Nov. 27, 1970.
28 United Nations, General Assembly Resolution, A/CONF. 48/P.C. –2/Add – May 7, 1970.
29 Khoshoo, T.N., *Environmental Concerns and Strategies* (New Delhi: Ashish, 1988), p. 5.
30 'Environment in the Indian Parliament: Rajya Sabha' Environmental Service Group – WWF-India, 1980, pp. 19–20.
31 State of Environment Report, 1976 Doc. UNEP/G.C./30 Chapter III, Para 54, Nairobi, 1970.
32 Carpenter, Richard A., *National System for the Environment: What Planners Need to Know?* (London: Macmillan, 1982), pp. 192–193.
33 UNEP/GC 6/9 Add.1 (Apr. 4, 1978), p. 3.
34 The Conference was held in Nairobi, from 29 August to September 1977. Under the Resolution of General Assembly No.3337 (XXIX) of 11 December 1974, preparatory work for the Conference was vested in the executive director of UNEP; and Res. 3511 (XXX) of 15 December 1975 entrusted the Governing Council to act in its capacity as the intergovernmental preparatory body for the Conference. The Conference attended by 95 states and other NGOs/IGOs adopted a plan of action to combat desertification.
35 UNEP/GC 7/9, 1982, p. 17.
36 UNEP/GCSS 1/2, 1982, p. 109.
37 Jaiswal, P.L. (ed.), *Desertification and Its Control* (New Delhi: Indian Council of Agricultural Research (ICAR), 1983).
38 Year Book of United Nations, UN: Publications Division, (1984), p. 689.
39 Government of India, Annual Report, Ministry of Environment and Forests, 1985-86, p. 4.
40 UNEP/GC. 55.1/2, Dec. 1, 1987, p. 45.
41 FAO, The State of Food and Agriculture, Rome, 1982, pp. 49-50.
42 Gadgil, M. and R. Guha, *Ecological History of India* (New Delhi: Oxford University Press, 1992).
43 Ibid., p. 32.
44 Valmik Thapar was the executive director of the Ranthambhore Foundation and South Asian chair of the Cat Specialist Group of IUCN – The World Conservation Union.
45 Valmik, Thaper, 'Millennium massacre', *Our Planet* 10(5) (2000), p. 31.
46 FAO, *Tropical Forests in the Areas and the Pacific* (Rome: FAO, 1976), p. 11.
47 Tolba, M.K., *Development without Destruction: Evolving Environmental Perception* (Dublin: Tycooly International, 1982), p. 13; FAO, *Forest Resource of Tropical Areas* (Rome: FAO, 1981), p. 81.
48 ESCAP, *Statistical Year Book for Asia and the Pacific* (Bangkok: ESCAP, 1992).
49 UNEP: Bangkok, UN Pub. 1984, p. 8.
50 FAO, op. cit., p. 88.

51 Eckholm, Erik P. *The Losing Ground: Environmental Stress and World Food Prospects* (New York: W.W. Norton and Company, Jan. 1, 1900).
52 FAO, *Forestry Paper: Forestry for Local Community Development* (Rome: FAO, 1987), pp. 5–7.
53 'State of India's Environment in 1984-85: "The Second Citizens" Report', New Delhi: CSE, 1985, and Bana, Sarus, 'Major Irrigation Projects Non-viable', *Financial Express*, 1 July 1987.
54 Paranjape, Vijay, 'Dams and Their Dangers', *Seminar* 333 (Feb. 1985), pp. 40–43.
55 Sawant, P.R., *River Dam Construction and Resettlement of Affected Villagers* (New Delhi: International India Publication, 1985).
56 Shatlwgna, M., 'Unrehabilitated Poor of Sri Sailam Project', *Economic and Political Weekly* 16(52) (26 December 1981), pp. 21–23.
57 Ibid., p. 24.
58 'Rehabilitation of Project Affected Person', Directorate of Information and Public Relations (Maharashtra), *Lok Rajya* 42(10) (Sept. 16, 1986), pp. 6-8.
59 Bana, Sarosh, 'Narmada Project Portends Disaster', *Indian Express*, 15 April 1987, p. 1.
60 Centre for Science and Environment, op. cit., p. 130.
61 Vegas, Philip and Geeta Menon, 'The Social Costs of Deforestation', *Social Action* 35(4) (Oct.-Dec. 1985), pp. 326–350.
62 Tiwari, S.D.N., 'Development of Forest in Narmada Basin', Paper presented at Seminar on Narmada, New Delhi, August 1987.
63 Kalpavriksha and the Hindu College Nature Club, 'The Narmada Valley Development Project, Development or Destruction', *The Ecologist* 15(5/6) (1986).
64 Rai, Usha, 'Dams Threaten Holy Places', *Times of India*, 23 December 1986, p. 17.
65 Paranjape, op. cit., p. 46.
66 'Siltation Threat to Hirakud Dam: Odisha seeks help from National Agencies', *Sambad English Bureau*, June 23, 2015. (https://sambadenglish.com)
67 Singh, N.K., 'Narmada Project – Charming Controversy', *India Today*, May 31, 1987, pp. 164–165.
68 UNEP/GCSS, 1/2, para 80, p. 33.
69 Ibid., para 82(e), p. 34.
70 ESCAP 'State of Environment in Asia and Pacific – UN', Bangkok, ECU/OES/MCEA/PM/4.1988.
71 UN. Doc. A/Con. 1977.
72 Ibid.
73 UNEP/GC, 55 1/2 para 80, p. 33.
74 Ibid., Paragraph 81, p. 34.
75 Annual Report 1988–1989 of Ministry of Environment and Forest. Government of India, pp. 42–43.
76 Agarwal, Anil and Sunita Narayan (eds.), *State of India's Environment: A Citizens' Report. 'Dying Wisdom, Rise, Fall and Potential of India's Traditional Water Harvesting Systems'* (New Delhi: CSE, 1997).
77 Khanna, op. cit., p. 130.
78 Agarwal, Anil, *State of the Environment Series, Slow Murder: The Deadly Story of Vehicular Pollution in India* (New Delhi: CSE, 1996).
79 Ibid., p. 55.
80 Agarwal, Anil and H.B. Mathur, 'Chowla, Peter and the Air Pollution control Unit, CSE', Special Report: The CNG Imperative, Down to Earth, vol.9, no 21, 31 March 2001, p. 28.
81 Ibid., p. 29.
82 United Nations General Assembly Resolution, 35/74 and 37/7 and Governing Council Resolution, 10/13 and 11/7 in Part VI.
83 Thaper, Valmik. 'Millennium Massacre' 'Our Planet': The magazine of the United Nations Environment Programme, vol.10, no.5, 1999, p. 32.

84 Khoshoo, T.N., 'A Case for Conservation', *The Hindu Survey of the Environment*, 1991, p. 125.
85 Ibid., p. 122.
86 Khanna, op. cit., p. 131.
87 Biodiversity is understood as the sum total of the species richness (e.g., plant, animals and microorganisms in a community or ecosystem).
88 Shamim, Mohammad and Jairaj Puri, 'Habitat: The Change That Destroys', *The Hindu Survey of the Environment*, 1991, p. 129.
89 Ibid., It provides the framework of strategy as well as programmes for the wildlife conservation.
90 Khoshoo, op. cit., p. 127.
91 Khanna, op. cit., p. 140.
92 Ibid., p. 141.
93 UNEP: GC 14/3 and GC 14/13 Apr. 1988.
94 UNEP/GCSS 1/3 of Dec. 1988, p. 23.

4
ENVIRONMENTAL GOVERNANCE AND SUSTAINABLE DEVELOPMENT IN SOUTH ASIA

A Study of Bangladesh, Nepal, Pakistan and Sri Lanka

Introduction

The South Asian Region: An Environmental Perspective

As nations develop, different sets of environmental concerns assume priority. Initially, prominence is given to issues associated with poverty alleviation, food security and development, natural resource management to prevent land degradation, providing adequate water supply, protection of forests and coastal areas from over exploitation and irreversible degradation. Attention to issues associated with increasing industrialization then follows, which include uncontrolled urbanization and infrastructure development, energy and transport expansion, the increased use of chemicals and waste production. More affluent societies focus on individual and global health and well-being, the intensity of resource use, heavy reliance on chemicals, and the impact of climate change and ozone destruction, as well as remaining vigilant on the long-term protection needs of natural resources.[1]

Progress through policy responses is often constrained in developing regions by weak institutions, insufficient human and financial resources, ineffective legislation, and a lack of effective monitoring and enforcement capabilities.[2] Sometimes, environmental institutions and regulations have been introduced at the request of external forces, such as international conventions and strategies, donor requirements, and structural adjustment programmes and are only internalized by countries. In the more developed regions of the world, experience with environmental management and conservation is extensive and of longer duration. Adequate safeguards in the initial stages were largely achieved through government-regulated command and control policies. Effective implementation of such policies relied on legislation and measures such as emission standards and limits as well as on maximum permitted rates of resource use.[3]

In the last many years, most countries in the South Asian region have undergone unparalleled social, political and economic transformations. Colonialism, which dominated most of the region of South Asia, was replaced by different political systems. Economies, which were largely agrarian, became industrialized, export-oriented and better integrated with global markets. Agriculture was intensified to increase production for home consumption and export. In order to have a better understanding of the environmental perspective of the South Asian region, the interlinkages of all these issues have to be understood.

Economic Growth

Rapid industrialization and economic growth have changed virtually every dimension of life. Yet, by many measures of education, health, nutrition as well as income, the quality of life within the region remains poor for most people. At least one in three Asians has no access to safe drinking water and at least one in two has no access to sanitation.[4] The advantages of rapid economic growth have not filtered down to all levels of society. Poverty remains a significant problem, particularly in South Asia, where more than 515 million of the region's 950 million people are poor[5] and around 39 per cent of the population is below the poverty line, with numbers still increasing. Environmental degradation in the region was largely due to poor farming methods, colonial expansionist land practices and mineral exploitation.

Population

Population densities in South Asia are among the highest in the world and there is great pressure on land resources throughout the region – in which some 60 per cent of the world population depends on 30 per cent of its land area.[6] The combination of rural poverty and population pressure has forced people to move to ecologically safer areas. In addition, the number of landless people is increasing.[7] The significant growth in population is one of the causes of environmental degradation and pollution. The combination of high population density and growth, rapid industrialization and urbanization, and poverty has taken its toll on the region's natural resource base, accelerated environmental degradation led to a substantial increase in air and water pollution. Other significant environmental problems include land degradation caused by deforestation and inappropriate agricultural practices, waterloss and mangrove clearance for aquaculture. Estimates of the economic costs of environmental degradation in Asia range from 1 to 9 per cent of national GNS.[8]

Natural Disasters

Cyclones, floods, storms, earthquakes, droughts, landslides, cloud bursts and volcanic eruptions, wildfires, forest fires and heat waves affect many countries in the region, causing great loss of life and extensive damage to property and infrastructure. These disasters seriously affect the pace of development. One cyclone in Bangladesh in November 1970 caused almost half a million casualties with colossal damage to property and infrastructure. Trends in disaster events show that during the period 1900–1991, there have been more than 3,500 disasters, roughly 40 a year and they have killed more than 24 million people.[9]

Forests

Average per capita forest cover for the region was 0.17 ha in 1995, considerably lower than the world average of 0.61 ha. Though there are large variations within the regions, the 555 million ha of forests that remained in 1995 seem incapable of satisfying the needs of the population, and domestic wood shortage is emerging as a challenge in South Asia. In Nepal, for example, nearly 90 per cent of all the energy consumed is still in the form of traditional fuel.[10] Continued development of urban and industrial infrastructure in forested areas may increase opportunities for forest exploitation by providing easy access for logging and encroachment.[11] Forest fires are another emerging challenge, contributing significantly to forest destruction, and will continue doing so unless major efforts are made to stop them.[12]

Biodiversity

The region includes parts of three of the world's eight bio-geographic divisions, namely the Palae arctic, Indo-Malayan and Oceania realms. The region also includes the world's highest mountain system, the Himalayas, the second largest rain forest and more than half the world's coral reefs. The South Asian sub-region is noted as the centre of diversity of wild and domestic cereals and fruit species.[13] Increased habitat fragmentation in South Asia has depleted the wide variety of forest products that used to be the main source of food, medicine and income for indigenous people.[14]

A major concern in South Asia, particularly in the Indian subcontinent, has been the loss of biodiversity brought about by compounding the long-term pressures on grasslands with rapid growth in human and livestock populations.[15] Hunting, poaching and illegal trade in endangered species have a widespread impact on biodiversity in many countries. In Pakistan, falcons are smuggled to the Middle East, lizards and snakes are killed for their skins, and crocodile hunting is still a popular sport and recreational activity.[16] The Chakaria Sundarbans, the oldest mangrove forest in Eastern Bangladesh, have been almost completely cleared for aquaculture,[17] and Thailand's mangrove forests and more than half of the total area (some 208,2000 ha) disappeared between 1961 and 1993.[18] The Asian Wetland Bureau has estimated that 15 per cent of all wetland habitat in South Asia is afforded some legal protection but only 10 per cent is totally protected. Further, the degree of protection in South Asia is greater than in any other region.[19]

Renewable Freshwater Resources

Freshwater withdrawal from rivers, lakes, reservoirs, underground aquifers and other sources increased more in Asia during the past century than in other parts of the world, from 600 cubic km in 1900 to approximately 5,000 cubic km by the mid-1980s.[20]

Agriculture, mainly irrigation, accounts for the major part of water withdrawals. In the more industrialized countries, agriculture accounts for up to 50 per cent of withdrawals, but this rises to more than 90 per cent in all South Asian countries except Bhutan and reaches 99 per cent in Afghanistan.[21] In common with other parts of the world the exploitation of water resources in the region has caused major disruption to hydrological cycles. Water development programmes for hydropower and to meet domestic and industrial requirements, coupled with deforestation in important watersheds, have reduced river levels and depleted wetlands. In addition, the mismanagement of water resources and increased irrigation have used groundwater reserves faster than they can be replenished; other activities, including the removal of vegetation from stream bank and flood control channelling, have changed the natural character of water course and estuaries.[22] South Asia's record with regard to safe water supply is poor. One in three Asians has no access to safe drinking water source that operates within 200 m of the home.[23] Access to safe drinking water is worse in South Asia. Almost one in two Asians has no access to sanitation services and only 10 per cent of sewage is treated at the primary level.[24] Dirty water and poor sanitation cause more than 500,000 infant deaths a year in the region, as well as a huge burden of illness and disability.[25] According to WHO, diarrhoea associated with contaminated water poses the most serious threat to health in the region, which accounted for about 40 per cent of the total global diarrhoea episodes in the under-fives.

A number of toxic pollutants also affect human health. For example, Asia's surface water contains 20 times more lead than surface waters in OECD countries, mainly from industrial effluents.[26] The worst lead contamination in the region is in South Asia.[27] Bangladesh and some parts of India suffer from arsenic contamination of groundwater, having disastrous effects on the local population. Agrochemical inputs, including fertilizers and pesticides, and animal wastes from livestock are a growing source of freshwater pollution. Excessive levels of nitrates from agricultural run-offs are a major cause of eutrophication throughout the region.[28] Demand for water will increase throughout the region into the next century. By 2025, India is expected to be water stressed with per capita water availability decreased to some 800 m^3. While agriculture will continue to use most water, freshwater demand is growing fastest in the urban and industrial sectors.

Oil and Air Pollution

Oil pollution is a significant problem along major shipping routes, and an increasing number of accidents have occurred in recent years. In the part of Chittagong in

Bangladesh, about 6,000 tonnes of crude oil are spilled a year, and crude oil residue and wastewater effluents from land-based refineries amount to about 50,000 tonnes a year.[29]

Atmospheric pollution increased significantly in much of the region, largely as a result of escalating energy consumption due to economic growth and greater use of motor vehicles. The use of poor-quality fuels with high sulphur content such as coal, inefficient methods of energy production and use, traffic congestion, poor automobile and road conditions, and inappropriate mining methods have exacerbated the situation. Forest fires are also contributing significantly to air pollution. Transportation contributes the largest share of air pollution to the urban environment. The total number of vehicles in the region in 1996 was about 127 million, 4.24 per cent more than in the previous year.[30] Lead pollution is a particular problem in megacities of South-east and South Asia. Ten of Asia's 11 megacities exceed WHO guidelines for particulate matter (particle pollution in the form of solids or liquids in the air) by a factor of at least 3[31] levels of smoke and dust, a major cause of respiratory diseases, are thrice the world average and more than five times as high as in industrial countries and Latin America.[32] Recent studies show that smoke and dust particles can significantly damage human health. According to WHO estimates, Bangladesh, India and Nepal together account for about 40 per cent of the global mortality in young children caused by pneumonia.[33] By 2020, SO_2 (sulphurdioxide) emissions from coal burning in Asia have surpassed the emissions of North America and Europe combined,[34] and if, current trends continue, they will more than triple within the next few years. A study of Nepal, for instance, estimates that total emissions will increase fivefold, about two-thirds of which will come from the transport sector.[35] While the region's contribution to the greenhouse effect and total world emissions of atmospheric pollutants are currently limited, both are increasing fast. Air quality is proving detrimental to human health in many parts of the region. Levels of urbanization in the South Asian region are relatively low. Some 23.6 per cent of Asians lived in urban areas in 1975, increasing to 34 per cent in 1995, less than half the levels found in North America, Europe and Latin America. However, with rapid economic development, particularly over the past 30 years, urban populations have increased fast with most of the urban population concentrated in a few cities. The impacts of rapid urbanization include encroachment on agricultural and forest lands, urban air and water pollution, unavailability of safe drinking water and the overexploitation of groundwater, increasing traffic congestion, noise pollution, and significant increases in municipal and industrial wastes.[36] India and Pakistan are home to the largest and fastest-growing cities in the subregion. Karachi and Mumbai are growing at 4.2 per cent per annum, followed by Delhi at 3.8 per cent per annum. Slums are growing in many cities. In Colombo, for instance, some 50 per cent of the population resides in slums and squatter areas,[37] while a similar percentage applies to the entire urban population in Bangladesh. Growing population has frequently outpaced the development of urban infrastructure. Access to sanitation ranges from 652 per cent in Pakistan to 100 per cent in Maldives.[38]

South Asia's environmental crisis is mainly a result of market and policy failures, neglect and institutional weaknesses.[39] The most important external force shaping the region's future is the increased integration with the world economy. The present scenario suggests that rapid economic growth and industrialization may result in further environmental damage, and Asia's particular style of urbanization, towards megacities, has exacerbated environmental and social stresses.[40]

The Asia Pacific region is responsible for around 75 per cent of all human-induced salinization in arid, semi-arid and dry sub-humid areas, the susceptible drylands of the world.[41] Irrigated agriculture has degraded existing arable lands and resulted in vast expanses of salinized and waterlogged soils. Pakistan and India could alone account for about 50 per cent, 30 million hectares, of the world's irrigated land damaged by salinization.[42] In Pakistan salt build-up in the soil is known to reduce crop yields by 30 per cent.[43]

Deforestation

Excessive cutting and clearing for agriculture, including commercial plantation crops, have been the two major direct causes of deforestation.[44] Illegal and unmonitored logging is also a significant cause of deforestation. Fuel-wood harvesting, irrigation and hydroelectric power development, mining, the expansions of urban and industrial infrastructure and railways, invasive species and cyclones have also contributed significantly to deforestation. Forest fires have also had a substantial impact.[45] In the light of rising population growth rates, particularly in the environmentally endangered regions of South Asia, one of the vital problems to be solved is that of an environmental consolidation and a sustainable future management of renewable resources.[46]

Economic development or the lack of it has a bigger impact on the environment than environmental actions per se. Effective environmental strategies exert a strong influence, if they are an integral part of development planning and policy-making. The National Conservation Strategies of Bangladesh and Sri Lanka were developed in parallel to development planning, without any tangible links to the development planning systems of either country. As a result, their success has been limited. The strategies of Nepal and Pakistan are one step closer to achieving integration with development planning. These strategies have included environmental sustainability components in development plans, and even corporate strategies and programmes in each country have established some form of link with the development planning process. This has given them credibility within key economic and planning agencies; as a result, they have succeeded in influencing major policies and plans. The National Conservation Strategy and many of the state's strategies undertaken were planned in close collaboration with the economic planning unit in the Prime Minister's Office and the economic planning units in the states. The Nepal National Conservation Strategy (NCS) was developed and implemented under the umbrella organization of the National Planning Commission. Pakistan has developed a formal link with the commissions, in the form of an environment unit, to help implement the NCS. Nepal has achieved the greatest degree of internalization within government, and a solid working relationship between the National Planning Commission (NPC) and sectoral ministries has been built up. The Nepal NCS has made significant contributions in the fields of environmental impact assessment, pollution control, environmental law, environmental planning, heritage conservation and environmental education. With top-level government and political support, the NCS has successfully been adopted in Pakistan as the environmental agenda. All the countries of South Asia have emphasized fitting these initiatives with the requirements of agencies such as the World Bank, United Nations Commission on Environment and Development, UNEP and Agenda 21.

Bangladesh

Bangladesh is the youngest state in South Asia. It has already undergone a number of political vicissitudes, moving from one-party dominance to one-party rule, to military regimes to a democratic transition. Unlike Pakistan, where it proved to be a torturous and time-consuming process, constitution-making was comparatively easy in Bangladesh primarily because of the relatively higher degree of ethnic and linguistic homogeneity.[47] In Bangladesh, an increasing concern for environmental issues, mostly at the state level, has been heightened by the two consecutive, devastating floods of 1987 and 1988, the fear of global warming, deforestation, desertification and also alarm caused by the serious depletion of flora and fauna necessary for maintaining the balance between environment and human existence compelled the government to take people as the central concern of Bangladesh's ecological future.[48] It is argued that the ecological problems and other related issues are significantly linked with the depletion of resources, land degradation, deforestation, pollution and extinction of various life forms, which is causing ecological menace and is leading Bangladesh to an 'ecological inferno'.[49]

Ecological Issues: Major River Systems

Ecological issues in Bangladesh are very much linked with geo-physical and climatic conditions, population pressures and also the availability of land resources, which are used by man for building and which also constitute the principal means of destruction of three major river systems, the Ganges-Padma, the Brahmaputra-Yamuna and the Meghna-Surma River, and these river systems combine to carry the world's biggest sediment load, thus forming the world's largest delta. The constantly active rivers in the delta change their flow, making the area vulnerable to deltaic behaviour. Heavy monsoon rainfall is a characteristic of Bangladesh, with abnormal rainfall causing floods. The flood-vulnerable area is very large, totalling some 82,088 sq. km, or 58 per cent of the total area of Bangladesh. The shift in river courses and riverbank erosion during the monsoon and flood dislocating the rural population is a normal feature in Bangladesh. It has been estimated that riverbank erosion dislocates an estimated one million people. The ecological, demographic and socio-economic consequences of the problem are far-reaching and often enormous.[50]

Cyclones, Population and Poverty, Deforestation

The general human development conditions are quite dismal, and more than half of the population survives in absolute poverty, the highest in South Asia. One of the greatest assets of Bangladesh is the vitality of its civil society. Some of the best-known NGOs in the world operate in Bangladesh, including the world-renowned Grameen Bank and Bangladesh Rural Advancement Committee (BRAC). The vitality and vigour of civil society in Bangladesh demonstrate a good deal of progress in human development which can be achieved at a fairly low cost. Bangladesh is also subject to several tropical cyclones which generally originate in the Bay of Bengal. In the last two decades, Bangladesh was struck by 33 cyclones, 7 of which were most severe.[51] These cyclonic storms claimed more than half a million human lives. Environmental issues in Bangladesh cannot be properly understood without taking the population dynamics into consideration. It has been rightly said that 'population growth' is the most important, constantly changing factor in the ecological equation affecting the demands on the natural resources of Bangladesh.[52]

Bangladesh, which once was considered a natural greenhouse, is now facing deforestation and environmental degradation. Currently 9 per cent of the total land area is covered by different types of forest and includes both natural forest (4%) and degraded natural forest areas (4–6%). The tropical forest-associated biodiversity in Bangladesh is the most threatened, and only 6 per cent of its original terrestrial habitat remains.[53] The situation has further deteriorated by an increase in demand for forest products such as fuel wood and timber. Although at the state level, there is awareness, Bangladesh still is characterized as a country, where there is almost no understanding and awareness of environmental problems.[54] While the institutional commitment to environmental concerns is at a formative stage, public awareness regarding problems is almost non-existent.[55]

Programme of Environmental Awareness: National Conservation Strategy

State-level concern has been demonstrated by the establishment of the Ministry of Environment and Forest and the declaration of 1990 as the 'Year of the Environment' and 1991–2000 as the 'Decade of the Environment'.[56] It is important that environmental concern is brought from the Secretariat building and conference rooms of five-star hotels to the people's backyard and to the far-flung corners of the country. Over the past several years in Bangladesh, pressure from multilateral banks and bilateral aid agencies to undertake a range of environment plans has resulted in a series of environmental strategy documents and other reports. They include a National Conservation Strategy, two National Environmental Management Plans and several Natural Resource Status Reports. They suggest that what was needed initially was a strong programme of environmental awareness to build the conceptual base for environmental governance institutions through support for strategy development.[57]

The NCS has to identify the obstacles to conservation and sustainable development and prescribe action to overcome them in an integrated, cost-effective manner. The terms of reference called for the establishment of links between conservation and the national development goals of Bangladesh. The strategy covers the following sectors: agriculture, conservation of genetic resources, cultural heritage, energy and minerals environmental education and awareness, environmental pollution, fishery, forestry and forest conservation, health and sanitation, human settlement and urban development, wildlife management and protected areas.[58] With the assistance of IUCN and UNEP, interest in the NCS was revived within the government and the development assistance community, which has emphasized the magnitude of conservation and development problems in Bangladesh and highlighted the need for strategic planning to address them.[59] Environment and development issues are also receiving a great deal of attention from donors and the government at the same time.

Many of Bangladesh's environmental problems are interdependent and closely related to its water resources. Bangladesh's water problems are aggravated by the introduction of green revolution technologies, population growth and inadequate management of resource utilization. All environmental and socio-economic crises are interdependent; whether it is water problem, poverty, fuel crisis, rising population, environmental degradation or difficulty in securing employment, they are not separate problems but interrelated[60] They are part and parcel of one general economic and ecological problem facing Bangladesh. Environmental administration in Bangladesh is characterized by a very high degree of centralization. A large number of ministries, departments and directorates, and statutory bodies are engaged in policies, programmes and actions that directly or indirectly affect the environment.[61] The Department of Environment and Forest was created in 1989 to enhance the capability of the government to effectively deal with environmental issues. The department was formed by renaming the Department of Pollution Control and was placed under the Ministry of Environment and Forestry.

The Environmental Protection Programme of the Department of Environment (DoE) has the following objectives:

1 Long-term sustainable development of the country;
2 Conservation of natural heritage in terms of natural resources;
3 A healthy and meaningful living environment for all citizens, including the poor. Environmental management in Bangladesh still revolves around pollution control.[62]

Hasan provides an illustration of ineffectiveness of environmental governance in Bangladesh.[63] With too many ministries and departments running the environmental programmes, the environmental governance and administration in Bangladesh is not effective and resulted in expanding the bureaucratic machinery. The Department of Environment is given the responsibility to coordinate all the activities relating to environmental management, with different agencies and bodies having their respective chains of command, but because of the lack of coordination, it has become less effective. This state of environmental governance and management is reminiscent of the overall effectiveness and priorities of the Bangladesh agricultural research system.[64]

The poor state of environmental administration and governance in Bangladesh is symptomatic of the overall state of administration and governance. The rules followed by the bureaucrats are, in most cases, outdated, complicated, and ineffective and provide no scope for monitoring transparency and above all, accountability.[65] In Bangladesh, like the developing world, a common method of exploiting a position of public responsibility is the threat of obstruction and delay. Hence, corruption impedes the processes of decision-making and execution at all levels. Corruption introduces an element of irrationality in all planning and plan fulfilment by influencing the actual course of development in a way that deviates from the Plan.[66] Bangladesh's environmental governance is in poor shape. Numerous departments are directly or indirectly engaged in environmental management and protection. However, lack of coordination and understanding of the complexities surrounding the issues have rendered the environmental management system largely ineffective. Widespread corruption has further exacerbated the problem. Inadequate, ill-defined and inappropriate policies have led to environmental degradation and undermined sustainable resource use and the livelihoods of the poorer sections of the community in many parts of Bangladesh.

Integration of the SDGs and the Updated and Revised NCS (2016-2031)

The first significant event happened in 1980, when the World Conservation Strategy was formulated. The second significant event was the adoption of the Millennium Development Goals (MDGs 2000)[67] in the late 1990s. The MDGs had among its goals targets related to environmental integrity but this was rather understated. The last event was the adoption of the SDGs in 2015 by the UN which has resource conservation, efficiency in resource use and maintaining integrity of the environment in general, specific natural resources in particular as the main focus and process for attaining, particularly, Goals 1 and 2 on eradication of poverty and hunger.[68] That means increasingly nature has taken a centre stage in development discourse and practice. The updated draft of National Conservation Strategy (NSC) in Bangladesh[69] takes its cue from all these three events (WCS, MDGs, SDGs) as well as the general development discourse surrounding man's relationship with nature.

The Government of Bangladesh is committed to conserving its natural resources guided by policy, law, strategies, international treaties and conventions, and also as enshrined in its Constitution. The Bangladesh National Conservation Strategy has become a key government document to be the guideline for the purpose. After Bangladesh endorsed the 1980s World Conservation Strategy, the government started working on developing the National Conservation Strategy document for Bangladesh. Between 1980 and 1993, several task forces were set up by the government to draft the National Conservation Strategy for Bangladesh. An NCS document was presented to the Cabinet in 1993. Since then, the draft had been revised and presented to the Cabinet in 2013.[70]

After due consideration of the last updated version in 2013, the Cabinet directed the Ministry of Environment and Forests (MoEF) to revise the draft yet once again to consider changes and also incorporate some new areas for analysis and formulation of strategy. The Cabinet also formed a Ministerial Committee headed by the finance minister with the provision of necessary secretarial services by the MoEF. To facilitate the project, the MoEF in April 2013 formed an Expert Committee comprising nine members. Subsequently the Expert Committee decided formulating a study to incorporate the desired revisions and updating and extensions outlined by the Cabinet. The Bangladesh Forest Department (BFD), financed by the Bangladesh Climate Change Trust Fund in collaboration with IUCN Bangladesh Country Office, had finally approved the implementation of this new project on 5 November 2015.[71] The overarching goal of the NCS (revised and updated) is to promote development through the conservation, development and enhancement of natural resources in the country within the framework of sustainable development,[72] particularly as envisioned under the Sustainable Development Goals. The outcome of the project is expected to create a conducive policy environment and strategy for conservation, development and enhancement of natural resources in the country. Finally, a Bangladesh National Conservation Strategy for consideration of the Ministerial Committee and finally the Cabinet for approval and implementation emerged.[73]

Few Observations and Final Decision on NCS: 2015–2030

First, several sectors/activities appeared closely linked and indistinguishable, such as urbanization, urban development and public housing, and these were merged together and subsequently discussed with the Expert Committee, and a clear analysis of each within the composite sectoral report was highlighted. Second, a few more new sectors/activities, such as marine and coastal resources, institutional issues and legal aspects were added to the NCS.[74] An important issue that was still missing and deliberated upon in the NCS was the issue of natural resources and cultural heritage. While accepting the newly added sectors/activities, it was decided in the Expert Committee that rather than dealing with these issues separately, they will be considered in the final draft. Third, the time schedule for the formulation of the NCS appeared to be quite tight. Originally there were 18 sectors mentioned in the last draft of the NCS and the Cabinet added 7 new sectors and activities, raising the number to 25.[75] Another issue of concern was pollution and effluents which can degrade the quality of other natural resources and the sources of pollution, primarily, domestic waste, municipal waste, industrial effluents and run-offs from agricultural fields and natural contaminants, such as arsenic contamination of groundwater are subsumed under specific sectors.

Another issue that came up during the preparation of the background reports was how to treat the issue of climate change, and it was decided that the issue has to be treated in all relevant sectors as this is a cross-cutting issue.[76] Major revisions and almost new analyses for understanding their policy implications that require full reviews were emphasized. Since the previous NCS did not have a holistic view of the issues involved, particularly from the perspective of a largely resource-dependent yet poor country like Bangladesh, a newly developed and globally adopted sustainable development framework and its goals and a wholly new conceptual framework for the NCS for moulding each sectoral profile and evaluation accordingly was used and adopted.[77] One new addition is the evaluation of NCS documents of other countries in the region and elsewhere to understand globally the best practices and assess their suitability for application to Bangladesh.[78]

Integration of SDGs into the National Conservation Strategy

Of the 17 SDGs, the first 5 relate to ending poverty and hunger, ensuring healthy living, education for all and gender balance in development, which are termed as the overarching objectives of the SDGs. The next goals are related to the management of natural resource base for achieving the first five goals.[79] There are specific goals and targets for the following natural resources under SDGs. These are water (Goal 6), modern energy (Goal 7), marine resources (Goal 14), forestry (Goal 15), biodiversity (Goals 2 and 15), inland fisheries (Goal 2) and livestock (Goal 2). All these form the core resources for management under NCS.[80] SDGs also include goals and targets for resource using sectors practically all of which form part of the NCS. These are crop agriculture (Goal 2), industries (Goal 9), power (Goal 7), transport and communication (Goal 9) and urbanization (Goal 11). The implementation of NCS necessitates the use of several strategies, policies and actions.[81] Again, practically all are subsumed under different goals and targets of the SDG for its implementation. These are human resources (Goals 3,4,5), gender issues (Goal 4), health and sanitation (Goals 3 and 6), disasters (Goals 11 and 13), environmental and international treaties and obligations (Goal 13 and 15), information and communication technology (Goal 9), financing (Goal 17), monitoring and coordination (Goal 17), institutional framework (Goal 16), legal aspects (Goal 16) and, lastly, natural and cultural heritage (Goal 11). A non-exhaustive list of mapping of the NCS sectors as approved by the Cabinet Sub-committee and the Expert Committee was linked to the goals and targets of SDGs.[82] That means, the implementation of the SDGs depends crucially on the implementation of the NCS. Indeed, it suggests that the NCS and its implementation form the core of the SDG and its implementation.[83] The NCS details out the characteristics of the configuration of the natural resources in the country along with the sectors that use these resources to create value in the economy and ultimately lead to the attainment of the first two overarching goals under SDG, namely the eradication of poverty and banishing hunger. Time and again the UN resolution calls for the sustainable management of natural resources (in other words, the basis of present NCS) as the process to attain the overarching goals of economic and social well-being. Even the particular issue of natural and cultural heritage,[84] which was omitted under the NCS, has been included as a full-fledged target under Goal 11. Thus, the core issues of a development process, which are environmentally sustainable, economically efficient and socially desirable, under SDGs and the NCS are remarkably regarded as same and similar.[85]

It has been observed that in Bangladesh an effort has been made to integrate Sustainable Development Goals into the National Conservation Strategy, and in the process, an environmental governance architecture has started emerging and evolving.

Nepal

Nepal is one of the least developed countries in the world, with an extremely low income and very poor human development indicators. Nepal was governed under the Rana family autocracy for more than a hundred years, until 1951, when a democratic structure was established for the first time. However, the democratic experiment was short-lived and the panchayat system of local bodies was introduced in 1960. This system lasted for three decades before the constitution of 1990 established Nepal as a multiparty parliamentary system.[86] After the 2015 earthquakes and successful national-level elections, Nepal has steadily implemented federalism and achieved economic progress. Through promulgation of the new Constitution in 2015, successful local and legislative elections in 2017 and successful national elections in 2020 (marking a potential watershed moment in Nepal's political transformation), a process of institutional strengthening and of the decentralization of power both geographically and culturally is underway (Executive Summary: Nepal Country Report 2022).

Nepal is a landlocked country with an estimated population of 12 million. Its geographical location in the Himalayan region makes it vulnerable to two of its powerful neighbours: China and India. Nepal is mainly an agricultural society, with nine-tenths of its population in rural areas and nearly half of its GDP originating from this sector. Land distribution is lopsided, resulting in an uneven distribution of income, which was the main reason for the increase in poverty, from 31 per cent of the population in 1977–1978 to 40 per cent of 1984–1985.[87] Even though Nepal is endowed with abundant land and water resources, only 18 per cent of the land is arable, the rest being mountainous and covered with snow most of the year. Nepal has the potential for generating 25,000 MW of hydroelectric power and supplying electricity to the entire South Asian region. This potential remains largely untapped: actual production of electricity is only 237 MW, or less than 1 per cent of its potential.[88]

Deforestation

Accounts of the 1988 monsoon floods in Bangladesh, which inundated over 60 per cent of the landscape, linked that tragedy to the Himalayan deforestation.[89] The Himalayan environmental degradation holds that the dramatic population growth in the mountains has resulted in deforestation, which is causing catastrophic increases in surface erosion, mass wasting, flooding, down-river sedimentation and desertification.[90] The continuation of current processes threatens to destroy the fragile environmental and social equilibrium of not only the mountainous uplands but also the densely populated lowlands.[91] During the past several years both government policies and subsistence farmers have caused deforestation. Recent deforestation has had a relatively small but unspecified impact on erosion, flooding and sedimentation, and reforestation schemes are not going to solve the environmental problems of the Himalayas.[92] Himalayan deforestation has deep historical roots. Most of the country's upland forests were converted to agriculture and pasture lands before 1960. Contemporary methods and uses are degrading forests at an alarming rate. 'National Planning Commission' estimates that during 1960–1980 the forest cover decreased from 60 per cent to 19 per cent; Karan and Ilima claim that one-fourth of the forests have been cleared in the last decade.[93] The World Bank states that 25 per cent of the Himalayan Forest has been lost in the last 20 years (IBRD, 1979). Government policies of the 1950s and 1960s decreased the security of forest tenure and thereby increased rates of forest degradation. Recent legislative reforms, which reverse the entire history of the government's conservation policies, allow the transfer of large areas of degraded forests and shrublands to local control. Political and bureaucratic systems based on patronage and corruption prevent the mobilization of Nepal's human and material resources to remedy the nation's problems.

National Conservation Strategy

In 1982, Nepal embarked on a National Conservation Strategy, which was completed in 1987 and its implementation coordinated by the National Planning Commission began in 1989.[94] Nepal's experience with the NCS process is one of the longest-running in the world and has been a major force in nurturing innovation and reform for environmental protection in Nepal, despite difficult institutional barriers. Preparation of the NCS document included consultation at national and village levels throughout the country. Political unrest heading to a revolution in 1990 and regular staff changes within government administration made it difficult to maintain consistency in approach and commitment. The NCS implementation programme has concentrated on building the basic components and skills for an environmental governance administration in Nepal. It includes developing national systems of environmental impact assessment (EIA) and planning. The positive responses to these initiatives justify the focus on activities that are of widespread and immediate use and have a strong support base in government and society.[95] In 1992, following three years of implementation, an Environmental Protection Council (EPC) was established. One major contribution to building institutional capacity was the establishment of an Environmental Core Group of officials from various ministries. This helped ensure that awareness of the NCS was widespread throughout the government.

Objectives

The objectives of the NCS are as follows:

- To satisfy the material, spiritual and cultural needs of the present and future generations of Nepal;
- To ensure the sustainable use of Nepal's land and renewable resources;
- To preserve the biological diversity of Nepal in order to maintain the variety of wild species, both plant and animal, and improve the yields and the quality of crops and livestock;
- To maintain the essential ecological systems, such as soil regeneration, nutrient recycling, protection and cleansing of water and air.[96]

The strategy document was intended to provide a long-term perspective on natural resource management to meet the country's development needs and aims to link sustainable development with conservation. It further sets out the results of 18 sectoral and cross-sectoral analyses, on topics including population and human settlement, energy and industrial development, the role of women and biological diversity. Further, a seven-part Conservation Action Agenda covers institutions, conservation awareness, policy, organization and administration, research, inventory and directed studies, resource planning and a programme of integrated village resource management. The NCS process is coordinated by the NPC and managed through a Swiss-funded commission project, with IUCN providing technical backing. The NPC is the principal planning agency of government, chaired by the prime minister, with advisory, rather than implementation responsibilities. The NCS team was closely involved in formulating Nepal's Eighth National Development Plan. The Eighth Plan (1992–1997) includes a chapter on the environment sector and recognizes and applies the various conservation principles of the strategy document, especially with respect to managing natural and cultural resources.

Strong endorsement of the World Conservation Strategy (WCS) by Prince Gyanendra Bir Bikram Shah led to an initiative to formulate an NCS for Nepal. A task force was established in 1982, led by the NPC, the Department of National Parks and Wildlife Conservation, and the Department of Soil Conservation and Watershed Management (DSCWM). Chaired by the vice-chair of the NPC, the task force consisted of high-level officials from the fields of National Planning and Conservation and Environment, as well as IUCN advisors.[97]

Implementation

Implementation did not begin until 1989; the political situation was also uncertain, with four changes of government, between 1989 and 1991. The government asked

IUCN-Nepal to play a major role in promoting implementation. The implementation project is staffed by locally recruited experts and one IUCN expatriate advisor. There were 25 technical staff, most with expertise in ecology, environmental management and environmental engineering, plus 20 support staff.[98]

The NCS document recognizes that it is neither appropriate nor feasible for the government to be totally responsible for implementation. Instead, there is a strong emphasis on involving individual land users, village and district government institutions, and the private sector and non-governmental organizations.[99] The implementation project focused on loopholes in the existing responsibilities of government departments and has worked to stimulate inter-ministerial cooperation in areas such as pollution control, environmental impact assessment and environmental education. The development of an Environmental Core Group (ECG) has been at the heart of the secretariat's approach to implementation. This group has a network of more than 90 senior officials and technical specialists from 20 ministries and departments as well as from all divisions of the NPC. The ECG mandate is to develop, test and apply new environmental policies and procedures. Through the initiative of ECG members, environmental units have been set up in all main departments and ministries. The ECG has mobilized and consolidated support for the NCS process within government and acted as the principal engine for policy reform appropriate to local conditions. It further includes five programmes in the fields of environmental assessment, environmental planning, heritage conservation, education and public information.[100]

National guidelines for environmental impact assessment were developed by the ECG and endorsed by Cabinet in 1991. Sectoral EIA guidelines for industry, forestry and water resources have been drafted and are tested by the relevant departments. It had formulated a national system of policies and standards for environmental planning. The draft has facilitated the preparation and implementation of eight model village environmental plans in two districts. Public awareness includes community-level wall newspapers, a weekly radio programme and a resource centre focusing on environmental issues of concern to Nepal. A nationwide network of journalists, trained through a series of media workshops, is preparing individual and group reports on environmental topics for the local and national media. Apart from these programmed implementation activities, the NCS team is regularly called upon to contribute to the range of additional environmental initiatives. The NCS secretariat was closely involved in preparing the documentation of UNCED. In 1993, the secretariat also helped prepare a National Environmental Policy and Action Plan (NEPAP) at the behest of the World Bank.

To address recent environmental degradation, technical studies are being carried out, including a 25-year Kathmandu Valley Perspective Plan. The status of the NCS, as a project within government, but independent in terms of funding and staging, has had advantages during the political upheavals and subsequent institutional adjustments that have shaken Nepal from the last 1980 to 1992. IUCN's role in providing technical help and stimulation has been very important. IUCN's bond with Nepal stems from the country's long-standing state membership in the union, going back to more than 40 years. It ensured that the NCS always had strong institutional support, even when the government was distracted and administrative links with the strategy process became limited. The long-term commitment by the Swiss funding to NCS process was crucial in maintaining IUCN's involvement.

The main loophole in the NCS process has been the lack of a united and coordinated approach to all elements of strategy implementation by donors and the government. The select range of activities addressed by the NCS implementation project as essential ingredients of an environmental governance and administrative system was well coordinated and supported. However, mechanisms for giving practical effect to the NCS as the strategic framework for all development investment and actions were lacking. The government's slow approach to developing environmental regulatory measures has created delays in implementation and has made the whole process ineffective. UNDP and UNEP attempted to establish an environment coordinating group, but institutional rivalries and imperatives, particularly involving the World Bank, sank the initiative. IUCN's potentially important role, as an independent technical organization helping government effectively to assess its coordinating function, was limited due to lack of resources. A much more concerted effort by donors to broaden the base of support for the NCS in the long term that reinforces the central core of activities and institutional initiatives managed through the NPC is needed. Nepal still has a long way to go in restructuring its political system and in accelerating the pace of its economic growth, environmental preservation and human development.[101]

The National Strategic Framework (2015–2030)

National Conservation Strategy was enacted in 1988 in Nepal, and the NCS implementation was carried out from 1990 to 1997. Although the National Council for the Conservation of Natural and Cultural Resources (NCCNCR) envisaged by the NCS has not been formed, the NCS Implementation Review study (2013) revealed that its implementation has yielded many achievements.[102] Major achievements are highlighted as the mainstreaming of the concept of NCS started in 1990 when the issue of mobilization of natural resources and heritage was incorporated into the Constitution of the Kingdom of Nepal (1990), and the recent periodic plans became successful in enhancing people's awareness and orientating themselves towards environmental improvements and nature conservation. The NCS has had a conceptual impact on the policies and development documents that were formulated after its enforcement. Separate chapters on the environment and natural resource management were incorporated in the Eighth and Ninth Plans,[103] which were formulated thereafter. Incorporation of environmental education in school education, in the bachelor-level curriculum of the Education Faculty and in the training programmes of the governmental and non-governmental sectors and establishment of the Ministry of Population and Environment are taken as milestones in this regard.[104] The NCS (1988) included a series of recommendations, including the introduction of a Churia Conservation Policy, the establishment of a Crop Protection Section for regulating the application of pesticides, and joint pasture management by the Department of Forest and Livestock Management. In addition to these qualitative achievements, some of the direct and indirect quantitative achievements related to nature conservation took place following the implementation of the NCS (1988).[105]

On the other hand, the National Strategic Framework (2015–2030),[106] which became effective in 2015, has covered and highlighted issues like forest, biodiversity and environmental services, agriculture, land management, physical infrastructure development and environment, society, culture and education, climate change and energy, water resource and disaster risk management, environmental education, health and law, and policy provisions.

As the 'Nature Conservation National Strategic Framework for Sustainable Development' (2015–2030) has been formulated at a time when Nepal is recovering from a devastating natural disaster and Post-Millennium Sustainable Development Goals are a priority throughout the world, this framework has become highly topical and relevant.[107] Drafted for the first time in Nepal, this National Strategic Framework is not a strategy in itself but an umbrella framework, which emphasizes nature conservation, sustainable use of natural resources and equitable distribution of their benefits. The Government of Nepal is going to implement this framework through its periodic and annual plans and programmes over the next 15 years (National Strategic Framework for Sustainable Development from 2015–2030). Based on this document, the National Planning Commission plays the role of a facilitator in exploring measures for solving problems faced in nature conservation, as well as in overall development, facing emerging challenges and appropriately addressing the issues that have emerged, guiding the concerned sectors in taking appropriate steps. This has provided a basis for integrating nature conservation into sectoral development, guiding the various ministries in formulating their plans based on sectoral strategies and evaluating nature sensitivity of the programmes prepared by the sectoral ministries.[108]

Rapid industrialization, urbanization, electricity generation, transportation and construction of physical infrastructure have adversely affected agricultural land, forests, reservoirs and the environment, which has, in turn, caused a loss of biodiversity and ecosystems, as well as environmental pollution. This loss of biodiversity and core ecosystems has compelled international attainment of the Millennium Development Goals such as reduction of poverty, hunger and morbidity. An assessment of the millennium ecosystem services, carried out in 2005, corroborated that the increase in global population and concomitant increase in demand for food, water, cloth and energy have led to negative impacts on nature and individual ecosystems. The traditional models of development and concepts that emerged in the 1980s did not adequately address emergent and important environmental and ecosystem issues. After the Rio Convention in 1992, however, environmental and ecosystem concerns started emerging as integral components of socio-economic development. In 2000, of the eight MDGs adopted, only the seventh goal, regarding ensuring environmental sustainability,[109] is directly concerned with nature conservation, while the other goals are concerned with issues of poverty reduction, primary education and gender equity. As these goals, except the goal of ensuring environmental sustainability (Goal 7), are particularly focused on growth and income generation, they did not necessarily generate positive impacts on nature conservation. Sustainable development has faced a number of challenges, including increase in global temperature due to green gas emissions; decline in carbon capacity due to deforestation and forest degradation; emission of harmful chemicals into the atmosphere by industries, factories and atomic plants; and land erosion, siltation and salinization due to exploitation of non-renewable mineral resources over the past four to five decades. In this context, integration of nature conservation into development efforts has become imperative for achieving Sustainable Development Goals.[110]

Programmes such as green economy, green enterprise and ecosystem-based adaptation are coming into prominence, and consequently, Rio+20 has announced its decision to maintain sustainable development as a priority after 2015, when the MDGs (2000–2015) are scheduled to end. Of the 17 goals proposed by post-2015 development agenda, 6 are directly related to environment and nature conservation (i.e., zero hunger [ending hunger, achieve food security and improved nutrition and promoting agriculture]; sustainable management of water and sanitation for all; affordable and clean energy; climate action; sustainable use of the oceans, seas and marine resources; sustainable use of terrestrial ecosystems, sustainable management of forests, combatting desertification and halting land degradation and biodiversity loss), whereas the other 6 are indirectly related to environment and nature conservation (i.e., ending poverty in all its forms; ensuring healthy lives and promoting well-being for all for all ages; sustainable cities and communities; responsible consumption and production; promoting peaceful and inclusive societies, access to justice for all and inclusive institutions at all levels; strengthening global partnership for sustainable development). And 5 reaming SDG goals are: quality education and promoting lifelong learning opportunities for all; achieving gender equality; productive employment and decent work for all; industry, resilient infrastructure, and promoting inclusive and sustainable industrialization; and reduced inequalities within and among countries.[111] Therefore, as the goal of sustainable development after 2015 is to achieve a balance between development and nature conservation, integration of nature conservation in development interventions has become imperative. In 1980, the International Union for Conservation of Nature (IUCN) prepared and started implementing the World Conservation Strategy.[112] Being a member state of IUCN, Nepal also formulated and enforced the National Conservation Strategy in coordination with the National Planning Commission in 1988. As there were no other strategies at that time, NCS 1988 served as the overarching national strategy on nature conservation.[113] After the country's political shift in 1990, various sectoral strategies and master plans were formulated. Since then, these strategies and master plans have been guiding thematic and sectoral development programmes. In order to graduate from the category of least developed countries to that of a developing country by 2022 as per GoN's (Government of

124 Environmental Governance and Sustainable Development

Nepal) decision,[114] Nepal tries to achieve progress at a rapid pace, and this progress is realized through proper planning so as to reduce the impacts of future natural disasters like the earthquake that occurred in 2015. Through the efforts of a single sector, it will not be possible to achieve goals such as an 8 per cent GDP growth rate and other goals within the next few years; increasing irrigation coverage from 1,311,000 ha to 1,713,000 ha by 2027; rehabilitating 1.6 million ha of degraded land by 2033; increasing the proportion of population with access to electricity facility from the current 67 per cent to 87 per cent; achieving 'one house, one tree' and 'one village, one forest'; maintaining open space for every 25,000 persons in each municipality area; maintaining a playground in each electoral constituency; bringing down overall poverty from 23.8 per cent to 18 per cent by the end of the Thirteenth (2013–2016) and Fourteenth Plan (2016–2018) periods and rural poverty from 27 per cent to 10 per cent within 20 years; and other goals.[115] Further, as natural resources are used in the course of development, achieving balance between overall development and nature conservation has become imperative. Since thematic and sectoral strategies are centred on regional development, a single umbrella strategy[116] has been adopted to guide overall development, to play a coordinating role in their implementation in order to achieve the Sustainable Development Goals. Although the NCS has previously adopted such objectives, they could not be implemented effectively. Furthermore, as conceptual changes have taken place in the country in a number of areas, including nature conservation, since 1988, this strategic framework for nature conservation has been prepared for sustainable development to suit the contemporary context.

This National Strategic Framework, formulated in Nepal for the first time, is an umbrella strategy that emphasizes nature conservation, sustainable use of natural resources and equitable distribution of their benefits, and hence, it covers all other sectoral strategies related to nature conservation. The GoN will be implementing this framework through its periodic annual plans and programmes over the next 15 years. The NPC will play a coordinating role and will also guide the concerned sectors to take appropriate measures, in order to explore solutions to issues encountered in nature conservation and overall development, and will also address emerging challenges and concerns, based on this document. In this way, the framework has provided a basis for integrating nature conservation into sectoral development, guiding the various ministries in sectoral strategy-based planning processes and evaluating nature sensitivity of the programmes prepared by the thematic and sectoral ministries. This framework in six chapters[117] has analysed, introduced and presented a detailed account of how nature conservation is integrated into development efforts, highlighting arrangements for the implementation of the framework and assessing monitoring, evaluation and information management.

Limitations

Although the Nature Conservation National Strategic Framework identifies priority sectors and sub-sectors, it does not encompass all the activities that are carried out by the thematic and sectoral agencies. Furthermore, the issues that directly or indirectly affect conservation are covered taking into consideration only the economic, social and environmental aspects. These priority sectors have been identified and prioritized based on the NCS (1988), its review, and emerging issues and challenges. It has been further revised and refined following interactions with stakeholders at various phases. Hence, although efforts have been made to cover all sectors concerned with nature conservation in this framework, it has not been possible to include many sectors, owing to a lack of periodic data on progress in the past.[118] Nevertheless, efforts have been made to review the actual situation by carrying out, among other assessments, interactions on common issues among various stakeholders and reviewing available data.

It has been concluded, with these observations, that the strategy has included environmental sustainability components in development plans, and even corporate strategies and programmes have established some form of link with the development planning process. This has given them credibility within key economic and planning agencies, and as a result, they have succeeded in influencing major policies and plans. The National Conservation Strategy and framework and many of the state's strategies undertaken were planned in close collaboration with the economic planning unit in the Prime Minister's Office, and the economic planning units in the states. The Nepal National Conservation Strategy and National Conservation Framework were developed and implemented under the umbrella organization of the National Planning Commission. Nepal has achieved the greatest degree of internalization within government, and a solid working relationship between the National Planning Commission and sectoral ministries has been built up. The Nepal NCS and National Conservation Framework have not only made significant contributions in the fields of environmental impact assessment, pollution control, environmental law, environmental planning, heritage conservation and environmental education but also played a major role in strengthening the environmental governance institutions.

Pakistan

The divergence between economic growth and human development is greater in Pakistan than in other South Asian countries. Economic growth has created the potential of the upliftment of human lives. Yet widespread poverty prevails because of powerful interlocking feudal industrial interests, which pre-empt most of the gains of development.[119] Pakistan has gone through a turbulent period of nationhood in the 75 years of its existence. During this period three separate constitutions were promulgated in 1956, 1962 and 1973. It took Pakistan 23 years after independence to hold its first general elections to Parliament in 1970 on the basis of adult franchise. For nearly half of its existence, Pakistan has been ruled by the military. And despite the democratic transition since 1988, the influence of the army over the country's politics remains strong.[120]

Pakistan presents a fascinating combination of many contradictions. The country has one of the lowest literacy rates in the world; yet some of its highly educated people have dominated many international forums. The country treats its women very poorly, with some of the lowest indicators of gender development. And yet it elected the first-ever woman prime minister in a Muslim country, not just once but twice within five years. Fascinating contradictions abound in a country of weak institutions and strong individuals, of economic growth without human development, of private greed and lack of social compassion, of election rituals without real democracy.[121] Concern over the pollution danger and other threats to the environment has grown considerably in the last two to three decades. While the protective sentiments for the ecosystem have galvanized major community movements in the advanced countries, it is yet to create popular involvement in developing countries like Pakistan.

National Conservation Strategy

Pakistan's National Conservation Strategy (NCS) began with a two-year start-up phase, followed by three years of preparation; during this time a strategy document was prepared, viewed, revised and submitted for cabinet approval and was approved in 1992.[122] The main implementation phase was launched with a donors' conference in January 1993, although some implementation began in 1991, with allocations in the federal budgets for 1991–1992 and 1992–1993. The implementation plan emphasized institutional strengthening and capacity building and recognized that projects are important because they are the most visible evidence of implementation.[123] The strategy is extended into the provinces, with the development of provincial conservation strategies, for example, the Sarhad Provincial Conservation Strategy in the North-west Frontier Province.[124]

The Pakistan NCS shows the value of high-level support for strategy process from the national government. The NCS is a strategy to conserve the environment of Pakistan, maintain its resource base and ensure that its development efforts are environmentally sound. It further assesses the state of Pakistan's environment, including land, freshwater and marine environments and resources, biodiversity, minerals, energy supply and demand, and cultural heritage. In addition, it reviews governmental, non-governmental and community institutions and organizations concerned with environment and development and also analyses the environmental implications of Pakistan's economic structure and policies/sectoral policies and laws.

In the second section, it proposes elements of a strategy covering agriculture, forest management, land rehabilitation, livestocks management, water resources, marine and coastal resource management and so on and outlines supporting policies, measures on population, education, communications, research and technology, women in development, training and environmental information system.

The third section of the strategy sets out a ten-year action agenda and implementation plan which identifies 14 core programmes, such as, maintaining soils in croplands, increasing irrigation efficiency, protecting watersheds, supporting forestry and plantations, restoring rangelands and improving livestock, protecting water bodies and sustaining fisheries, conserving biodiversity, increasing energy efficiency, developing and deploying renewable resources, preventing pollution, managing urban wastes, supporting institutions for common resources, integrating population and environment programmes and preserving the country's cultural heritage.[125] For each of these programmes, the strategy specifies long-term goals, expected outputs and critical tasks for the federal and provincial leadership, government departments, districts, communities, individuals and non-governmental organizations. It further proposes building institutions to support the action agenda and implementation plan, paying particular attention to federal provincial leadership, increasing inter-agency cooperation, enhancing departmental capacities, improving district level coordination, involving the corporate sector, and cooperating with communities and NGOs. The aims of the NCS are highlighted as to introduce fundamental changes in work and lifestyle to protect the interests of present and future generations.

- To facilitate the integration of environmental considerations into the daily economic, social and physical decisions of individuals, communities, companies and government;
- To facilitate the incorporation of environmental policy considerations into government's economic, social and physical development process;
- To revitalize community-based management of the sustainable use of common resources and infrastructure.

Objectives

The NCS objectives are defined according to conservation, sustainable development and efficiency to maintain essential ecological processes, to preserve biodiversity and to restore degraded natural resources cost-effectively. Principles governing the NCS process relate to public participation, integration of environment and economics, and quality of life. Ensuring greater public participation in development and environmental management to achieve better public awareness of environmental concerns and promoting participation by strengthening grassroots institutions are emphasized. A merger of environment and economic development in decision-making by allocating environmental responsibility to economic decision-makers and installing environmental monitoring systems is highlighted. Constituent programmes of the implementation plan have been incorporated in the Eighth Five Year Plan, which also includes a chapter on environment, based on the NCS document. An environment section has been established in the planning and development division within the Ministry of Planning to provide environment-economic policy analysis that integrates environmental considerations into planning, which will review and promote sustainable development projects. This is expected to further ensure the integration of NCS programmes into existing planning and decision-making.

The Government of Pakistan asked IUCN for a National Conservation Strategy, and IUCN prepared a document outlining the rationale and scope of an NCS, a rough budget and the necessary organizational arrangements. On the recommendation of IUCN, the Environment and Urban Affairs Division of the Planning and Development Department was given the responsibility to implement the strategy. The prospectus and 12 sectoral papers were discussed in a national workshop sponsored by IUCN in 1986, which was attended by some 50 national experts who reached a consensus that the country's environmental problems were serious and rapidly getting worse and that they have to be addressed in a cross-sectoral manner. The proceedings of the workshop were jointly published in 1987 (GOP, IUCN and CIDA: Towards a National Conservation Strategy for Pakistan: Proceedings of the Pakistan Workshop, Asian Art Press, Lahore). The NCS secretariat was responsible for preparing the strategy document, which took three years, wherein more than 3,000 agencies and individuals were consulted, including schools and colleges in Islamabad, Karachi and Lahore.

Review of information, consideration of options and reaching an agreement involved:

- Provincial and sectoral workshops to review draft sector papers and generate options;
- Drafting sub-committees to correct sectoral recommendations and policy options;
- Comments on first, second and third drafts of the NCS document from all divisions of the federal government and all four provincial governments;
- To consider principles, work plans and option drafts, and to review and approve the NCS document.[126]

The draft NCS document was approved by the Steering Committee in July 1991 was further, approved in a special one-agenda session of Cabinet in March 1992. IUCN-Pakistan played a key role in facilitating the strategy, whereas National Development Plan and UNEP played a supporting role in funding sector workshops, ensuring that the strategy was locally relevant.

Implementation

Under its contract with IUCN, CIDA retained responsibility for monitoring and evaluating the strategy's preparation. Four monitoring missions were undertaken during this phase which facilitated a participatory evaluation by associated individuals (a Canadian consultant). Participants considered the NCS an excellent technical and policy framework for sustainable development in Pakistan. The action plan in the NCS document was not specific regarding an implementation programme. Therefore, IUCN conducted nine implementation design workshops, involving research, federal, provincial, education and training, judicial, urban, and business, communications, and community development institutions and constituencies. Cabinet approved the strategy by setting up an NCS Implementation Committee, headed by the Minister of Environment, including other key ministers. The Implementation Committee approved an action plan with four components:

- Strengthening technical, regulatory and planning institutions, local participatory institutions to coordinate the NCS, improving federal and provincial environmental protection agencies, strengthening technical institutions;
- Formulating a coherent broad-based communications campaign for mass environmental awareness;
- Creating a supportive framework of regulations and economic incentives;
- Implementing projects in NCS priority areas.[127]

The first three components demonstrate a clear emphasis on capacity building, with the recognition that the strategic design must not be buried under a pile of individual projects. Of the 14 core programmes of the strategy document, 8 were prioritized to be implemented:

- Increasing irrigation efficiency;
- Protecting watersheds;
- Restoring rangelands and improving live stocks;
- Protecting water bodies and sustaining fisheries;
- Conserving biodiversity;
- Increasing energy efficiency;
- Preventing and abetting pollution;
- Managing urban wastes.[128]

The projects are divided into three categories: enhancing human welfare, increasing economic efficiency and promoting nature conservation. Other objectives are pointed out:

- Roundtables established to involve the corporate sector;
- Increased participation by women;
- Training in environmental assessment;
- Setting up of a Sustainable Development Policy Institute.[129]

The NCS is extended into the provinces through the development of provincial conservation strategies. An environment sector, which has as its counterpart the Sarhad Provincial Conservation Strategy (SPCS) unit, has been set up in the Planning and Development Department of the Government of the North-west Frontier Province. The task of the environment sector is to ensure that environmental issues and concerns are fully reflected in all public planning. The SPCS is developed with the approval of the chief minister and is overseen by a Steering Committee chaired by the additional chief secretary, who heads the Planning Environment and Development Department. Steering Committee members include secretaries and heads of government departments, politicians and representatives of NGOs, the media, the chamber of commerce and industries with IUCN providing technical assistance, including the coordinating facilities.

Creation of the cabinet Implementation Committee immediately after approval of drafting of the implementation plan and the broad constituency of support for the NCS were regarded as key factors favouring implementation. The NCS has set a course, but it has to be extended beyond the federal government to provincial and local governments, business NGOs and communities, which emerged as a big challenge. The strategy has made great progress, but some significant challenges and limitations surfaced, for example, institutional development with continuing and expanded political will, additional human and financial resources, greater environmental awareness and attention to social equity. Remaining problems include lack of education, entrenchment of vested interests and a lack of good governance, impacting further strategy development.

The relatively broad-based Steering Committee has been replaced by a Cabinet Committee of Federal Government, where the decisions have the force of the Cabinet. The strategy has benefited in two ways to drive it forward and keeping it on track. One was the Steering Committee and NCS secretariat, and the other was IUCN-Pakistan. Communicating the strategy, its process and its messages is key to its success. Although a fairly sophisticated communication programme existed, the NCS team realized that specialized formal and informal communication skills are needed to reach communities and other potential participants. The NCS has been criticized for its inadequate emphasis on equity and poverty, the basic needs of the majority of people and population. The first cycle of the strategy set a clear and strong direction and is somewhat modest in scope, increasing the issues it addresses in later cycles as progress is made. Similarly, the strategy has been criticized for concentrating on the federal government without fully involving departmental staff and for not involving provincial governments, communities, and the corporate sector. Greater participation is needed but this too can be achieved through successive cycles of the strategy.

Pakistan and the 2030 Agenda for Sustainable Development

Pakistan is committed to the 2030 Agenda for Sustainable Development right from its inception, in 2015. In February 2016, it became the first country in the world to adopt the Sustainable Development Goals as part of its national development agenda through a National Assembly Resolution.[130] Learning from the experience of the Millennium Development Goals, Pakistan's national and provincial assemblies established SDG Task Forces to oversee progress on the goals. In fact, Pakistan started a conversation around the post-2015 Agenda as early as 2013, when nationwide consultations helped identify the priorities that were weaved into Pakistan's national development perspective. These developments reflect Pakistan's commitment to the SDGs.[131] More recently, the 12th Five Year Plan and provincial medium-term development strategies are all aligned with the 2030

Agenda. Pakistan's political commitment to these priorities, such as reducing poverty and child stunting, improving transparency and accountability, and promoting gender equality and women's empowerment, supports the 2030 Agenda.[132] To improve vertical and horizontal coordination among different tiers of government and non-governmental stakeholders, seven SDG Support Units have been established at the federal, provincial and federally administered area levels. These units, guided by the federal Ministry of Planning, Development & Reform (Mo PD &R) and provincial and administrative area Planning & Development Departments (P&DDs), have been instrumental in collating Pakistan's first Voluntary National Review (VNR).[133] The review process encompassed several comprehensive and inclusive stakeholder consultations, spread over months, focusing on seven predetermined themes. Pakistan has designed a comprehensive National SDG Framework, which was approved by the National Economic Council (NEC), the country's highest economic policy-making forum, in March 2018.[134] This framework sets baselines and targets for SDG indicators and feeds into the SDGs' Monitoring and Evaluation Framework. The framework is now guiding the provinces and federally administered areas to determine their development priorities, based on local needs. To bolster the implementation of the SDGs, the provinces have instituted Technical Committees and Thematic Clusters. The nomination of focal persons at all levels of government, down to the districts, is helping them to align their development priorities with the 2030 Agenda. This institutional arrangement has been instrumental in guiding the alignment of federal and provincial national policies, sectoral plans and growth strategies with the 2030 Agenda.[135] National data collection tools have been modified to improve data availability, with a focus on the inclusivity, equity and sustainability aspects of the SDGs. Transparency is a major hallmark of monitoring and evaluation architecture.[136] The Local Government Summit in 2017 and several other events have been arranged to raise awareness among grassroots-level public officials and parliamentarians, to prioritize the SDGs in response to local needs.[137] Civil society and academia are fully supportive of the government in terms of achieving these targets. Pakistan is also working to implement the inclusive nature of the 2030 Agenda by developing communication platforms that cater to cultural, linguistic and geographic diversity, to ensure that 'no one is left behind'. Since 2016, several policies and laws have been approved and promulgated. A greatest number of legislative frameworks relate to SDG 16 ('Peace, Justice and Strong Institutions'), SDG 8 ('Decent Work and Economic Growth') and SDG 4 ('Quality Education').[138] Commitment to poverty alleviation remains a key focus. Through Pakistan's multi-sectoral poverty reduction strategy and targeted interventions, progress has been made despite persistent challenges. Over the past ten years (Voluntary National Review – Pakistan 2019), the poverty headcount has fallen by 26 percentage and multi-dimensional poverty by 16 percentage.[139] The national poverty alleviation programme was launched in 2019 to expand social protection, safety nets and support human capital development throughout the country. The National Socioeconomic Registry is updated to target the poorest more effectively and to ensure that no one is left behind.[140] Despite its miniscule carbon footprint, Pakistan faces the enormous impacts of global climate change. Therefore, climate adaptation is imperative for the country. Pakistan has initiated actions to protect the environment and contribute to minimizing the adverse impacts of climate change. Both climate adaptation and mitigation are reflected in the country's policy and implementation approach. After the successful completion of Pakistan's Billion Tree planting drive across 350,000 hectares – the first Bonn Challenge pledge to meet and surpass its target – Pakistan has scaled up the initiative to the 10 billion Tree Tsunami.[141] This five-year, country-wide tree planting drive aims to restore depleted forests and mitigate climate change. With the launch of the Clean and Green Pakistan and Recharge Pakistan initiative, the country has taken the lead in 'nature-based solutions for ecosystem restoration' among developing countries, with the added benefits of safeguarding biodiversity and generating livelihood opportunities.

Objectives of the Voluntary National Review

Pakistan is committed to the 2030 Agenda for Sustainable Development and its 17 Sustainable Development Goals (SDGs) in October 2015. Pakistan's first Voluntary National Review[142] outlines the country's level of preparedness for achieving the SDGs, reports progress on several SDGs and puts forth future plans, which hinges on multi-stakeholder engagement, institutional mechanisms, the allocation of financial resources and the streamlining of policies. Although growth and social development go side by side, growth in Pakistan has not contributed to balanced social development in recent decades, primarily due to high levels of population growth. As a result, many of Pakistan's social indicators do not match significant levels of economic growth, averaging 4 per cent per year. This makes the Sustainable Development Goals a key priority for the country.[143]

After committing to the 2030 Agenda for Sustainable Development in 2015, Pakistan became the first country in the world to adopt the SDGs as its own national development goals through a National Assembly Resolution in February 2016.[144] At the same time, Pakistan's Parliament became the first to establish an SDG Unit dedicated solely to the 17 goals.[145] Pakistan began working on the SDGs as early as 2013, when the United Nations selected Pakistan as one of the countries to conduct consultations on the post-2015 development agenda. The key development priorities were identified during consultations. These included peace and security, governance, inclusive economic growth, the rule of law, social development, gender equality and women's empowerment, sustainable low-cost energy, disaster response and preparedness, and the much-needed broader role of the developed world. These priorities were incorporated in Pakistan's long-term perspective development document.[146] In 2014, the National Assembly aligned its long-term Strategic Plan 2014–2018 with the post-2015 agenda.[147] In 2017, National and Provincial Parliamentary Task Forces were created to focus on the SDGs during parliamentary work.[148]

Pakistan's local government system is in place to bring the government closer to the people. This is enshrined in the Constitution, which empowers each province to set up its own local governments. The local government system in Pakistan is a three-tier system, integrated through a bottom-up planning approach.[149] It is supported by the provinces' local government departments, with the Federal Ministry of Inter-Provincial Coordination ensuring national-level coordination. Effective implementation of the 2030 Agenda in Pakistan hinges on the effectiveness of the local government system – a potentially viable tool for embedding the SDGs at the grassroots level. The goals' achievement ultimately depends on the ability of local and provincial governments to promote integrated, inclusive and sustainable development.[150] Since the adoption of SDGs, the country has worked to mainstream the SDGs in all its policies, plans and strategies. Pakistan's long-term development agenda, provincial development strategies and five-year plans are all aligned with the SDGs. All tiers of government are actively participating in the SDGs' implementation.[151] In 2017, the first Local Government Summit on the SDGs identified education, employment, energy, water, and peace and governance as major issues to address. The Public Sector Development Programme (PSDP) has increased spending on energy, law and order, and security at the federal level. In tandem, education, health, and water and sanitation receive higher share of provincial budgets. Based on the priorities highlighted by the public during consultations on the post-2015 agenda, and the debates undertaken during the Local Government Summit in 2017, an objective criterion, encompassing seven dimensions, was developed to prioritize national requirements. This seven-dimensional criterion[152] guided inter-provincial discussions on identifying national priorities. Based on the outcomes of these discussions, a framework was devised to prioritize the SDGs in the Pakistani context. Accordingly, a National SDG Framework was prepared. The National Economic Council, chaired by the prime minister, is Pakistan's highest forum for the approval of plans for the implementation of policies. In 2018, the National Economic Council approved the National SDGs Framework.[153] The framework prioritizes the Global Goals into three categories. While all goals will be worked on simultaneously, Category 1 goals are those that require immediate attention to achieve rapid results which will pave the way for achieving the remaining goals. The framework's categories prioritize the goals as follows:

- Category 1 – SDG 2 ('No Hunger'), SDG 3 ('Good Health and Well-Being'), SDG 4 ('Quality Education'), SDG 6 ('Clean Water and Sanitation'), SDG 7 ('Affordable and Clean Energy'), SDG 8 ('Decent Work and Economic Growth)' and SDG 16 ('Peace, Justice and Strong Institutions');
- Category 2 – SDG 1 ('No Poverty'), SDG 5 ('Gender Equality'), SDG 9 ('Industry, Innovation and Infrastructure'), SDG 10 ('Reduced Inequalities'), SDG 11 ('Sustainable Cities and Communities') and SDG 17 ('Partnerships for the Goals');
- Category 3 – SDG 12 ("Responsible Consumption and Production'), SDG 13 ('Climate Action'), SDG 14 ('Life below Sea') and SDG 15 ('Life on Land').[154]

Targets for Sustainable Development to be sustainable, Pakistan recognizes that the three core dimensions of development must be connected: social, economic and environmental. Working towards these three main dimensions of the SDGs compels the country to seek for interconnectivity when devising policy frameworks. Lessons learned from the MDGs in Pakistan speak of the need to devise integrated policies that connect these three dimensions of development. For instance, poverty reduction requires strong policies in terms of employment, social protection, better health and nutrition and a clean environment. Similarly, ending stunting requires a clean environment, good hygiene within and outside of the home, proper nutrition and better healthcare and health education. Following an integrated approach, Pakistan has made considerable progress on several fronts, including reducing poverty, reducing stunting among children, increasing school enrolment and promoting gender equality by reforming policies and introducing legislation to empower women.[155] To achieve Pakistan's sustainable development targets, effective coordination is emphasized among all the stakeholders, including the government, private sector, civil society and academia, in terms of devising and effectively implementing policies. To address financing and governance issues, the government has engaged a group of experts to identify solutions. Academia and think tanks have established special SDG Units and SDG Centers of Excellence to spearhead research on different goals. On the climate front, Pakistan's target is to further minimize its carbon footprint and take steps to safeguard the environment, such as large-scale tree planting campaigns and extending the country's forest cover.[156]

Since adopting the 2030 Agenda in 2016, the Government of Pakistan has taken three important steps to mainstreaming the SDGs in public planning: (1) institutionalization, where provincial and administrative area growth strategies reflect each federating unit's respective development priorities (2) project preparation and (3) Localization plans. These are a useful guide, offering a prioritization tool to rank the SDGs. This tool has been used to review different policies, their theoretical underpinnings and empirical findings in three broad stages: (a) mapping targets in five exclusive themes, (b) identifying and quantifying relative criteria at the goal-level and (c) prioritizing targets through systematic numeric ranking.[157] A comprehensive and technical methodology was used to devise Pakistan's National SDG Framework, which guides future development strategies for the 2030 Agenda. The SDG targets were prioritized using a seven-dimensional[158] criterion, the dimensions being width, depth, multiplier, urgency, requirement of lower structural change, the need for lower finances and relevance for all provinces. The framework is based on five critical pathways (CPW)[159] that can converge to reduce regional inequalities by fostering inclusive, sustainable development:

- CPW1: Improve governance and security;
- CPW2: Increase access to quality social and municipal services;
- CPW3: Increase investment, employment and productivity in key sectors and improve economic growth;
- CPW4: Improve environmental stewardship and climate action;
- CPW5: Reduce inequalities and improve social cohesion.

The targets in the framework were ranked as 'high', 'medium-high', 'medium-low' and 'low'. Rankings were then mapped back to each SDG. The goals were further classified into three categories, while all goals will be worked on simultaneously, Category 1 goals, like SDG 2 ('no hunger), SDG 3 (good health and well-being), SDG 4 (quality education), SDG 6 (clean water and sanitation), SDG 7 (affordable and clean energy), SDG-8 (decent work and economic growth) and SDG 16 (peace, justice and strong institutions) are those that require immediate attention to achieve rapid results which will pave the way for achieving the remaining goals that are covered in the Categories 2 and 3 goals. The Category 2 goals are highlighted as, SDG 1 (no poverty), SDG 5 (gender equality), SDG 9 (industry, innovation and infrastructure), SDG 10 (reduced inequalities), SDG 11 (sustainable cities and communities) and SDG 17 (partnership for the goals). Category 3 goals are highlighted as, SDG 12 (responsible consumption and production), SDG 13 (climate action), SDG 14 (life below sea) and SDG 15 (life on land), each requiring continuous policy and budgetary support, prioritizing the goals.[160] Provincial SDG prioritization frameworks also follow the five CPWs as their building blocks.

Limitations and Challenges

While Pakistan is well prepared to achieve the SDGs, several challenges remain. Financing the SDGs in a slow growth environment is going to be a trying task, compounded by the knowledge and technology gap in developing local solutions and improving efficiency through improved governance.[161] Exploring innovative financing, developing a Responsible Business Framework as well as engaging local universities in devising local solutions for local problems is the strategy that Pakistan is pursuing. Pakistan is advancing towards its commitment to the 2030 Agenda by working to strengthen institutional mechanisms, enhancing awareness, creating productive partnerships and improving coordination.[162]

Several challenges still limit Pakistan's progress on the SDGs. A lack of efficient coordination has deterred implementation, led to the duplication of efforts and made it difficult to build synergies. Coordination challenges also limit stakeholder participation in different types of consultations. A lack of awareness of, and knowledge on, policy coherence and the interlinkages among the SDGs is a major hurdle to developing an appropriate policy mix to achieve the SDGs.[163] Limited awareness at the grassroots level is another important challenge to implementing programmes and projects related to the SDGs. To achieve the 2030 agenda through the implementation of Pakistan's national development priorities, the government is committed to ensuring that institutional arrangements and policies to catalyse growth and sustainable development while addressing current challenges.[164] Resource gaps are addressed through innovative financing modes; by building synergies and clearly defining roles and responsibilities at the federal, provincial and local levels; building robust partnerships among all stakeholders; and seeking technology transfer from developed economies. Pakistan's large population often dispersed across sizeable geographic areas and a lack of financial resources pose important challenges to achieving nutrition-related targets.

Notwithstanding economic and financial challenges, Pakistan continues to work towards achieving the SDGs through innovative, targeted and focused implementation strategies in the social, economic and environmental spheres. The government maintains the current momentum through consistency in plans, policies and the institutional strengthening process. Moving forward, the Government of Pakistan stands firm in its commitment to the 2030 Agenda and the momentum generated by putting institutional support mechanisms in place and aligning the SDGs with the country's short- and long-term development priorities. In addition to a continued focus on alleviating poverty in all its forms and eliminating hunger, the government focuses on fostering growth, achieving sustainable development and transforming Pakistan into an industrialized economy led by an innovative, healthy population, and making full use of modern knowledge and technology. Pakistan enhances the implementation capacity of its institutions through the transfer of technical knowledge from global experts in the fields of environmental sustainability, responsible consumption and production, and innovation. For effective reporting on the 2030 Agenda, all tiers of government are also working to reduce data variability across the country. Pakistan partners with international experts to benefit from successful models and, through adaptation, develop local solutions of the 2030 Agenda in Pakistan.[165]

Nevertheless, several challenges remain; the task of planning and implementing the 2030 Agenda for Pakistan's rapidly growing population must incorporate diverse local contexts, build local capacities and strengthen institutions. The gigantic challenge of climate change requires intensive community mobilization efforts. A cornerstone of implementing the SDG goals is building on existing alliances and forging new partnerships, leveraging technology and mobilizing innovative sources of finance.[166] Hence, partnerships with a broad array of stakeholders, including the private sector and civil society, supported by international community, continue to guide this process. The most important challenge is to harness the potential of youth, by leveraging the opportunities through innovative financing and making use of technologies, partnerships and cross-sectoral innovations to ensure that the country delivers on commitments and creates a solid foundation that enables sustainable national development, regional growth and global prosperity.

Sri Lanka

Sir Lanka poses a baffling dilemma. Its human development indicators are among the highest in the world, often surpassing those achieved in more prosperous regions of the developing countries, and sometimes even the human progress made in industrialized countries. Sri Lanka is often cited as a model of human development, yet the country has had to face serious ethnic tensions.[167] Currently, it has a population growth rate of 1.5 per cent compared to the average of 2–3 per cent for South Asia; its adult literacy rate, at 90 per cent, is one of the highest in the developing world; primary and secondary school enrolments are 99 per cent and 74 per cent, respectively, with virtually no gender differences in either literacy or school enrolments. Basic health facilities are available to 93 per cent of the population; life expectancy at 72 is 11 years longer than the South Asian average of 61 years.[168] These impressive figures are the result of a conscious policy effort of successive governments in Sri Lanka to invest in social development over the past decades. They show that even a relatively poor country with a per capita income of around US$600 can achieve success in providing all the basic social services to its people, given unwavering political commitment and if a determined policy effort is there.[169]

Being a tropical country, Sri Lanka is richly endowed with forests, natural water and coastal resources. In recent years, land degradation has become a major environmental problem. Forest cover has been depleted from an estimated 24 per cent of the total land area in 1989 to 20 per cent in 1992; at the present rate of deforestation forest cover is projected to shrink to 10 per cent by 2000–2020.[170] With a high rainfall of 2,000 mm per year, Sri Lanka is rich in freshwater resources, but the supply of freshwater is constrained by the rapid rise in the demand for water for irrigation, power generation and domestic and industrial use. In the coastal zone, erosion and mining of coral reefs to obtain lime for the construction industry have become major environmental hazards. Even though the level of industrialization in Sri Lanka is relatively low by international standards, a 1989 survey on industrial pollution revealed that out of 7,610 industries, 291 had a high polluting potential, and 1,900 a medium polluting potential.[171] The major sources of air pollution are the burning of fuel wood and industrial emissions from power plants. Solid waste management is another growing environmental problem. In keeping with international trends, Sri Lanka has placed increased emphasis on environmental management. Some of the steps taken to improve environmental management include the creation of a cabinet rank ministry for the environment, strengthening of the Central Environmental Authority (CEA) and technical support to create environmental awareness in development-oriented ministries and institutions. In recent years increased resources have been allocated by the central government for environmental improvement activities. In 1991, a National Environment Act (NEA) introduced environmental standards for industries. District law enforcement committees have been established to regulate industrial pollution. All development projects are subject to a mandatory environmental impact assessment in the preparatory stage.

National Conservation Strategy

Sri Lanka's strategy process has been dominated by the preparation of documents. An NCS document was completed in 1988; it was followed in 1990 by a Draft Action Plan setting out a large number of activities to implement the NCS. After the World Bank intervention, an Environment Action Plan was prepared in 1991 to identify high-priority actions. The government felt that this was too narrowly focused and prepared a NEAP, for the period 1992–1996, which the Cabinet approved in 1991.[172] The Sri Lankan National Report to UNCED, published in

1991, describes the links between environment and development.[173] The UNCED report and the NEAP serve as the blueprint for environment and development programmes in Sri Lanka.

Objectives

The major objective of the NCS is to ensure that environmental protection is incorporated into the development strategy of the country. The NCS document is divided into six parts: environmental profile, management of ecosystems for sustainable development, human activities involving the environment, constraints to environment conservation and sustainable development, the strategy policy and implementation.[174] The NEAP is the programme of action for implementing the NCS. It identifies 12 programme areas: land resources, water resources, mineral resources, coastal resources, forestry, biodiversity and wildlife, urban pollution, industrial pollution, energy, environmental education, culture and institutional capacity.[175]

The government first showed interest in the NCS following a consultation with IUCN in 1977. Strategy development began in 1982 when the president of Sri Lanka appointed a task force to prepare the NCS as Sri Lanka's response to the World Conservation Strategy. In 1988, the National Environment Act was revised to give control and regulatory functions to the Central Environment Authority (CEA). The same year, the CEA prepared a drat Action Plan (DAP). A Steering Committee was established under the chair of the CEA to oversee the process. Ministries, departments, corporations, universities and NGOs were consulted. Workshops and informal meetings were held to discuss the revision of the DAP, which was published in November 1990.

Also in 1990, the World Bank and the newly established Ministry of Environment and Parliamentary Affairs (MEPA) prepared an Environmental Action Plan (EAP). In 1991, therefore, MEPA prepared a NEAP. Cabinet regards it as the definitive national action plan for the environment for the period 1992–1996. The Ministry of Environment is responsible for catalysing and coordinating implementation. Government adoption of the NCS and the NEAP is a sign of its commitment to developing a practical national environmental strategy linked to the economic development and planning system. The improved political climate for addressing environmental issues includes the creation of the Cabinet Committee on Environment, the strengthening of the National Environment Act and the establishment of a National Environmental Steering Committee. NESC is responsible for monitoring and evaluating the progress of the NEAP. It consists of representatives of different ministries and is co-chaired by the secretaries of the Ministry of Policy and Planning and Implementation and the Ministry of Environment and Parliamentary Affairs.

The succession of strategy documents from 1982 to 1991 culminated in a more focused definition of what needed to be done; each document was an improvement over its predecessor in this respect. Nonetheless, results would have come more quickly if greater priority had been given to action rather than documentation. There are several obstacles to the development and implementation of environmental strategies in Sri Lanka.

The Government of Sri Lanka finds it increasingly difficult to finance the priority actions set out in the NEAP. A high level of donor support is necessary to implement the recommended actions and programmes. Usually, the government lacks the resources to assume the costs, and the more a strategy relies on donors for funding, the more likely it is to break up into unrelated projects and lose strategic coherence.

There is serious shortage of trained personnel, particularly at the professional level in MEPA and CEA. There is little capacity within the private sector to handle environmental matters such as pollution abatement.

The main institutional weaknesses are as follows:

- Lack of clear institutional mandates and responsibilities for environmental policy-making at the national level;
- Fragmentation and lack of coordination among the many agencies with responsibilities for land, water or other natural resource management;
- Lack of capacity for pollution control and monitoring.[176]

Sustainable Sri Lanka 2030 Vision and Strategic Path

At the 1992 Earth Summit in Rio, the world community adopted Agenda 21, containing over 2,500 wide-ranging recommendations for sustainable development (SD), as well as the UN Framework Convention on Climate Change (UNFCCC). Ten years after, at the 2002 Johannesburg World Summit on Sustainable Development (WSSD), world leaders adopted follow-up proposals, including endorsement of the 8 Millennium Development Goals launched in 2000. Twenty years later, at the 2012 Rio+20 Earth Summit, nations agreed to work towards a broad-ranging post-2015 agenda. This process eventually led to the universal acceptance of the 17 Sustainable Development Goals and 2030 Agenda by all UN member states, in 2015. Within this framework, all countries are expected to set out their vision for 2030 and to submit annual national reports on progress towards the SDG. In 2015, UN member states, including Sri Lanka, universally accepted the 17 SDGs and 2030 Agenda.[177] The Sri Lanka National 2030 Vision[178] is a strategic document that seeks to interpret the presidential vision which was outlined on 2 January 2017, when he gave the mandate to a Presidential Expert Committee (PEC) to prepare the report. In that vision, by 2030 Sri Lanka hopes to become a sustainable, upper-middle-income nation with an economy that is prosperous, competitive and advanced, an environment that is green and flourishing, and a society that is inclusive, harmonious, peaceful and just. Expanding on the president's vision, this report also sets out a practical pathway to reach such a sustainable future by 2030, including sustainability perspectives in the intermediate years of 2020 and 2025. The contributors to this report, including PEC members and others, who have followed an inclusive process in preparing the report, recommended wider systematic country-wide consultation of all relevant stakeholders, thereby ensuring wide ownership of the emerging path towards the vision of Sri Lanka in 2030.

The PEC report seeks to set out Sri Lanka's current country profile and status, key issues and opportunities relating to sustainable development, future priorities, and new initiatives and options to achieve ambitious goals by 2030. Because the long-term future is uncertain, the focus is on a sustainable path for Sri Lanka, a balanced, inclusive green growth, or BIGG, pathway,[179] which identifies strategic actions on this path and emphasizes that future adjustments have to be made along the way keeping in mind unforeseen changes and new emerging facts during the country's progress towards 2025 and 2030. Besides empowering Sustainable Sri Lanka 2030 Vision and Strategic Path, it is further visualized that a balanced, inclusive green growth path will enable Sri Lanka to become a world leader of sustainability by 2030.[180] The BIGG path follows the Sri Lankan tradition of the middle path, which is people-oriented, and open socio-economic system, which is based on democratic and pluralistic institutions; able to protect Sri Lanka's ethical values, heritage and environment; and built on respect for freedom, justice, equal opportunities and human rights. Inspired by the ennobling elements of the country's rich past, the three major groups (government, private sector and civil society) are disciplined enough to play balanced, cooperative and effective roles within this framework.[181] The strategic national sustainable path is simple and clear in language, which empowers people and provides guidance to the government, as well as civil society and the business community. The core framework seeks to harmonize the economic, social and environmental dimensions of the SD triangle.[182]

Limitations

It is also important to set out the limitations imposed by the mandate: time and resource constraints. First, the task is to produce a strategic document, within a one-year time frame, focusing on priority sustainable development issues. Therefore, this report is not intended to be a detailed national development plan but to avoid duplicating the work of other branches of government that are already carrying out that task. Existing national, sub-national, sectoral and other plans and data, fitted within a consistent and comprehensive long-term conceptual framework, were used. Second, some of those other planning documents are not necessarily fully consistent with the BIGG path which is not within the PEC's mandate to revise those plans, but key recommendations provide guidance on how other national, sub-national and sectoral plans are brought into alignment within the BIGG path. Such detailed plans and policies are updated and revised to achieve the 2030 vision, by transforming values, mindsets, institutions, behaviours, processes, methods, tools, technologies, projects and policies.[183] The transformation processes required within the government bureaucracy, civil society, business and industry are worked out through inclusive consultations. Third, the lack of success in executing past strategies and policies is examined that had come up with some new ideas like, constant changes in policy by successive governments, inadequate implementation frameworks, bureaucratic bottlenecks, lack of accountability, skill limitations, poor resource deployment or other related factors. The Vision report identifies ongoing and future policies and plans, and recommends policies and measures to improve implementation. Fourth, a simple baseline projection, without extensive macro-analysis, was done and that assumes a 5–6 per cent GNP growth rate and 1 per cent population growth rate up to 2030, and the results are projected as robust over a range of values around this baseline. Furthermore, some key scenarios are explored at the sector level (e.g., worst case disasters affecting agriculture), to provide strategic guidance on risk management.[184]

The 2030 Vision strategy is proposed to be updated at periodic intervals in the future, based on new information and experience gained along the way. Such a process ensures a robust and flexible strategy based on a multi-focal vision aimed at the short- (2020), medium- (2025), and long-run (2030) future.[185] The implementation of the 2030 Vision also requires substantive improvements in national statistical capacity, especially strengthening disaggregated data gathering and analysis. The nationwide, inclusive, consultative process that was recommended builds a consensus and wide ownership, ultimately facilitating successful implementation. Further, the strategic recommendations in the PEC report are aligned with more detailed plans and policies of other ministries and institutions. Improved governance is essential for effective implementation,[186] and secretaries of key ministries are suggested to lead the effort and to achieve this goal, dynamic interaction with Parliament, the judiciary and ongoing constitutional initiatives, besides the vital role played by the business community and civil society are highlighted. The problems humanity faces due to the overconsumption of natural resources, pollution, and inequality have been recognized, and the manner in which these can be changed to ensure prosperity for all while protecting the planet has been outlined in the 2030 Agenda for Sustainable Development. The corresponding 17 SDGs are globally accepted as the solution because they are comprehensive, holistic and inclusive. Sustainable Sri Lanka 2030 Vision and Strategic Path has acknowledged the role played by youth, who are considered a fundamental investment for the successful implementation of the SDGs.[187] This paradigm shift which tries to harness the youth of Sri Lanka to lead a national strategy for sustainable development is going to be a dynamic and productive force for the achievement of the SDGs.

South-Asia Co-operative Environmental Programme (SACEP)

In the Asia and Pacific region, a South Asia Co-operative Environment Programme (SACEP), with headquarters in Colombo, Sri Lanka, has been the focus for coordinated regional programme. Afghanistan, Bangladesh, Bhutan, India, Maldives, Nepal, Pakistan and Sri Lanka participate as members in SACEP, whose council has emphasized the need to integrate environmental protection with economic development. A Regional Seas Programme for South Asian seas is coordinated by SACEP, with support from UNEP. All these eight countries of South Asia, also participate in the South Asia Co-operative Environment Programme, which prepares a report on the state of the region's environment with a special focus on natural disasters, and this includes a programme for bilateral and multilateral cooperation.

Unity in Diversity

At the regional level, the SACEP region displays a great deal of unity amidst diversity. Thus, it affords a unique opportunity for a collaborative approach towards the protection and management of the regional environment. The establishment of the South Asia Co-operative Environment Programme constitutes an important landmark in regional cooperation in South Asia.[188] As the very name suggests, SACEP is a regional grouping of countries located in South Asia with quite a number of them having common shoreline washed by the Indian Ocean. The very concept of a regional grouping on the subject of the environment is unique. There are regional groupings, which are political, economic, religious and social. But this is the first occasion on which South Asian countries have been brought together solely on the subject of environment.

SACEP was created out of the initiative taken by the regional office for Asia and the Pacific of the United Nations Environment Programme located in Bangkok, which convened an inter-governmental expert group meeting at the Regional Centre for Technology Transfer in Bangalore, from the 10th to 15th of March 1980.[189] The purpose of the meeting was to consider the feasibility of establishing a South Asia Co-operative Environment Programme with eight countries of the South Asia sub region.

The inter-governmental expert group meeting was of the unanimous opinion that the South Asia sub-region, stretching from the Himalayan chain at one end to the Indian Ocean on the other, with a diversity of landforms, water bodies, climates, soil, natural vegetation and human settlements, afforded a unique opportunity for a collaborative approach towards the protection and management of the regional environment. Amidst such diversity, there existed a great deal of unity in ecological characteristics, the way of life and the problems of development common to the region. The deliberations were held under the headings 'environmental management', 'management of natural resources', 'desertification', 'regional seas programme', 'energy and environment', 'education and training' and 'the establishment of a South-Asia Co-operative Environment Programme'.

The meeting unanimously adopted that it would be mutually beneficial to establish a regional organization of these member countries enjoying equal status and named the organization the South Asia Co-operative Environment Programme. The meeting also unanimously resolved that the Secretariat of the Coordinating Committee of SACEP be established in Sri Lanka. With the approval of the Cabinet, the Secretariat of SACEP was established. The Secretariat reviews the progress of the various programmes and activities that are undertaken under SACEP, and in particular, the activities of the focal points are identified for specific subject areas. The Secretariat identifies additional areas and activities that are of concern to SACEP from time to time including activities that the countries may wish to undertake directly or between two or more of them. SACEP also secures technical and other assistance and arranges meetings of the member countries for formation of programmes.

Articles of Association

It was also decided at the meeting that the Regional Office of UNEP should convene a high-level meeting of government officials to work out 'articles of association', programme modalities and functions of each focal point of the coordinating committee

The meeting also adopted the establishment of several focal points which serve as the primary institutional points for carrying forward the agreed programme components on the specialized areas identified. Some of the key focal points identified so far are: India for conservation of biodiversity, legislation, education and training; Maldives for energy, wildlife and environmental quality, and coral island system; Bangladesh for corals, island ecosystems, mangroves, deltas and coastal areas; Nepal for tourism and participatory forest management; Pakistan for mountain ecosystem, social forestry and wildlife; Sri Lanka for environmental impact assessment, cost-benefit analysis, environment and development; SACEP for regional seas programme and so on.[190] These focal points serve as the receiving and disseminating centres for organized exchange of materials and information and ongoing field activities in the participating countries. Such focal points are linked to each other and to the Secretariat.

The principal objective of this new organization is to promote and support the protection, management and enhancement of the environment, individually, collectively and cooperatively. It is hoped that SACEP will be in a position to make a more meaningful contribution towards the judicious use of the resources of the environment and thereby help achieve the goals of development of the member countries, such as removal of poverty, reduction of socio-economic disparity and enhancement of the quality of life of people.

The launching of SACEP is an initiative of indisputable significance for South Asian countries as well as for the world community as a whole. All the countries in the sub-region are confronted with the problems of rapidly growing populations. They also face the irreversible problem of the depletion of natural resources and the rapid degradation of the environment. In the development process, the environment – that is, the life support system of the society – has to be enhanced. That is why, SACEP attempts to harmonize the relationship between environment and development, both at the national level and at the regional level.[191] The member countries were also mindful of the fact that there was a wealth of knowledge and expertise which could be shared for mutual benefit. Moreover, being developing countries, some of them realized that solutions to the problems of the environment could emerge from the region itself rather than the application of imported solutions from the industrialized world.

The institutional arrangements comprised the Governing Council, the Consultative Committee and the Secretariat. All members of SACEP are represented in the Governing Council at ministerial level. It is the principal review and deliberative body of SACEP which is responsible for determining policies and programmes. The Governing Council meets once a year.

The Consultative Committee is the executive arm of SACEP. Its headquarters is in Colombo, and it is represented by all member governments through permanent representatives nominated by the respective governments. The Consultative Committee meets once in three months and gives guidelines to the Secretariat on the implementation of the different programmes and projects and monitors the progress of the activities of SACEP. The most significant innovative feature in the institutional framework of SACEP is the establishment of 15 subject area focal points, with each country assuming responsibility for one or two focal points. Each focal point is expected to coordinate all activities pertaining to the subject area assigned in its own country as well as with the related agencies in the other countries of the region with the full knowledge and concurrence of the Secretariat.[192]

The Secretariat consists of a director and a number of other officials appointed by the Governing Council from among the nominees of member governments as well as other supporting staff. The Secretariat identifies areas of critical concern, formulates programmes and projects and obtains the assistance of UNEP as well as other multilateral and bilateral donor agencies in the implementation.

The deliberations, both in Bangalore and Colombo, clearly showed that each country had made considerable progress in tackling the common problems of the environment in one way or other. Nepal, for instance, has to its credit a solar water heating device designed, manufactured and marketed in that country. Sri Lanka has made considerable headway in the design and manufacture of a windmill with a water pump. India has made great strides in research and development in several fields of renewable energy such as biogas, solar and wind energy.[193] All these developments were bound by the common thread of appropriate technology, low costs and use of local materials as well as local skills. Thus, a regional arrangement of this nature results in a sharing of the research effort, transfer of technology and the fostering of bilateral trade among these countries.

An Assessment of SACEP (1982–2030)

South Asia Co-operative Environment Programme is an inter-governmental organization, established in 1982 by eight member governments of South Asia to promote and support protection, management and enhancement of the environment in the region. South Asia is one of the most diverse regions in the world. Surrounded by the Himalayas to the North and Indian Ocean to the South, the South Asian region, covers a diversity of ecosystems from lush tropical forests to harsh, dry deserts. It is also one of the most populous regions, with over one billion people living in India alone. The movements of peoples over thousands of years have resulted in strong commonalities between cultures and yet there remains a huge diversity of languages, religions and outlooks across the subcontinent. Most of the South Asian nations share many similar environmental problems, stemming from poverty and its consequences on natural resources. According to the World Bank, during the past decade, South Asia has been the second fastest economically growing region in the world, and their efforts at increased production have put increasing pressure on natural resources and the environment.[194] Significant natural resource concerns of the region include depletion of water quality and quantity, dwindling forests and coastal resources, and soil degradation resulting from depletion and salinization. Many countries of the region have taken actions for the protection and management of the environment. SACEP supports national government's efforts for environmental protection and sustainable development. Since its creation, SACEP has implemented a number of projects and programmes in the areas of environment education, environment legislation, biodiversity, air pollution and the protection and management of the coastal environment. SACEP is also Secretariat for the South Asian Seas Programme.[195] The Malé Declaration on control and prevention of air pollution, and its likely transboundary effects for South Asia, is another significant effort which encourages intergovernmental cooperation to combat the transboundary air pollution problem.[196]

Here is an environmental profile of South Asia:

- The region covers almost one-twentieth of the earth's surface and provides a home to about one-fifth of the world population;
- The degree of urbanization in 1999 ranged from 7 per cent in Bhutan to 33 per cent in Pakistan. Mumbai, Calcutta, Delhi, Karachi and Dhaka are fast-growing cities, with population more than 10 million;
- Over 30 per cent of the population earns less than one dollar per day and the per capita GNP for 1998 ranged from US$ 210 to 130. It is US$ 210 in Nepal to 1,130 in Maldives.[197]

Although the economies of the countries are primarily agricultural, industrialization has increased during the past decade.

- South Asia is home to 14 per cent of the world's remaining mangrove forests, and the Sundarbans found between Bangladesh and India is one of the largest continuous mangrove stretch in the world;
- Six per cent of the world's coral reefs are in the South Asian seas. The atolls of Maldives and Lakshadweep islands of the region are biodiversity-rich marine habitats;
- Hindu Kush Himalayan belt is home to over 25,000 major plant species, comprising 10 per cent of the world's flora;

146 Environmental Governance and Sustainable Development

- The region is prone to natural disasters such as cyclones, floods and landslides, earthquakes and heat waves, and the region accounted for over 60 per cent of disaster-related deaths worldwide.

A healthy environment, resilient society and regional prosperity for the present and future generations are the vision of SACEP through the 2020–2030 decade.[198] In 1982, almost four decades ago, SACEP was created to fulfil a vision based on the following three assumptions:

- Recognition of environmental degradation caused by factors like poverty, overpopulation, overconsumption and wasteful production threatening economic development and human survival;
- Integration of environment and development as essential prerequisites to Sustainable Development;
- Importance of cooperative action in the South Asian region, where many ecological and development problems transcend national and administrative boundaries.[199]

The most important goals and objectives of SACEP are articulated as promoting regional cooperation in South Asia in the field of environment, both natural and human in the context of sustainable development and on issues of economic and social development which also impinge on the environment and vice versa; also supporting conservation and management of natural resources of the region and working closely with all national, regional, and international institutions, governmental and non-governmental, as well as experts and groups engaged in such cooperation and conservation efforts.

Post-2015 South Asia Development Agenda

In 1992, Earth Summit at Rio, member countries adopted Agenda 21, a blueprint to rethink economic growth, advance social equity and ensure environmental protection. Twenty years after this, during 20–22 June 2012, the United Nations Conference on Sustainable Development (Rio+20), which took place in Rio de Janeiro, Brazil, resulted in a focused political outcome document, which contains clear and practical measures for implementing sustainable development. This outcome of the Rio+20 Conference initiated an inclusive intergovernmental process to prepare a set of Sustainable Development Goals.[200] The post-2015 development agenda[201] builds on the progress achieved through the MDGs: eight goals established after the Millennium Summit of the United Nations in 2000. At the same time, it also addresses persistent issues and new challenges facing people and the planet. Instead of addressing the dimensions of sustainable development separately, the SDGs have adopted an approach that fully integrates the social, economic and environmental dimensions of sustainable development in these countries of South Asia.

Regarding the 2030 Agenda for Sustainable Development, the UN Sustainable Development Summit held on 25 September 2015, at UN Headquarters in New York, adopted the post-2015 development agenda, which includes a set of 17 Sustainable Development Goals to end poverty, fight inequality and injustice, and to tackle climate change by 2030. The SDGs, otherwise known as the Global Goals,[202] build on the Millennium Development Goals, eight anti-poverty targets that the world is committed to achieving by 2015. The new Global Goals, and the broader sustainability agenda, go much further than the MDGs, addressing the root causes of poverty and the universal need for development that works for all people.[203] The summit was convened as a High-level Plenary meeting of the UN General Assembly and co-chaired by the presidencies of the 69th and 70th sessions, Uganda and Denmark, respectively.[204] Approximately 160 heads of state or government and 30 ministers attended the Sustainable Development Summit, along with over 9,000 delegates and around 3,000 accredited journalists, and all the South Asian countries are committed to abide by these goals.

Biodiversity Conservation

Biodiversity conservation[205] is another area where regional cooperation between and among the countries of South Asia is evident. South Asia region, not only is famous for its spectacular natural beauty and biological wealth but also is home to one of the oldest civilizations of the world. The region's geographical expanse and topography contain several diverse marine and terrestrial ecosystems that harbour a rich variety of faunal and floral species. As biodiversity has no geopolitical boundaries and it is connected among all member countries of South Asian region by land or sea, the regional approach is necessary for conservation of biodiversity in a holistic manner.[206] The regional cooperation will help build harmonization, facilitate to mitigate adverse impacts on biodiversity in the region, and build capacity through exchange of information, technology and knowledge. SACEP, the regional intergovernmental environment body for South Asia, is fulfilling this function. Biodiversity conservation is one of the priority areas of SACEP since its establishment in 1982. Major concerns[207] are as follows:

- Capacity development;
- Technical assistance to SACEP member countries for development of National CHMs;
- Assistance to National Biodiversity Strategy and Action Plan Updating;
- South Asia Regional Biodiversity Clearing House Mechanism;
- One of the most important SACEP GC decisions on biodiversity conservation is the 12th SACEP GC resolution on 'South Asia's Biodiversity Beyond 2010'.

Another significant area on which SACEP is focusing attention is the concept of sustainable consumption and production,[208] which is recognized in the Johannesburg Plan of Implementation adopted in 2002 at the World Summit on Sustainable

Development (WSSD). It was acknowledged that sustainable consumption and production forms one of the three overarching objectives of, and essential requirements for, sustainable development, along with poverty eradication and the management of natural resources in order to promote economic and social development. It was recognized that fundamental changes in the way societies produce and consume are indispensable for achieving global sustainable development and called for all countries to promote sustainable consumption and production patterns. It also called for governments, relevant international organizations, the private sector and all major groups to play an active role in changing unsustainable consumption and production patterns. Following on the implementation of the 2013 SACEP Governing Council decision to promote SCP within policy-making mechanisms of South Asian countries, as well as the Asia-Pacific Road map of the Rio+20, a 10-Year Framework of Programmes on Sustainable Consumption and Production (10YFP), SACEP is undertaking a number of activities/projects to promote SCP in the region. Preparatory meeting for the establishment of the South Asia Forum on Sustainable Consumption and Production (SCP)[209] led to the establishment of South Asia Forum on Sustainable Consumption and Production,[210] which is a significant development in this context. Another area of major focus by SACEP is associated with waste management in South Asia.

Waste Management

The South Asian region, with a current population of 1.6 billion, is experiencing rapid urban growth. Increasing population, urbanization, industrialization and changing consumption patterns are resulting in the generation of increasing amounts of solid waste and diversification of the type of solid waste generated. The region is reported to generate approximately 70 million tonnes of solid waste per year, with per capita values ranging from 0.12 to 5.1 kg/person/day and an average of 0.45 kg/capita/day[211] (World Bank). There are various factors that attribute to poor solid waste management, such as lack of public awareness, unplanned city growth, high waste generation and non-functioning of existing systems. Rate of urbanization, scavenger role for recyclable separation and the capacities of existing municipalities for solid waste management are also important factors that are considered regarding waste management. Increased solid waste generation creates more environmental problems in this region, as many cities are not able to manage it due to institutional, regulatory, financial, technical and public participation shortcomings. As decided at the Ninth Governing Council held in August 2005, waste management[212] is identified as one of the priority areas in the SACEP work programme. Activities of SACEP related to waste management in South Asia[213] are highlighted as follows:

- SACEP is the sub-regional Secretariat for International Partnership for expanding waste management services of local authorities (IPLA), a dynamic knowledge platform as well as a decentralized network established to address the needs of local authorities in achieving sustainable waste management;

- SACEP conducted a scoping exercise on e-waste management in South Asia 'in New Delhi, India in 2007 in collaboration with the Development Alternatives' (DA);
- SACEP organized the 'South Asian Games Waste Management Programme' at the South Asian Games held in Colombo, Sri Lanka, in 2006;
- SACEP supported UNCRD (United Nations Centre for Regional Development) in organizing the first National 3R workshop for South Asia held in Dhaka, Bangladesh, 2007;
- Prepared the 'Framework for Marine Litter Management in South Asia';
- Since 2006 SACEP has been organizing many activities to commemorate International Coastal Cleanup Day;
- Introduced the Blue Flag Beach Conservation Programme to the South Asian Seas Region in collaboration with Foundation for Environment Education (FEE) and UNEP and organized National Workshops in Bangladesh, India, Maldives and Sri Lanka in February–March 2010 with the assistance of the National Focal Points. These workshops provided a platform to further develop national programmes;
- In response to the mandate of the 9GC-SACEP on Waste Management, with the support of the member countries, COP8 of the Basel Convention recommended establishing a Basel Convention Regional Centre for South Asia at SACEP Secretariat, and 10GC-SACEP held in January 2007 endorsed same.

Climate Change

Climate change is one of the most important environmental and socio-economic issues facing the world today. In Asia and the Pacific region, and particularly in the South Asia sub-region, prominent increases in the intensity and/or frequency of many extreme events have become a regular occurrence. Despite growing efforts to reduce greenhouse gas emissions, some impacts such as higher temperatures, more intense floods, droughts, wildfires, earthquakes, heat waves and rising sea levels are becoming inevitable that challenge decision-making at all levels. The 9th meeting of Governing Council (GC) of SACEP held in August 2005 at Thimphu, Bhutan, identified the 'adaptation to climate change'[214] as one of the key areas concerning the region, and since then, it has been in the SACEP's work programme as a priority area. Adaptation refers to adjustments in ecological, social or economic systems in response to actual or expected climatic stimuli and their impacts; and changes in processes, practices and structures to moderate potential damages to benefit from opportunities associated with climate change. Some significant activities carried out/organized by SACEP are given:

1   Regional Workshop on Lessons Learnt in Strategy Implementation on Climate Change Adaptation (CCA) in water sector, which was organized from 9 to 10 September 2014 in Sri Lanka in collaboration with Global Water Partnership-South Asia (GWP-SAS), provided a platform to share Change Adaptation in Water Sector.[215]

2 9GC-SACEP, held in August 2005, identified adaptation to climate change as one of the key areas concerned in the South Asian region. SACEP entered into collaboration with NBRI (National Botanical Research Institute), a Centre of Excellence for the South Asian region, to develop a programme on adaptation issues related to climate change, and a regional workshop, 'Climate Change and Its Impact on Flora of the South Asian region', was conducted from 9 to 12 March 2008. A proposal on adaptation to climate change developed by SACEP and NBRI was discussed and finalized during the workshop.[216]

3 SACEP conducted a School Environment Awareness Programme in collaboration with CEE (Centre for Environment Education, Gujarat), India, Ministry of Education, Ministry of Environment and Natural Resources, Sri Lanka and the South Asia Youth Network. A series of school competitions (Essay, Drawing and Painting, Posters and Photography) from January to July 2007 for different age categories were held, and the main theme of these competitions was Adaptation to Climate Change and Waste Management.[217]

4 A MoU with CEE-India (Ahmedabad) was signed in October 2006. Under this CEE-SACEP initiative,[218] 'A Scoping Exercise on Adaptation to Climate Change' was conducted on 10 January 2007, in which government representatives and experts participated to consider a regional status of the priorities on adaptation to Climate Change under a consultative process. The report was finalized and submitted to the 10GC-SACEP.

Regional Seas Programme

The establishment of the South Asia Co-operative Environment Programme (SACEP) in 1982 as the regional environmental hub facilitated the introduction of the Regional Seas Programme in South Asia. As a result, the 11th Governing Council of the United Nations Environment Programme, held in 1983, recommended the development of a Regional Seas Programme for South Asia[219] in close collaboration with SACEP and the governments in the region. As the first follow-up activity to this recommendation, SACEP called up a meeting of the National Focal Points of its five maritime states (in March 1984), where the countries committed themselves to the development of an action plan to protect and manage the marine environment of the South Asian seas. The South Asian Seas Programme (SASP)[220] is one of 18 such programmes and the South Asian Seas Action Plan (SASAP)[221] was adopted in March 1995 by the region's five maritime countries (Bangladesh, India, Maldives, Pakistan and Sri Lanka). The South Asia Co-operative Environment Programme (SACEP) serves as the secretariat of SASP.

Instruments Governing SASP

SASAP follows existing global environmental and maritime conventions and considers the Law of the Sea as its umbrella convention. SACEP is the Secretariat for the South Asian Seas Programme, which facilitates the development and implementation of SASAP, which is agreed upon by the member countries at the Intergovernmental Meetings of Ministers (IMM).[222] The Consultative Committee provides the Secretariat with policy guidance on the implementation of the decisions taken at the IMMs. Each country has designated a National Focal Point for coordinating SASP activities. The costs of running the Programme are provided by the member countries as annual contribution and donors' support is obtained to implement specific projects.[223]

The SAS Region

The South Asian Seas Region[224] is categorized into two distinct geographical regions (Mainland and island nations), while Maldives and Sri Lanka are categorized as island nations, Bangladesh, India and Pakistan are situated on the Asian mainland. The SAS Region is comprised of the marine and coastal waters of Bangladesh, India, Maldives, Pakistan and Sri Lanka and is physically divided by the Indian subcontinent into three distinctive areas: two large marine ecosystems – the Arabian Sea in the west and the Bay of Bengal in the east; and a large area of the Indian Ocean to the South of India and Sri Lanka. This region encompassing mangroves, sea grass beds and coral reefs support some of the richest concentrations of biodiversity in the world. The vast flood plains of Ganges, Brahmaputra and Meghna River deltas between Bangladesh and India harbour the Sundarbans,[225] which is the largest continuous mangrove forest in the world (around 6,000 km). The SAS Region contributes to over 6 per cent of the world's coral reef area, which include all three major reef types (atoll, fringing and barrier)[226] that are represented to varying degrees. The Lakshadweep, along with the Maldives and the Chagos of the British Indian Ocean Territory, form an interrupted chain of coral atolls and reefs on a contiguous submarine bank covering a distance of over 2,000 km.[227] There are also extensive reefs around the Andaman and Nicobar Islands as well as in the Gulf of Mannar between India and Sri Lanka. These coastal and marine ecosystems represent some of the most important habitat for the survival of more than hundred globally threatened species.[228] Further, five out of the seven marine turtle species that are globally threatened feed in the open marine waters and the single largest breeding ground of olive ridley is found in Orissa on the east coast of India, where mass breeding occurs in three nesting beaches; Gahirmatha, Deviriver mouth and Rushikulya. The nesting at Gahirmata at the mouth of the river Maipura near Dhamra is the largest sea turtle rockery in the world with 100,000–500,000 turtles nesting there each year.[229] The entire population of Maldives and in Sri Lanka, where more than 32 per cent of the population resides within the coastal belt are dependent on the sea resources to generate their livelihoods and therefore, have great economic, social and cultural importance to individual countries and to the region as a whole.

The overall objective[230] of the South Asian Seas Action Plan is to protect and manage the marine environment and related coastal ecosystems of the region in an environmentally sound and sustainable manner. The specific objectives[231] are as follows:

- Establish and enhance consultations and technical cooperation among states within the region;
- Highlight the economic and social importance of the resources of the marine and coastal environment;
- Establish a regional cooperative network of activities concerning subjects/projects of mutual interest for the whole region.

The Action Plan for the South Asian Seas Programme was formally adopted by five South Asian countries, namely Bangladesh, India, Maldives, Pakistan and Sri Lanka, at a Meeting of Plenipotentiaries of the concerned countries held in New Delhi, on 24 March 1995.[232] The action plan helps the member countries in protecting the marine environment of the region, besides specifying the needs under the main components of environmental assessment, environmental management, environmental legislation and institutional and financial arrangements, and identified the areas where priority activities are developed for implementation. These priority activities are in the following four[233] specific areas:

- Integrated Coastal Zone Management (ICZM);
- Protection of Marine Environment from land-based activities;
- Human resources development through Regional Centers of Excellence;
- Development of national and regional oil and chemical spill contingency plan.

Another area of significant regional cooperation between and among the countries of South Asia is seen in the efforts of SACEP to free rivers and seas from plastic pollution in recent years.

Plastic-Free Rivers and Seas for South Asia

The fifteenth meeting of the Governing Council of SACEP, held on 6 November 2019 in Dhaka, Bangladesh, endorsed the SACEP's collaboration with the World Bank and their development partners to formulate and implement a regional project, 'Plastic-free Rivers and Seas[234] for South Asia', involving SACEP as an implementing agency for the benefit of the South Asian region. The Project Development Objective (PDO)[235] of the Plastic-Free Rivers and Seas for South Asia (PRS) project is to catalyse actions that reduce the flow of plastic pollution into South Asian seas. The project consists of four main components totalling US$ 40 million from International Development Association to be implemented over a period of five years in all eight countries in South Asia region. The project has three components.[236] These significant supportive efforts are listed:

1 **Supporting Competitive Block Grant Investments to Reduce Plastic Waste:**

   i   Supporting circular plastic economy solutions to reduce plastic waste by implementing a programme of regional competitive block grant investments, providing Regional Competitive Block Grants (RBGs) to eligible organizations in South Asia (Eligible RBG Beneficiaries);

   ii facilitating exchange of circular plastic economy knowledge between Eligible RBG Beneficiaries and selected South Asian countries and promoting awareness-raising activities.

2 **Leveraging Public and Private Sector Engagement and Solutions:**

   i   Supporting development of strategies, action plans, policies, and standards to harmonize plastic pollution mitigation measures through: (a) developing and implementing a multi-year plastic policy support programme, working with leading universities and organizations; (b) developing a database for lifecycle analysis, data collection, and modelling related to plastic across selected industry value chains; and (c) supporting communication activities;

   ii Supporting circular use of plastic in the economy through regional public-private collaboration and engagement in South Asia, including designing and organizing annual and more frequent meetings of representatives from public and private sectors.

3 **Strengthening Regional Integration of Institutions**

   i   Supporting the construction of SACEP's new headquarters and providing technical assistance, and building capacity of, SACEP to discharge its functions, including coordination with relevant regional organizations and uniform collection, analysis and interpretation of pollution data;

   ii Supporting the development of a fund for sustainability of existing activities and accelerating circular plastic economy solutions (Sustainability Fund).

So, what SACEP has achieved is something unique; where there was no organization, no machinery for cooperation, the countries of South Asia have decided to set up the machinery for it. It is hoped that through the non-controversial subject of the environment, a stronger and more viable United Nations of South Asia would emerge for further collaboration in the fields of political, economic and social development. These countries of South Asia are committed to the objectives and goals of SACEP and UNEP, and even though they are not in a position to make financial contributions of any meaningful magnitude, they still assist SACEP and UNEP with their well-developed manpower resources and reassure them (both SACEP and UNEP) that they would make every effort to discharge the duties and responsibilities as elected members of the UN Environment Assembly of UNEP and the Governing Council of SACEP to the best of their ability.

Summary and Conclusion

This chapter has analysed and explored the role of SACEP in the management of environmental issues and challenges in the South Asian region. The chapter began with an analysis of environmental perspective of the South Asian region, highlighting various environmental challenges specific to the region, such as, rapid industrialization and economic growth; rising population; frequent natural disasters; deforestation; biodiversity loss; air, water and oil pollution; land degradation; and related issues. In all these South Asian countries the role of national conservation strategies in not only creating environmental awareness but also strengthening environmental governance institutions, their objectives and implementation have been separately and briefly discussed. Further, the contributions of National Conservation Strategies have been analysed, revised and updated covering SDGs and MDGs, which are integrated into the development planning of the South Asian countries under study. In the process, an attempt is made to update facts and data related to the environmental perspectives in South Asia and also with regard to National Conservation Strategies that emerged as sustainable development strategies in all the South Asian countries (under study) covering the period till 2030. A unique part of this chapter is the role of SACEP in the management of environmental issues in the South Asian region, which is also updated to cover the period till 2030, titled, An assessment of SACEP:1982–2030.

It has been observed that effective environmental strategies must exert a strong influence on development if they are an integral part of development planning and policy-making. All the countries of South Asia have insisted that the requirements of agencies such as the World Bank, United Nations Commission on Environment and Development, UNEP and Agenda 21 must fit in with all these initiatives. It has been further observed in this chapter that there is a growing feeling among the nations that environment problems can be solved through regional and sub-regional cooperative efforts. Further, this chapter concludes with the observations that solutions to the problems of the environment could emerge from the region itself rather than the application of imported solutions from the industrialized world. The governments must show greater commitment. Environment is no longer seen in the narrow sense of pollution control. Today it is much more than a concern for health and sanitation. Developing countries in particular, especially through the efforts of SACEP and UNEP, have resolved themselves to the position that environmental considerations rest heavily on resource management. Another healthy feature vis-à-vis environment in many developing countries is the establishment of central authorities in charge of the overall subject of the environment at the highest national level. A significant feature in the establishment of these authorities or ministries is the fact that there is neither centralization nor curtailment of the planning, implementation and regulatory powers enjoyed by the organizations and specialized agencies. These institutions play a catalytic role by assisting in the formulation of overall policies and coordinating different programmes and projects, thus preventing overlapping and guiding different agencies towards a common goal through a process of consultation and participation. UNEP has set an example and given the lead by playing a similar role in the international sphere.

Notes

1 UNEP, *Global Environment Outlook 1: Executive Summary: Overview of Regional Status and Trends* (Nairobi: UNEP, 1999), p. 3, https://www.unep.org/unep/era/geo1/exsum/ex3.
2 UNEP, *Global Environment Out Look 1: Executive Summary: Overview of Regional Policy Responses* (Nairobi: UNEP, 1999), p. 2, https://www.unep.org/unep/era/geo1/exsum/ex4.
3 Ibid., p. 4.
4 Asian Development Bank, *Emerging Asia: Changes and Challenges: Asian Development Plan* (Manila, Philippines: Asian Development Bank, 1997).
5 UNDP, *Human Development Report 1997* (New York: Oxford University Press, 1997).
6 UNESCAP/ADB, *State of the Environment in Asia and the Pacific. (United Nations Economic and Social Commission for Asia and the Pacific and Asian Development Bank)* (New York: United Nations, 1995).
7 7WRI, UNEP, UNDP & WB 1998 Complied by UNEP GRID Geneva from United Population Division, 1996.
8 ADB (1997), op. cit.
9 CRED, *Centre for Research on the Epidemiology of Disasters, Disaster Events Database* (Brussels, Belgium: CRED Disasters in the World, Nov. 1991).
10 WRI, UNEP, UNDP and WB 1995. Complied by UNEP, GRID Geneva from United Nations Population Division, 1996.
11 ADB, *The Environment Programme: Past, Present and Future* (Manila, Philippines: Asian Development Bank, 1994).
12 Ibid.
13 ASEAN, First ASEAN State of the Environment Report. ASEAN Secretariat, Jakarta, Indonesia. 1997.
14 MRC/UNEP.D Making River Basin Diagnostic Study: Final Report Mekong River Commission, Bangkok, Thailand, 1997.
15 WCMC, *Global Biodiversity: Status of Earth's Living Resources* (Cambridge, UK: World Conservation Monitoring Center, 1992).
16 Government of Pakistan. Country Report and State of Environment in Pakistan, presented at the Regional Meeting on the State of the Environment in Asia and Pacific, Myanmar, 1994.
17 UNESCAP/ADB, op. cit.
18 GESAMP, *Impact of Oil and Related Chemicals and Wastes on the Marine Environment*, GESAMP Reports and Studies No.50 (London, UK: IMO, 1993).
19 Singh, Samar, 'The Biological Value of the Asia-Pacific Region. Biodiversity Conservation in Asia-Pacific: Constraints and Opportunities', Proceedings of a Regional Conference, 1994.
20 Da Cunha, 'Sustainable Development of Water Resources', International Symposium on Integrated Approach to Water Pollution, 1989.
21 Compiled by UNEP Grid Geneva from WRI, UNEP, UNDP and WB 1998.
22 Ibid.
23 ADB (1997), op. cit.

24 Ibid.
25 WHO? Our Health. Report or the WHO commission on Health and the Environment. WHO Geneva, Switzerland 1992.
26 ADB 1997, op. cit.
27 Ibid.
28 UNESCAP/ADB, State of the Environment in Asia and the Pacific 1995 (United National Economic and Social Commission on Asia and the Pacific, and Asian Development Bank). United Nations, New York, United States. 1995.
29 Khan, M.A., *Problems and Prospects of Sustainable Development: Management of Urban Water Bodies in the Asia and Pacific Region* (Bangkok, Thailand, 1993).
30 International Road Federation, (ed.), *World Road Statistics 1997* (Geneva Switzerland, and Washington, DC: URF, 1997).
31 WHO and UNEP? Urban Air Pollution in Megacities of World. United Kingdom: Blackwell, 1992.
32 ADB 1997, op. cit.
33 WHO, The Work of WHO in the South-East Asia Region, 1 July 1991–30 June 1993. WHO New Delhi, India. 1993.
34 World Bank, *Environment Matters: Towards Environmentally and Socially Sustainable Development* (The World Bank: Washington, DC, 1997).
35 Shrestha, R.M. and S. Malla, 'Air Pollution from Energy Use in a Developing Country City: The Case of Kathmandu Valley', *Nepal, The International Journal on Energy* 21(9) (1996), pp. 785–794.
36 Compiled by UNEP Grid Geneva from United Nations Population Division 1997, and WRI, UNEP, UNDP, and WB 1998.
37 Government of Sri Lanka, State of Environment of Sri Lanka, Ministry of Environment and Parliamentary Affairs, Colombo, Sri Lanka. 1994.
38 WRI, UNEP, UNDP and WB 1996, Government of Maldives: Statistical Yearbook of Maldives 1997, Male, Republic of Maldives, 1998. ASEAN, 'First ASEAN State of the Environment Report', ASEAN Secretariat, Jakarta, Indonesia, 1997.
39 ADB (1997), op. cit.
40 Ibid.
41 UNEP, *World Atlas of Desertification*, 2nd edition (London: Arnold, 1997).
42 Postel, S., *Water for Agriculture: Facing the Limits*, World Watch Paper. 93 (Washington, DC: World Watch Institute, 1989).
43 World Watch Institute 1997. op. cit.
44 FAO 1997, UNESCAP/ADB (1995). State of the world's Forest FAO, Rome Italy, 'State of the Environment in Asia and the Pacific', 1995 (United Nations Economic and Social Commission for Asia and Pacific and Asian Development Bank). United Nations, New York, United States. 1997.
45 Gadgil, Madhav and Ramachandra Guha, *Ecological History of India* (New Delhi: Oxford University Press, 1992).
46 Klaus, Seeland, *Man in the Forest: Local Knowledge and Sustainable Management of Forests and Natural Resources in Tribal Communities in India* (New Delhi: D.K. Printworld, 2000).
47 Human Development in South Asia: Country Profiles: Bangladesh, 1997, p. 42.
48 Wolfgang, Werner L. (ed.), *Aspects of Ecological Problems and Environmental Awareness in South Asia* (New Delhi: Manohar Publisher, 1993), p. 3.
49 Biswas, Asit K. and S.B.C. Aggarwal, *Environmental Impact Assessment in Developing Countries* (Oxford: Reed International Books, 1992), p. 16.
50 Nurul Alum, S.M., *Perception of Ecological Problems and Its implications for Bangladesh's Ecological Future* (Dhaka: Jahangirnagar University, 1990), p. 41.
51 Hussain, Azfar and Iqbal Shailo, 'Deforestation in Bangladesh: Towards an Ecological Inferno', *ADAB News* XV (Mar.-Apr. 1989), pp. 30–46.
52 Hussain, Azfar, *Floods in Bangladesh: Recurrent Disasters and Peoples Survival* (Dhaka: University Research Centre, 1987), p. 19.

53. Reis, Newsletter no.1, Department of Geography, Savar, Dhaka: Jahangirnagar University, 1984.
54. Nazem, Nurul Islam, 'Management of Environmental Disaster in South Asia: A Regional Approach', *Bliss Journal* 9 (1988), p. 357.
55. Dean, P.B. and Trey go, Wit, 'The Environment and Development in Bangladesh: An Overview and Strategy for the Future', A Report prepared for the Bangladesh Programmes, Asia Branch, CIDA, Oct. 1989, p. 3.
56. Malcom, Janis and Johnson Nels, 'Conservation of Biological Diversity in Bangladesh: Status, Trends and Recommended Responses', A Report Submitted to USAID, Bangladesh, 1989.
57. Jeremy, Carew Reid (ed.), *Strategies for Sustainability: Asia* (London: Earthscan Publications, IUCN, WCU, 1997), p. 17.
58. Ibid., p. 19.
59. Ibid., p. 18.
60. Awareness refers to realization of or consciousness on many issues and problems by the people. Awareness reflects a more explicit and concrete manifestation of a problem. Perception means people's thinking or ideas about a particular event or issue.
61. The state-level concern has been expressed by various speeches, statements and messages of the president of Bangladesh, ministers and government officials. President Ershad has declared 1990 as the Year of Environment and 1991–2000 as the Decade of Environmental Problems at the Non-aligned Summit, at Belgrade and at the meeting of commonwealth heads at Kuala Lumpur.
62. WCED, *'Our Common Future', World Commission on Environment and Development* (Oxford: Oxford University Press, 1987).
63. Hasan, S., 'Environmental Governance in Bangladesh: Problems and Issues', in M. Allaudin and S. Hasan (eds.), *Bangladesh: Economy, People, and the Environment* (Australia: University of Queensland, 1996), Economic Conference Monograph Series-1, p. 234.
64. Ibid., p. 236.
65. Ibid., pp. 237–238.
66. Allaudin, M. and C.A. Tisdell, 'Bangladesh and International Agricultural Research: Administrative and Economic Issues', *Agricultural Administration* (1986), pp. 1–20.
67. Bangladesh National Conservation Strategy (NCS): 2016–2031 (Executive Summary, Part-I), IUCN Bangladesh, 2016 (103 pages).
68. Transforming Our World: The 2030 agenda for Sustainable Development: A/Res/70/1 United Nations 2018 (Website: Sustainabledevelopment.un.org).
69. Bangladesh NCS, op. cit., p. 1.
70. Ibid., p. 2.
71. Ibid., p. 1.
72. Ibid. (During the preparation of NCS emphasis was given on a unified and holistic sustainable development framework rather than standalone sectoral issues.) pp. 15–18.
73. Bangladesh NCS, op. cit., p. 12.
74. The Inception Report, op. cit., p. 13.
75. Ibid., p. 14.
76. Ibid., p. 13.
77. Bangladesh NCS, An SDG Framework for NCS, pp. 18–20.
78. Ibid. (Few Observations), pp. 13–14.
79. An SDG Framework for NCS, op. cit., p. 18.
80. Ibid., para 2, p. 18.
81. Ibid., para 3, p. 18.
82. Ibid., para 4, p. 18.
83. Ibid., p. 15.
84. Ibid., pp. 15–18.
85. United Nations Transforming Our World: 2030 Agenda for Sustainable Development, A/Res/70/1, 2015.

86 Human Development in South Asia: Country Profiles: Nepal, 1997, p. 50.
87 Ibid., p. 49.
88 Ibid., p. 51.
89 Zafarullah, H., 'Towards Good Governance in Bangladesh: External Intervention, Bureaucratic Inertia and Political Inaction', in Allaudin and Hassan (eds.), 1996, pp. 146–162; Subhan, R., *Bangladesh: Problems of Governance* (Dhaka: University Press Limited, 1993).
90 Myrdal, G., *The Challenge of World Poverty* (Harmondsworth, UK: Penguin, 1971), p. 237.
91 Smith, William E., 'A Country under Water', *Time*, Sept. 19, 1988, pp. 30–31.
92 Galloway, Joseph L., 'Nature and Man torture Bangladesh', *U.S. News and World Report*, 19 September 1988, p. 35.
93 Ives, Jack D. and Messerli Bruno, *The Himalayan Dilemma: Reconciling Development and Conservation* (London: Routledge, 1989).
94 Myers, Norman, 'Environmental Repercussions of Deforestation in the Himalayas', *Journal of World Forest Resource Management* 2 (1986), pp. 63–72.
95 Kadkar, Ram, 'Nepal: National Conservation Strategy', in Jeremy Carew Reid (ed.), *The Book Strategies for Sustainability: Asia* (London: IUCN, WCU Earthscan Publications, 1997), p. 41.
96 Metz, John J., 'A Reassessment of the Causes and Severity of Nepal's Environmental Crisis', *World Development* 19(7) (1991), p. 806.
97 Kadkar, op. cit., p. 42.
98 Ibid., p. 43.
99 Claiborne, William, 'Erosion Is Laying Waste to Life in Shadow of the Himalayas', *The Washington Post*, July 18, 1983, p. 415.
100 Kadkar, op. cit., p. 51.
101 Ibid., p. 52.
102 Nature Conservation National Strategic Framework for Sustainable Development (2015–2030), Government of Nepal: National Planning Commission, Kathmandu, 2015, p. 10 (see major achievements).
103 Eighth and Ninth Plans were launched after the environment conservation policy incorporated into the constitution started adopting the policies of conservation. In the Eighth Plan programmes such as environment and resource conservation programmes, institutional development environment impact assessment, conservation of natural and cultural resources, mass awareness building implementing suitable programmes were highlighted, whereas in the Ninth Plan environment and nature conservation received greater priority.
104 Nature Conservation NCS (2015–2030) op. cit., para 2, p. 10.
105 Ibid. (see Chapter 2), pp. 5–15.
106 National Strategic Framework for Sustainable Development (164 pages).
107 Ibid., p. 16.
108 Ibid. (see Chapter 2: Contextual Analysis of Nature Conservation), p. 5–20.
109 Ibid. (see Chapter 3: Integration of Nature Conservation into Development Efforts), pp. 35–44.
110 Ibid., pp. 35–38.
111 Ibid., Introduction (Chapter 1), pp. 1–3.
112 Ibid., para 6, p. 1.
113 Ibid., p. 1–2.
114 The United Nations has determined three indicator s(to be elevated to the rank of a developing country), namely per capita average gross national income, Human Assets Index (HAI) and their standards. To be elevated to the rank of developing countries, it is mandatory to fulfil at least two of the three indicators. Since production and income play major role in achieving this goal, the achievement of this goal is also directly associated with nature conservation (in Nepal).

Environmental Governance and Sustainable Development 159

115 Some of the major dimensions of 13th Plan are highlighted as integrating environment conservation into physical infrastructure development, economic and social development programmes, documenting biodiversity and indigenous knowledge and skills for conservation, participatory management of forest heritage, community-based, inclusive, equitable and poverty reduction-oriented programmes in sustainable and balanced conservation, as climate change has emerged as a global concern according greater importance to environmental conservation and promotion.
116 Op. cit., pp. 1–5.
117 National Strategy Framework for Sustainable Development, p. 164.
118 Ibid., (See Limitations), p. 3.
119 Human Development in South Asia: Country Profiles: Pakistan, 1997, p. 37.
120 Ibid., p. 39.
121 Ibid., p. 38.
122 Karbraj, Aban Marker, G.M. Khattak, and Syed Ayub Qutub, *Pakistan: NCS' in the Book Strategies for Sustainability: Asia* (ed.), Jeremy Crew Reid (London: Earthscan Publications, 1997), p. 55.
123 Ibid., p. 56.
124 Carew-Reid, Jeremy (ed.), *Strategies for Sustainability: Asia', IUCN The World Conservation Union* (London: Earthscan Publications, 1997), p. 41.
125 Ibid., p. 42.
126 Ibid., p. 47.
127 Hassan, Shaukat, *Environmental Issues and Security in South Asia*, Adelphi Paper No.263 (London: International Institute for Strategic Studies, 1991).
128 Carew-Reid, op. cit., p. 55.
129 Karbraji, op. cit., p. 56.
130 Devwatch Newsletter Jan–March 2018, Issue 1, National Assembly of Pakistan.
131 Pakistan Millennium Development Goals Report 2013: Planning Commission, Government of Pakistan.
132 Pakistan's Implementation of the 2030 agenda for Sustainable Development: Voluntary National Review, Government of Pakistan, 2019.
133 Ibid., Voluntary National Review, see Introduction, pp. 12–14.
134 Ibid., Voluntary National Review, National Priorities, p. 13.
135 Ibid., p. 14.
136 Ibid. (Introduction).
137 Ibid. (Policy and Enabling Environment), Chapter 3, p. 19.
138 Ibid., p. 20.
139 National Poverty Report (2015–2016), Planning Commission, Pakistan.
140 Multidimensional Poverty in Pakistan (2016), Planning Commission.
141 Op. cit., Voluntary National Review ('Progress Review on Selected Goals: Key Initiatives'), p. 48.
142 Ibid. (VNR, Government of Pakistan, 2019), p. 11.
143 Ibid., p. 12.
144 Ibid., para 2, p. 12.
145 Ibid., para 3, p. 12.
146 Op. cit. Introduction, pp. 12–13.
147 Ibid., pp. 12–14.
148 Ibid.
149 Ibid., para 4, 12.
150 Ibid., para 2, 13.
151 VNR, National Priorities, op. cit., p. 13.
152 Ibid., para 2, p. 13.
153 Ibid. (VNR), National SDGs Framework, para 4, p. 13.
154 Op. cit., VNR, 'Targets for Sustainable Development', pp. 13–14.
155 Ibid., p. 14.

156 Op. cit., VNR, Progress Review on Selected Goals, Key Initiatives, pp. 48–50.
157 Ibid., VNR, Mainstreaming and Institutionalizing SDGs in Public Planning, p. 62.
158 Ibid., Introduction and Chapter 5, pp. 12–13.
159 Op. cit., Mainstreaming and Institutionalizing, SDGs, p. 62.
160 Ibid., p. 63.
161 Op. cit. (VNR, Critical Challenges), p. 67.
162 Ibid., Critical Challenges, para 2, p. 67.
163 Ibid., Way Forward, para 3, p. 68.
164 Ibid., SDG Target 17.3 (Mobilize additional financial resources for developing countries from multiple sources).
165 Ibid., Voluntary National Review (Pakistan, 2019), Way forward, p. 68.
166 Ibid., p. 67.
167 Human Development in South Asia: Country Profiles: Sri Lanka 1997, p. 51.
168 Ibid., p. 53.
169 Ibid., p. 54.
170 Ibid., p. 56.
171 Ibid., p. 61.
172 Ranatunga, M.S., L. Wijesinghe, and A. Ranjit Wijewansa, *Sri Lanka: National Conservation Strategy* (London: Earthscan Publications, 1997), p. 63.
173 Ibid., p. 64.
174 Ranatunga, M.S., *Sri Lanka: Natural Resources and the Environment, EIU Country Profile* (London: The Economist Intelligence Unit, 1997), p. 7.
175 Ibid., second paragraph.
176 Carew-Reid, op. cit., p. 85.
177 Sustainable Sri Lanka 2030 Vision and Strategic Path: Presidential Expert Committee (Jan. 2019) (315 pages). See 'Introduction and Overall Integration', p. 2.
178 Ibid. See Introduction and Overview, p. 38.
179 Ibid., para 2, p. 38.
180 Ibid., pp. 38–39.
181 Ibid., Vision Document, Executive Summary, Cross Cutting Themes, p. 20.
182 Ibid., 2030 Vision and Summary, pp. 32–35.
183 Sustainable Development Vision and Balanced Inclusive Green Growth (BIGG) Path, pp. 41–42.
184 Ministry of SD and Wildlife, National SDG Report Card, Dec. 2017.
185 M. Munasinghe, 'Environmental Economics and Sustainable Development', Paper Presented at the 1992 Earth Summit, Rio De Janeiro, and reproduced as Environment Paper No.3, The World Bank, Washington, DC.
186 Sri Lanka Voluntary National Review on the Status of implementing the Sustainable Development Goals: Government of the Democratic Socialist Republic of Sri Lanka (published by Ministry of Sustainable Development, Colombo, 2018), pp. 16–21.
187 Ibid., VNR, pp. 22–40.
188 Ibid., p. 86.
189 Ibid., p. 87.
190 Ibid., p. 88.
191 UNEP, Governing Council Session, 'Economics, Environment and Development', Nairobi, 13th to 16th May 1981, UNEP/GCS/9, p. 21.
192 National Seminar on Marine Environment and Related Eco-Systems: Sri Lanka Foundation Institute, Colombo, 22 and 23 July 1980, p. 17.
193 Ibid., p. 18.
194 The South Asia Co-operative Environment Programme, An Overview. Accessed internet on 6 May 2020.
195 SACEP, About Us: An Overview Accessed Internet on 6 May 2020.
196 SACEP, About Us: An Overview 'Male Declaration of 1998'. Accessed internet on 5 May 2020.

197 SACEP Regional Overview, An Environmental Profile of South Asia. Accessed internet on 7 May 2020.
198 SACEP, About Us: Vision. Accessed internet on 7 May 2020.
199 Ibid.
200 www.sacep.org/programmes & activities. Accessed internet on 9 May 2022.
201 The SACEP post-2015 development agenda highlights the key challenges and critically analyses the means of implementation of Rio+20 Outcome and the policy programme responses in South Asia in addressing environment priorities identified in Rio outcome document 'Our Common Future' as well as their linkages to poverty reduction and development.
202 Sustainable Development Goals (SDGs) are also known as Global Goals. See www.globalgoals.org.
203 These new Global Goals and the sustainability agenda go beyond MDGs and identify the common and transboundary priorities of South Asian countries to achieve sustainable development.
204 The United Nations Summit for the adoption of the post-2015 development agenda was held from 25 to 27 September 2015, in New York, and convened as a high-level plenary meeting of the General Assembly (A/69/L-85 – Draft Outcome Document).
205 https://www.sacep.org/about-us Programmes and activities of SACEP. Accessed on 6 June 2020.
206 Ibid., Programmes and activities of SACEP. Accessed internet on 7 June 2020.
207 Programmes and activities of SACEP: Biodiversity Convention, 'Major concerns' Accessed internet on 11 June 2020.
208 www.sacept.org/ Programmes and activities: Sustainable Consumption and Production (SCP). Accessed internet on 25 June 2020.
209 The 13th Meeting of Governing Council of SACEP promotes SCP within policy-making mechanisms of South Asian countries as well as the Asia-Pacific Road map of the Rio+20 10-Year Framework of programmes on Sustainable Consumption and Production. SACEP has been promoting a number of activities in support of SCP in the region.
210 www.SACEP.org/ Programmes and activities: South Asia Forum of Sustainable Consumption and Production (SAFSCP). Accessed internet on 27 June 2020.
211 www.SACEP.org/ Programmes and activities: Waste Management. Accessed internet on 1 July 2020.
212 In 2005, the 9th Governing Council identified waste management as one of priority areas in SACEP, which is the sub-regional secretariat for international partnership for authorities (IPLA) in South Asian region.
213 www.SACEP.org, Programmes and activities: Activities of SACEP related to waste management in South Asia. Accessed internet on 9 May 2022.
214 www.sacep.org/ Programmes and activities: Adaptation to climate change. Accessed internet on 15 May 2022.
215 www.sacep.org.com/ Programmes and activities. Accessed internet on 22 May 2022.
216 Ibid.
217 www.sacep.org.com/ Programmes and activities – Accessed internet on 23 May 2022.
218 www.sacep.org.com/ Programmes and activities – Accessed internet on 24 May 2022.
219 www.sacep.org.com Regional Overview: Regional Seas Programme, Accessed internet on 23 May 2022.
220 SASP is one of the 18th Regional Seas Programme of the United Nations Environment Programme (UNEP). The priority activities are highlighted in four specific areas: integrated coastal zone management, development of national and regional contingency planning, human resources development and regional centre's protection of marine environment from land-based activities.
221 www.sacep/org.com/SASAP. Accessed internet on 28 May 2022.

222 Ibid., SASAP, Accessed internet on 28 May 2022.
223 www.sacep.org.com/ Programmes and activities: The South Asian Seas Region, Accessed internet on 29 May 2022.
224 Ibid.
225 Ibid., SASR. Accessed internet on 29 May 2022.
226 www.sacep.org.com/ The South Asian Seas Region. Accessed internet on 30 May 2022.
227 SACEP: SASR. Accessed internet on 31 May 2022.
228 Ibid., SASR. Accessed internet on 31 May 2022.
229 Accessed internet on 1 June 2022.
230 Ibid.
231 www.sacep.org.com/South Asian Seas Action Plan (SASAP): Specific Objectives. Accessed internet on 3 June 2022.
232 SASAP: Priority areas. Accessed internet on 3 June 2022.
233 www.sacep.org.com/South Asian Seas Programme Action Plan: Priority activities. Accessed internet on 3 June 2022.
234 SACEP and World Bank are collaborating to formulate and implement a US$ 40 million, regional project 'Plastic Free Rivers and Seas for South Asia'. www.sacep.org.com/Milestones. Accessed internet on 15 June 2022.
235 Milestones: Plastic Free Rivers and Seas for South Asia: Projects Development Objectives. Accessed internet on 18 June 2022.
236 Plastic Free Rivers and Seas for South Asia: Components of the Project. Accessed internet on 18 June 2022.

5
UNEP
An Appraisal

Introduction: Positive Factors and Developments

The United Nations Conference on Human Environment marked the culmination of efforts to place the protection of the biosphere on the official agenda of the United Nations.[1] Specific aspects of the environment have been the object of international negotiations and arrangements, but the concept of the collective responsibility of nations for the quality and protection of the earth as a whole did not gain political recognition until the years immediately preceding the Stockholm Conference. The Stockholm recommendation for a United Nations Environment Programme was implemented by the UN General Assembly on 15 December 1972, through Resolution 2997, which established the necessary institutional and financial arrangements.[2] The principal accomplishments of the Stockholm Conference were twofold: the official recognition of the environment as a subject of general international concern and the institutionalization of that concept in the United Nations Environment Programme (UNEP).[3]

There is no doubt in recognizing the fact that UNEP has contributed significantly and positively in various ways in the protection and preservation of global environment. It plays an important role in developing scientific knowledge on the state of the global environment by enabling and encouraging innovative solutions to the management of a variety of the complex environmental problems on a wide range of scales, from local, to regional, to global.[4]

UNEP shares its knowledge and wisdom with United Nations agencies that are preoccupied with economic and sustainable development issues, such as, United Nations Development Programme and specialized agencies such as Food and Agricultural Organisation, the World Health Organization and the United Nations Children's Fund (UNICEF).[5] UNEP also nurtures partnerships with major segments of civil society such as non-governmental organizations, the business community,

DOI: 10.4324/9781003370246-5

women's organizations, consumer and sports associations. UNEP's promotion and synthesis of environmental information have generated many 'state of the environment reports' which created worldwide awareness of emerging environmental problems. It recognizes that learning and training are fundamental for achieving a sustainable future. Today, global communities face changes that touch the very core of their value systems, involving changes in attitudes and perspectives, and in this enterprise, the central task is to assist decision-makers and planners to cultivate a new mindset with the goal of incorporating the imperatives of environmental conservation and sustainable development into the decision-making process.[6]

UNEP acts as the institutional framework to service this enterprise of learning, training and skills exchange. UNEP assists in the development of new analytical tools, such as national environmental profiles, impact assessment, environmental accounting and environmental audits.[7] It also acts as the sponsor for a variety of international conferences, programmes, plans and agreements covering such diverse agendas as water resources management, transboundary movement of hazardous wastes, biological diversity, land degradation and environmental law.

UNEP provides an integrative mechanism through which a large number of separate efforts, intergovernmental and non-governmental, national, regional and local are reinforced and interrelated. The policies of UNEP are set by the Governing Council, which receives reports and recommendations from the executive director. Its membership, first set at 27 by the Prep com, was increased to 58 by the General Assembly, thus giving the developing countries a clear majority. From the viewpoint of some observers in developed countries, UNEP appears to be one more forum for voicing of disappointments of ex-colonial states and yet another channel for development assistance. These misgivings have been partially relieved as the developing countries' majority, under the leadership of the Secretariat, has come to recognize the seriousness of environmental issues.

The First Three Sessions of the Governing Council: The Priority Areas of Developed and Developing Countries

The first three sessions of the Governing Council (1973–1975) were formative and set the course of UNEP to the present date. The first session of the Governing Council met in Geneva from 12 to 22 June 1973.[8] It was at this session that the consolidation of the political interests and numerical strength of the developing countries enabled them to influence the policies of UNEP. Development-related priorities predominated. One of the prime interests of the developed nations was Earth watch, a worldwide monitoring programme to detect significant changes in critical environmental conditions.

The developing nations rejected the proposed priority areas put forward by Maurice Strong, placing excessive emphasis on environmental problems caused by industrial states. They appointed their own drafting group and substituted their own priorities. The priority areas[9] adopted were human settlements, land, water and desertification, trade, the transfer of technology, oceans, and conservation of nature, wildlife, genetic resources and energy.

The second session held at Nairobi from 11 to 22nd March 1974 was described as harmonious. The council reaffirmed its previously designed priority areas and assigned 'functional' priority to the creation of Earth watch and its major components: the Global Environmental Monitoring System (GEMS) and the International Referral System (INFOTERRA).[10] Executive Director Maurice Strong, in his report to the 29th session of the General Assembly, explained that the information referral and monitoring systems were to be developed with an emphasis on assistance and training to meet the needs of the developing countries.[11]

The third session of the Governing Council, which was held from 17 April to 21 May 1975, adopted over 25 major decisions that were consistent with established objectives and priorities.[12] The developing nations maintained their dominance and suggested a balance between the interests of the developed and developing nations. According to Maurice Strong: 'As long as measures to protect and improve the environmental capital is not counted as a cost, economic incentives will continue to run counter to environmental interests.'[13]

During these initial sessions of UNEP, no consensus was reached on the responsibilities of the developed nations for developing nations' environmental problems. Industries based in developed countries contributed to the environmental problems of the developing world, such as timbering, mining and trade in endangered species; tourism; and mono-crop agriculture. Third World leadership often took positions, which in the developed countries appeared to be contradictory.

Environment Assembly of UNEP and the First Three Sessions

Earlier, UNEP was administered by a 58-member Governing Council, but at the Rio+20 UN Conference on Sustainable Development in 2012, 'universal membership' was accepted for UNEP, to include the full 193 member states of the UN. This is a significant development in the evolution of environmental governance, which replaced UNEP's Governing Council with the UN Environmental Assembly (UNEA) of UNEP, composed of the full 193 UN members.[14] UNEA acts as a parliament of the environment and sets UNEP's agenda. The first meeting of UNEA (UNEA-1) was held in June 2014.[15]

The first three sessions of the UNEA along with the fifth session (which commemorates the 50th anniversary of UNEP) are briefly discussed here. The first session of the UN Environment Assembly discussed and adopted resolutions on major issues of illegal trade in wildlife, air quality and environmental rule of law, financing the Green Economy, the Sustainable Development Goals and 'delivering on the environmental dimension of the 2030 Agenda for Sustainable Development'.[16] Further, it was regarded as a success, with the adoption of a 'Ministerial Outcome Document' (2014), which called for the achievement of 'an ambitious, universal implementable and realizable Post-2015 Development Agenda' that integrated all the dimensions of sustainable development for 'the protection of the environment and the promotion of inclusive social and economic development in harmony with nature'.[17] The second session of the United Nations Environment Assembly of the United Nations Environment Programme (UNEA-2) was held from 23 to 27 May 2016 in Nairobi, Kenya. The meeting brought together over 2,500 delegates from 174 countries, and the Assembly was preceded by the Science and Policy Forum, which convened more than 250 representatives from the science and policy-making communities from over 100 countries. For two days, current issues of human interaction with the environment were discussed, with sessions covering the SDGs, the GEO-6 assessments, a range of frontier and merging issues with a perspective of strengthening the science-policy interface.[18] The third session of the UN Environment Assembly (UNEA-3) took place from 4 to 6 December 2017 in Nairobi, Kenya. It considered the overarching 'theme of pollution'.[19] Outcomes of the session included a political declaration on pollution, linked to the Sustainable Development Goals (SDGs); resolutions and decisions adopted by member states to address specific dimensions of pollution; voluntary commitments by governments, private sector entities and civil society organizations to clean up the planet; and the Clean Planet Pledge, a collection of individual commitments to take personal action to end pollution in all its forms.[20] In addition to the Open-Ended Committee of Permanent Representatives (OECPR) meeting, which was held from 29 November to 1 December, UNEA-3 was preceded by these meetings: Global Major Groups and Stakeholders Forum, Science Policy Forum and Sustainable Innovation Expo.

The most recent UNEA session was held during the raging pandemic. The overall theme of UNEA-5 is 'Strengthening Actions for Nature to Achieve the Sustainable Development Goals'.[21] The theme calls for strengthened action to protect and restore nature and nature-based solutions to achieve the SDGs in its three complementary dimensions (social, economic and environmental). UNEA-5 provides member states and stakeholders with a platform for sharing and implementing successful approaches that contribute to the achievement of the environmental dimension of the 2030 Agenda and the SDGs, including the goals related to the eradication of poverty and sustainable patterns of consumption and production.[22] UNEA-5 also provides an opportunity to take ambitious steps towards building back a better and greener world by ensuring that investments in economic recovery after the COVID-19 pandemic contribute to sustainable development. In light of the restrictions related to the COVID-19 pandemic, member states and stakeholders have decided that UNEA-5 take place in a two-step approach. The first session (UNEA-5.1) was held online on 22–23 February 2021 with a revised and streamlined agenda that focused on urgent and procedural decisions. Substantive matters that require in-depth negotiations are deferred to a resumed in-person session of UNEA-5 during 28 February–2 March 2022 (UNEA-5.2).[23] UNEA-5.2 was followed by a Special Session of the UN Environment Assembly (UNEA-SS), held on 3–4 March 2022, which devoted to commemorate the 50th anniversary of the creation of UNEP in 1972.[24] UNEA-5 is an opportunity for member states to share best practices for sustainability. It creates a momentum for governments to build back better after the COVID-19 pandemic through green and sustainable recovery plans.

United Nations Environment Programme's accomplishments and achievements are recounted from year to year in the introductory reports of the executive director to the annual sessions of the Governing Council. For the developing nations, two aspects emerged as crucial: (1) the emphasis on development as a vehicle for raising the quality of the environment and (2) the location of the main office of UNEP, in a developing country.

Critical Problems

UNEP experienced mixed fortunes in its first decade. It had what Sandbrook describes as 'perhaps one of the most difficult jobs in the entire U.N. system'.[25] The critical points are that it was not designed as an executive body with the same kinds of powers as FAO, UNESCO and other specialized agencies. It had to coordinate the work of others. Some UN agencies promote policy initiatives with others and provide information to others. It had a huge constituency, effectively the entire natural environment and paltry finances. It was compelled to involve itself in multifarious activities and concerns and often had to take action based on the chance emergence of opportunity rather than on any carefully considered long-term plans. It had no power of positive interference in the management of local and regional environmental matters. Its problems fell into four main categories: financial, managerial, political and constitutional.

Financial

The strength of UNEP had ultimately to be measured in terms of the finances it had available. It made a promising start, with US$60 million pledged to the Environment Fund in just 60 months; the target was US$100 million in five years. Governments made a voluntary contribution to the UNEP Environment Fund, varying in scale from that of the United States – which, during the first eight years of UNEP's existence, contributed US$40 million, 36 per cent of the funds, to the 64 countries, such as Zambia, Jordan and Paraguay, which made no contributions at all between 1978 and 1981. In the first five years of its operation, UNEP planned to raise US$20 million per year, a target that it very nearly achieved. The complications of short falls in income were aggravated by contributions arriving later or at the end of the financial year or in non-convertible currency (currency that can only be spent in the donor country, commonly in the Eastern bloc). Such problems made forward planning difficult.

Managerial

UNEP's second major handicap lay in its own internal management. It was accused by its critics of having a limited perspective on global problems, of being inefficient, of failing to delineate its priorities, of being too centralized in the person of the executive director and of trying to do too much with too little to show in the end. The responsibility for these problems was seen to lie partly in UNEP's own management policy and partly in the limitations inherent in UNEP's constituency. Maurice Strong, despite his role in Stockholm and in the foundation of UNEP, was new to environmental matters, so, on his appointment as first executive director, he drew on the areas he knew best – business, politics and international public service – when appointing senior staff to launch UNEP.[26] The long-term result was that UNEP faced internal management problems based more on bureaucratic procedures than on professional approaches to the problems of the environment. There was little improvement in the first four-year term of Strong's successor, Dr. Mostafa Tolba.[27] By 1980, personnel issues had reached such a low ebb that the permanent representatives to UNEP of the European Community Countries compiled a short report outlining what they saw as the key problems: autocratic management with no delegation of authority, all decisions centralized in the person of the executive director, low staff morale and high turnover, and the unhealthy state of the Environment Fund, which could be blamed partly on dissatisfaction with the performance of UNEP and its executive director.[28] An unsuccessful attempt was made in 1980 to replace Dr. Tolba at the end of his first term in office. The impermanence of its headquarters arrangements and the postponement in 1980 of new building plans made the management problem no easier. Only in 1983, after being based temporarily in several different offices, UNEP finally moved into permanent office buildings on the outskirts of Nairobi.

Political

The third source of problems was UNEP's location in Nairobi. It helped redress the imbalance in the location of UN agency headquarters, and it helped broaden the minds of European and North American environmentalists by drawing attention to the situation in developing countries. But it also isolated UNEP from the industrialized countries which held the balance of world economic and political power and, hence, made many of the key decisions affecting the global environment. UNEP itself felt that the location made it difficult to recruit highly qualified staff, and demanded that UNEP create a new infrastructure, whereas if it had been located within an existing UN agency, it would have been able to draw from a common pool of personnel while directing its resources at substantive action.[29] The location in Kenya also tended to leave UNEP divided between the more developed world, which supported emphasis on global problems such as global warming, ozone depletion, conservation of biodiversity and climate change, and the developing world, which adopted UNEP as its own UN 'agency', emphasizing issues of environmentally sound development, and hoped UNEP to be particularly sympathetic to the needs of the developing countries in any conflict of interest with industrialized countries. This reached such a point where the Governing Council and later Environment Assembly were accused of rejecting project suggestions from the Secretariat because these were not sufficiently Third World oriented.[30] The division in turn helped sour relations between the Governing Council and the Secretariat.

Constitutional

The fourth source of problems was constitutional. UNEP headquarters was divided between the Programme Proper and the Fund. Thus, the first division was between the Programme and the Fund. The former was responsible for devising programmes which may or may not have been put into effect in the form of projects. The latter was responsible for managing finances for these projects. This division demanded that all UNEP activities be approved by both the Fund and the Programme and involve staff from both sides in formulation and execution. The second division was between the Secretariat and the Governing Council. The council was drawn from 58 member governments, representatives of whom were voted onto the council for three-year terms by the UN General Assembly. It was charted with a number of basic responsibilities: to promote international environmental cooperation, to guide policy and review progress in UN agencies, to draw the attention of governments to emerging global environmental problems, to ensure that development in developing countries was not handicapped by environmental management policies and to manage the Environment Fund. Critics charged that the council was successful in the last of these responsibilities, but not in the others. Martin Holgate, president of the 11th Session of the Governing Council in 1984, was concerned that during 1983, procedural matters had become so dominant that they tended to obscure the real justification of UNEP: 'that people all over the world have environmental needs that the United Nations should meet'. Too much time was spent by the plenary session 'on discussions that were unlikely to help resolve these appalling problems'.[31]

The most critical constitutional problem lay in the nature of UNEP's relationship with other UN agencies. Its task of persuading other agencies to execute programmes was hampered by the fact that it had few incentives to offer and no means of enforcing its policies.[32] Furthermore, some of UNEP's slow progress reflected the fact that the environment had proved more complex, and more costly to monitor, than Stockholm delegates realized.[33] The action plan was widely regarded as too broad, too vague about priorities and above all too lacking in means of implementation. Unlike WHO and FAO, observed Dr. Mostafa Tolba, UNEP could not 'say at the end of the day that it had eradicated a disease here or planted so many thousand hectares of rice fields there'.[34]

In 1976, the UN General Assembly reviewed its institutional arrangements for the environment and concluded that they were inadequate, the UNEP itself admitted their shortcomings.[35] The incompatibility between the desire for a small secretariat and the need to maintain the wider ranging aims of the programme was a difficult task which had still not been resolved. Perhaps the most fundamental problem was that UNEP's role was consistently and widely misunderstood.[36] UNEP was never intended to be a full UN-specialized agency but simply the environmental programme of the United Nations system. Its role was catalytic. The first three areas were those agreed on before Stockholm: the Global Environmental Monitoring System, Marine Pollution, and INFOTERRA. As its own activities evolved, so UNEP's credibility and capacity to do the job of coordination and compilation of a truly system-wide programme increased.[37]

Major Achievements in the First Decade

In 1982, a new exercise, the System-Wide Medium Term Environment Programme (SWMTEP), was introduced which was designed to define precisely what each UN agency would do in the period from 1984 to 1989. Most of the description of action to be taken by each agency has been dismissed as bland and highly generalized; SWMTEP furthermore outlined areas of interest but spoke only in broad terms of appropriate action.[38] It was so broad that nearly every initiative on the environment could be fitted into one or another of its proposals. Rather than allowing UNEP to do a few things well, it ran the danger of further dividing the UNEP agenda. Yet ten years after Stockholm, UNEP confidently believed that its relations with other agencies had improved as the incidence of joint programming had grown.[39]

Often regarded as one of UNEP's major successes, if not its only real success in its first decade, was its Regional Seas Programme (RSP), in which UNEP successfully brought 120 countries and 14 UN agencies together to confront shared problems of pollution and coastal degradation in shared seas. Stockholm recommended action to end all significant marine pollution within three to four years, particularly in enclosed and semi-enclosed seas.[40]

The RSP was initiated in 1974 with a strategy carefully designed to respect national sensitivities particularly over data on pollution and to leave the actual implementation of each regional plan firmly in the hands of the coastal states concerned. The RSP involved most of the major UN bodies and illustrated exactly the kind of catalytic and executive role that UNEP could play.[41] The model for the RSP was the Mediterranean Action Plan (MAP), adopted by 16 Mediterranean states in Barcelona in January 1975. However, the Programme's success was limited by a number of fundamental problems, and despite these problems, the 'Regional Seas Programme' was widely regarded as UNEP's most effective undertaking in its first decade of operation. UNEP attributed the Programme's success to the fact that it was regional rather than global in nature, involving nations in urgent common problems that could only be solved by mutual action.[42]

UNEP was the most tangible result of the Stockholm Conference: Although imperfect and a decade later, still insufficient in many ways, it was the best form of institution possible given the limitations imposed by other UN agencies and by the low level of funding from governments. But it had severe obstacles placed in its path from the outset. It has had too little money, too few staff and too much to do. Eck Holm argues UNEP has simply not become the powerful global force that some once dreamed of.[43] Ten years after Stockholm, UNEP itself could claim only 'as a mixed record of achievement, fair-to-good progress in implementing some of the elements of the action plan', while for other elements, progress has been very slow.[44]

Critical Perceptions and Challenges

Mark Imber, in his book *Environmental Security and UN Reform*, analysing the role and performance of UNEP, has divided the problems faced by UNEP into three parts.[45] *First*, there is incompatibility between its role as conceptualizer of environmental policy and its commitment to action.

A *second* difficulty has been the relationship between UNEP and other United Nations' agencies. At the time of UNEP's establishment, the specialized agencies were resistant to UNEP's role of coordination and leadership. Within the United Nations Secretariat, UNEP has had the closest relations with UNDP. A declaration in 1980 by UNEP, UNDP, the World Bank and several regional development institutions promised the incorporation of environmental considerations in development policies and provide a rationale for closer collaboration in the future.[46]

UNEP is more often perceived as representing a special sectoral interest to which appropriate attention should be paid, rather than as a coordinating body intended to ensure the presence of environmental sensitivity throughout the United Nations system. It is important to the objectives of the Stockholm Declaration and Action Plan that UNEP play an effective role of leadership and coordination. The action plan distributed functions among the specialized agencies, non-governmental organizations and national governments 'so if there were a spurt in to systematic effort and coherent programming from within the system', it would have to come through the initiatives of the UNEP.

Since UNEP's establishment, its role has moderated and it has generally been the decisive factor in the incorporation of an environmental dimension into specialized agency developmental activities. Further, UNESCO, FAO, WHO and WMO point out environmental initiatives of their own which are necessary but insufficient for carrying out the Stockholm recommendations. UNEP's coordinative influence is essential to bringing their interrelating aspects into a coherent and concerted relationship. A *third* difficulty is that of developing effective leadership within UNEP and among nation-states, in the task of infusing development action with an informed environmental sensitivity. This is the purpose for which UNEP was established, which in essence is to assist the translation of the eco-development concept into operational reality.

An Assessment of UNEP

In this section of the chapter, an analysis of the views and opinions expressed by leading environmental activists (from the Centre for Science and Environment) as well as officials conducting UNEP programmes (an Interview method was followed) on the role and functions and an assessment of UNEP was highlighted after interviewing some officials – Veena Jha (project coordinator of UNCTAD in New Delhi), Mr. Nirmal Andrews (UNEP's regional director at Bangkok, Thailand), Sudarshan Rodrigues (consultant, Environment and GEF Division at UNDP Office at New Delhi), and Dr. Pradeep Monga (assistant resident representative, Environment and GEF Division) – as they represent the major UN functionaries at the implementation level. And an analysis of their views and opinions is presented in the following paragraphs.

Veena Jha,[47] discussing about the diminished role of UNEP, has highlighted several factors that contributed to the role and they are as follows:

1   There is a lack of commitment on the part of the developed countries of the North to implement the agendas on environmental issues (Rio and Stockholm agendas);
2   (i) Creation of the Commission on Sustainable Development has eroded the role of UNEP, (ii) UNEP is regarded as the global talk shop on environmental issues and (iii) environmental issues and environment have been mainstreamed into economic agencies and hence it has become difficult to differentiate UNEP's functions from other agencies;
3   Problems of resources and polarization of political issues have also eroded UNEP's credibility;
4   Asia and South Asia is a region where UNEP should be even more active. It is active in Africa compared to Asia;
5   The monitoring and stock-taking facilities and functions and role of other agencies/institutions like Asian Development Bank (ADB), United Nations Development Programme (UNDP), United Nations Industrial and Development Organisation (UNIDO) and ESCAP have eroded the role of UNEP.

Addressing the workshop 'UNEP/SACEP: South Asian Regional Forum on Environmental Co-operation between Government and the Private Sector' in New Delhi on 13–14 July 2000, Nirmal Andrews, the regional director of UNEP,[48] highlighted three most important priority areas for UNEP, and according to him, they are air pollution, water resources management and implementation of environmental policies and action programmes. In addition, according to him some of the most important priority areas for successful environmental cooperation in South Asia are highlighted as implementation of regional action programmes, reviewing and monitoring by the ministries of environment, regional roundtable conferences with the support of UNEP and SACEP, participation by international financial institutions and donor countries, regional review meetings on environment and reviewing the government documents on which cooperation is emphasized.

Veena Jha, focusing on major objectives of regional environmental programmes, emphasized four crucial factors which play a very significant role in the South Asian region, namely food security and effective participation to strengthen national capacities in multilateral processes, effective regional integration, cooperation and capacity building processes. She, further, noted, 'Real and substantial change and progress on environmental issues, requires deeper involvement, understanding, commitment and strengthening the institutional capacity.'[49]

UNEP functions in collaboration with other international institutions like ILO, UNCTAD, UNDP, UNESCO and UNIDO, complementing the work of these institutions, not duplicating their work. Further, UNEP works through conferences, seminars, workshops, strategy documents, research and joint projects, training and education.[50] Sudarshan Rodrigues,[51] consultant, GEF Division, at the UNDP office in New Delhi, commenting on the functions and role of UNEP, emphasizes a lack of presence of a regional office in South Asia, since the headquarters of the UNEP is at Bangkok, which contributes to the lack of efficiency in implementing its policies in the South Asian region.

Mr. Pradeep Monga[52] highlighted that any society committed to preservation and protection of environmental issues will support the main features of the implementation of Agenda 21 and the management of development. He identifies some operational projects which are being financed by GEF in India: (1) paper and pulp industry, (2) olive ridley turtle conservation, (3) medicinal plants conservation and sustainable utilization where local community participation is emphasized, (4) development and use of natural dyes in textiles where viable eco-friendly processes are used, (5) technical support for eco-friendly neem products, (6) support for small hilly hydel project, (7) bio-methanation processes (harnessing energy from waste through high rate bio-methanation process), (8) biomass energy for rural India, (9) fuel-cell bus development in India (it is futuristic technology) and (10) coal bed methane and its commercial use. According to him, the emerging areas for environmental cooperation in South Asia are as follows:

1 Environment-friendly products;
2 Environmentally sound clean technologies;
3 Capacity building through training key stakeholders, professionals and experts.[53]

Criticizing the UN agencies, including UNEP for lack of synergy to meet and tackle the environmental issues of the local people, he points out to a holistic approach with regard to sustainable development. He stressed the importance of regional programmes/regional networks setting up of more and more UNDP, UNEP regional offices and sectoral expertise in tackling environmental issues. According to him, monumental failures in capacity building will be resolved with the involvement of good NGOs, good local, regional, national and international institutions which will always lead to good results. There is always light at the end of the tunnel.[54]

The main criticism from the New Delhi-based NGO Centre for Science and Environment has focused on the functioning of UNEP. It is underlined that while UNEP was created to become a centre for coordinating global environmental action, it is not able to discharge its role with efficiency. It has been disempowered and has become a puppet in the hands of the developed and developing countries. Thus, it is not able to play the coordinating role effectively. It has been said that the UNEP has no power and enjoys only delegated authority and real power is wielded by the World Bank. Thus, UNEP's role has become insignificant and ineffective and it has merely become a talkshop.[55]

Evolving Nature of Environmental Issues

Programmes set in motion since Stockholm, for example, trade in endangered species, environmental monitoring and regional seas, have built in momentum with institutional support. In its tenth anniversary report in 1982, UNEP pointed to an evolution of environmental issues growing out of changing perceptions of the environment.[56] A worldwide change and growth in environmental perception was cited from three basic sources: (1) decisions by United Nations bodies revealing governmental acceptance of new principles, (2) growing numbers of scientific findings concerning the environment that have been gaining wide public acceptance and (3) practical experience in dealing with environmental issues.

A growing public awareness of environmental problems and national commitments in principle to environmental protection suggest that national and international policy-making will almost certainly be concerned with environmental problems in the coming future. The social and political divisions of the world burden and delay the implementation of environmental policies to which nations have in principle agreed. International environmental agreements today are negotiated and implemented in the political context of antagonistic cooperation. The need for institutions for environmental governance, better fitted to the task, has been voiced repeatedly by environmental policy makers.[57] The building of consensus is proceeding in many ways, exemplified in its larger dimensions by the environmental education movement, the Man and Biosphere Programme, the World Conservation Strategy, networks of NGOs and multinational corporate bodies, which are coming closer to the evolving structure of international cooperation.[58]

Science assists the process of defining the tasks of international environmental action, but building a coherent and effective structure for global environmental collaboration is a collective work of art-no science provides a blue print. Humanity was never given an organizational manual for spaceship earth.[59] Defence of the earth is possible only in a world which is divided in many ways, yet united and cooperative to safeguard the survival of the human kind in many other ways.

Challenges of Implementation: From Vision to Action

When world leaders gathered in Rio de Janeiro, Brazil, in June 1992 to attend the United Nations Conference on Environment and Development (UNCED), their aim was to address the interlocking global crisis of ecological and economic decline. Among the key results of this conference were the 'Rio Declaration', which acknowledged 'precautionary' and 'polluter pays' principles, the need to change consumption patterns and the importance of addressing global poverty, and Agenda 21, which is a far-ranging, comprehensive global action programme approved by the international community for environmental conservation and sustainable development.[60] Wolfgang Sachs,[61] eminent expert on development issues from Germany, highlighted several issues confronted by the developing countries, like sustainability, tackling poverty and malnutrition, and the development models of Northern and Southern countries were blamed for this. He was critical of the

Rio Conference and found that 'the official discourse on sustainable development does not look at "conservation of nature" but instead tried to reinforce the idea that development means economic growth'.[62] His criticism of the Western model of development is significant. To him 'development has become a race between nations in an economic arena, where it has become imperative for those who are behind to catch up, and given the ecological crisis, this race leads to an abyss'.[63] According to him the issue of sustainability was not addressed adequately at the Rio Conference. He conceptualized a world that is sustainable and just, and such a world can be achieved only by 'greening of the mind'.

Building Consensus

After Rio, there has been a substantial follow-up on its recommendations, starting with the United Nations Environment Programme itself and proceeding through national governments, regional authorities and the civil society. Underlying the time-bound goals and action programmes of the UNEP is the message that not only the United Nations agencies but also the national governments, the business community, the local authorities, non-governmental organizations, and every man, woman and child have a role in conserving the environment.

Environmental degradation does not stop at national frontiers. Growing human numbers and rising levels of economic activity have produced consequences which are regional or even global in scope. For instance, no single country is responsible for the depletion of the ozone layer, for the pollution of marine resources and for the extinction of biological resources, and only international cooperation can create the conditions for their regeneration.[64] One of UNEP's priorities is to encourage responses to these environmental problems at the local, regional and global level by bringing governments together and by developing policies, programmes and negotiating agreements.

Chapter 38 of Agenda 21[65]

This chapter called up UNEP to take 'the leading role' in international environmental law. Since UNCED, UNEP has built on this foundation and also ensured that the existing agreements are not only better implemented but also oriented to the economic needs of the developing world. The following sections will discuss some of the agreements forged under the auspices of UNEP after UNCED.

The Convention on Biological Biodiversity

This came into force on 29 December 1993, with 161 members and the establishment of close linkage with other intergovernmental bodies such as the Commission on Sustainable Development, the Intergovernmental Panel on Forests, the Second Committee on the United National General Assembly and the Global Environment Facility (GEF). It has become an international treaty with universal participation.

To support the implementation of the convention, UNEP has launched a comprehensive Biodiversity Programme and Implementation Strategy.[66] UNEP, jointly with the World Conservation Union (IUCN) and World Resources Institute (WRI), prepared the Global Biodiversity Strategy to stimulate and implement local national and international action. It also launched the Global Marine Biological Diversity jointly with the Centre for Marine Conservation, IUCN and World Fund for Nature (WWF). This strategy emphasizes the distinctive conservation needs of life in oceans, coastal waters and estuaries to help leaders decide how to protect the living systems of the sea.

The Convention to Combat Desertification (CCD)[67]

It was adopted in June 1994. UNEP provided substantial financial and technical support to the CCD negotiations, supporting the Interim Secretariat, the International Panel of Experts, in the preparation of background documents and case studies, and the participation of NGOs and experts at preparatory meetings and negotiating sessions.

UNEP

UNEP is built on supporting the science through assessments carried out by the Inter-governmental Panel on Climate Change (IPCC), supporting information dissemination through the information unit for conventions and in building consensus through projects such as 'Activities Implemented Jointly' (AJJ).[68]

UNEP helped develop internationally agreed guidelines on climate impact assessment and mitigation strategies for the costing of greenhouse gases abatement. As part of the country studies on impacts of climate change, UNEP prepared, in cooperation with IPCC, a handbook on methods for climate change impacts and adaptations assessment, to complement the IPCC Technical Guidelines for Climate Change Impacts and Adaptations Assessment.

The UNEP Collaborating Centre on Energy and Environment (UNCCEE) developed a methodological framework and coordinated country studies in ten developing and industrialized countries, on greenhouse gas abatement costing. UNEP organized an international conference in Copenhagen on national actions to mitigate global climate change, to discuss the aims of IPCC and how they could be translated into practical action.

After UNCED, the Montreal Protocol on the Substances that Deplete the Ozone Layer Has Recorded Some Major Successes[69]

The continuum of negotiations from Montreal to London to Copenhagen to Vienna to San Jose has served not only to clarify several provisions of the Protocol and accelerate the phase-out of several ozone-depleting substances but also to put many ambitious work plans in place.

One of UNEP's More Meaningful Achievements Has Been the Finalization of the Lusaka Agreement

This provides for the establishment of an international task force for combating illegal trafficking in African wildlife. It is the only regional wild life treaty establishing a regional Interpol. It seeks to implement the provisions of CITES by conducting undercover investigations in close cooperation with established national law enforcement agencies in Africa. It also implements and reinforces the provisions of the Convention on Biological Diversity which aims at enhancing awareness on the conservation and sustainable use of biological diversity. UNEP provides guidance on environmental issues in implementing forest principles and assesses their implications for sustainable forest management practices. UNEP developed its Forest Policy and Proposed Action Programme for 1996–2000 and provided expertise on a substantial basis to the CSD's Inter-governmental Panel on Forests (IPF).[70]

Global Environmental Facility (GEF)

GEF was established and operates on the basis of collaboration and partnership among the GEF implementing agencies (UNDP, UNEP and World Bank), as a mechanism for international cooperation for the purpose of providing new and additional global environmental benefits in the following focal areas: biological diversity, climate change, international waters and ozone layer depletion.

UNEP's role in the GEF is spelt out in the Instrument for the Restructured Global Environment Facility, which calls for UNEP to:

- Play the primary role in catalysing the development of scientific and technical analysis and in advancing environmental management in GEF- financial activities;
- Provide guidance on relating the GEF-finance activities to global, regional and national environmental assessment, policy framework and plans and to international environmental agreements;
- Establish and support the Scientific and Technical Advisory Panel (STAP) as an advisory body to the GEF.[71]

Technology Transfer

Technology Transfer constitutes a crucial element in developmental concerns related to the environment. The key issues in technology transfer in the emerging global industrial system concerns:

- Access of developing countries to technologies and to the international networks in the developed world;
- Appropriateness of technologies;
- Integration of technology transfer with policies for environmental protection.[72]

UNEP works in partnership with businesses and industries to push the frontiers of environmental management. It aims at building bridges between leading companies and those who require expertise, and it facilitates technology transfer between developing and developed countries. While technology transfer is often a business-to-business transaction, governments in partnership with industry have the responsibility to develop the institutional capacity that forms the basis for effective technology transfer. As one of the implementing agencies of the Multilateral Fund of the Montreal Protocol to assist developing countries, UNEP's Ozone Action Programme gathers practical information, conducts workshops and training on technical and policy issues, and assists country programmes.

UNEP's International Environmental Technology Centre (IETC), established in 1993, has offices in Osaka and Shiga, Japan. The mandate is the transfer of environmentally sound technologies to large urban areas in developing countries and countries with economies in transition. It addresses problems such as sewage and solid waste management as well as the management of lakes and reservoir basins.[73] UNEP has actively tried to promote and implement the principles contained in the Rio Declaration on Environment and Development through the elaboration of environmental law regimes at the national, regional and global levels. A number of Rio principles, such as those of global partnership and common but differentiated responsibilities, have been implemented in various conventions and continue to inspire emerging instruments.

UNEP since UNCED

After Rio, the prime challenge before UNEP was not only to promote and implement its environmental agenda but also to integrate it strategically with the goals of economic development and social well-being. Since UNCED, UNEP has modified the way it works with an emphasis on achieving results, partnerships, decentralization and the cultivation of new constituencies while strengthening old ones.

One of UNEP's most meaningful steps in this direction was the adoption of an integrated approach[74] to the preparation and implementation of its programmes. UNEP's integrated approach focuses on four major environmental themes:

- Sustainable use and management of natural resources;
- Sustainable production and consumption;
- A better environment for human health and well-being;
- Globalization and the environment.

Regional Focus

Another significant step taken by UNEP after UNCED has been the enhancement of its regional presence. UNEP regards a strong regional architecture as central to its ability to manage and advance its larger global environmental agenda. UNEP's regional offices are now involved as facilitators and enhancers of cooperation not only with the governments and with national and sub-national institutions but with international, inter-governmental and non-governmental institutions. In disseminating information, advice and training and in providing forums for consultation and consensus-building, UNEP's regional offices are playing a more meaningful role, in partnership with headquarters. These are the regional offices:

1   Regional office for Africa (ROA);
2   Regional office for Latin America and the Caribbean (ROLAC);
3   Regional office for West Asia (ROWA);
4   Regional office for Asia and the Pacific (ROAP);
5   Regional office for Europe (ROE);
6   Regional office for North America (RONA).[75]

Post-Rio: Economic and Ecological Globalization and Challenges

The last 50 years have seen a virtual explosion of intergovernmental negotiations to formulate international environmental treaties. This 'ecological globalization' is the inevitable result of the ongoing processes of economic growth and economic globalization which not only integrates the world's economies together but also takes national production and consumption levels to a point that threatens the world's ecological system.[76] It is now widely recognized that there are many ecological problems that cannot be solved by one country acting alone; cooperation with other countries has become essential. As a result, the nations of the world are increasingly getting together to create systems of global economic and environmental governance. While the former is best represented by the creation of the World Trade Organization (WTO), a major step in the later area has been the creation of the United Nations Environment Programme. However, the two processes of globalization are not accompanied by any form of political globalization. As a result, no political leader has any interest to ensure that the emerging global ecological policy is managed in the best interest of the maximum number of people and on the basis of the principles of 'good governance', that is, equality, justice and democracy.

Further there is no clear and transparent mechanism to integrate the two processes of economic and ecological globalization. And environmental diplomacy has not turned into the establishment of fair and just global environmental governance systems. Delegations from developing countries usually argue that since they did not create the environmental problems in the first place, it is the industrialized countries which should take remedial action. But the developed countries argue that even if industrialized countries created the problem in the first place, developing countries will soon do so, as they follow the Western economic model. However, national political and economic interests weigh in first before global ecological interests or global equity and social justice. Southern political leaders have shown a lack of vision to recognize that environmental issues/treaties mainly deal with the sustainable sharing of the earth's ecological space-global goods like oceans and atmosphere – on which national economies depend. Therefore, what matters most is not aid or technology transfer but rights to the equitable sharing of the earth's ecological resources.

However, in this quest, there exist significant dichotomies and challenges. If these developments in global economic and ecological integration are taking place within a framework of democratic political globalization, the outcome will be in the best interest of all people. However, that is not always the case – there is no guarantee that national delegations, concerned primarily with national economic interests, will ensure outcomes that are in the best interest of global sustainable development.

Global environmental governance needs strong and reliable institutions to manage global resources and decide the rules as to how nations will live in an increasingly small world.[77]

Unlike economic globalization, managed by the all-powerful World Trade Organization (WTO), management of environmental globalization is the responsibility of several independent convention secretariats and multilateral agencies. Every environmental convention, such as the UN Framework Convention in Climate Change (FCCC), the Convention on Biological Diversity (CBD) and the UN Convention to Combat Desertification (CCD), has an independent Secretariat and an independent international institutional framework which guards its share of power and clout. These institutions create many forums but little leadership. As a result, while the WTO increasingly adjudicates on environmental issues through trade disputes, environmental concerns have not yet found an institution which is equally powerful to counter economic interests. 'While WTO which manages the world's trading system has become strong, environmental institutions have not become so in the years following the Rio Conference', remarked Klaus Töpfer, head of the UN Environment Programme (UNEP), at a conference organized by the WTO in Geneva in March 1998.[78]

World leaders have shown little vision over the design of a unifying institutional framework for the global environment. Unmindful of the high coordination costs of the institutional framework, Northern players have managed with others to set up environmental secretariats in their countries. Meanwhile, the concerns of Southern players have been twofold: dislodge/dismantle environmental institutions, and to push for the creation of new institutions in the face of 'emerging problems', instead of building upon existing institutions. UNEP was created in 1972 in the afterglow of the UN Conference on the Human Environment in Stockholm, to provide leadership and encourage partnership in caring for the environment by inspiring, informing and enabling nations and people to improve their quality of life without compromising that of future generations.[79]

The appointment of Klaus Töpfer, a former German minister as head of UNEP in 1998, was seen as an important step towards reforming the institution. Soon after his appointment, UN secretary-general Kofi Annan set up a task force on environment and human settlement under his jurisdiction to 'reinvigorate reform and operationalize' the UN agenda on environment. There is an ongoing battle for turf among heads of existing UN institutions to guard their territory and maintain status quo. For instance, Nitin Desai, undersecretary general in charge of the UN Department of Economic and Social Affairs and CSD, firmly distinguishes CSD from UNEP saying that CSD 'integrates environment and development', unlike UNEP, which is an environmental institution.

The issue of global environmental governance is a persistent one. When post-UNCED progress came up for review five years later at the 1997 UN General Assembly Special Session (UNGASS), several delegates were dissatisfied with the existing institutional infrastructure for global environmental governance. Germany, backed by Brazil, Singapore and South Africa, raised the possibility of a global environment organization (GEO), which would be an alternative to the existing scattered UN institutions on environment. In the words of Chancellor Helmut Kohl, 'it is important to create a global umbrella organization for environmental issues'.[80] In November 1998, while addressing the Congress of the World Conservation Union (IUCN), Jacques Chirac, president of France, called for a world authority on the environment. He stated, 'we must start by establishing an impartial and indisputable global centre for the evaluation of our environment'. Pointing to existing institutional loopholes in the global environmental governance infrastructure, he argues that 'there is a need for a single place that embodies the environmental conscience of the world'. In a call for bold action, he went on to demand that UNEP be given the job of 'federating the scattered secretariats of the global conventions, gradually establishing a world authority, based on a general convention that endows the world with a uniform doctrine'.[81] Proponents of free trade also raised the need for a global environmental organization. Renato Ruggiero, previous WTO director-general, called for the creation of an organization that could be 'the institutional and legal counterpart to the WTO that is a similar multilateral rule-based system for the environment a World Environment Organisation'. When the United Nations Conference on the Human Environment convened 50 years ago, a major shift was underway in how the world conceived environmental issues. At its root was the growing realization that a healthy environment was fundamental to human well-being and that people living on one side of the planet could be affected by how the environment was treated on another. It became clear that a concerted international effort was warranted, and UNEP was created. This is a time of interdependence. The planet is being interlinked by a global trading system, global information super highways, global financial systems, global communications and entertainment networks and global companies, and in the 50th year of its creation, the agenda for environment recognizes these trends.

While UNEP's major task has been to catalyse and coordinate environmental action, it has also served as a flagship, lending credence to the idea that there is an indispensable role for environmental institutions in the world. Building institutions, according to Elizabeth Dowdeswell (executive director), is clearly no longer a priority. Making sure that the ones we have are working in concert with each other is the priority. While circumstances have changed since 1972, there remains a strong sense that the world community needs an organization like UNEP to champion issues and act as an independent, objective and authoritative advocate for the global environment. The time is right to consider how UNEP can best adapt to the new circumstances and realize its potential as the United Nations voice for the environment.

In Elizabeth Dowd swell's view, there are three responsibilities that must fall on an organization such as UNEP by virtue of its unique position as the environmental body of the United Nations.[82] The first is the collection and assembly of facts into a picture of the planet's health. The second is the evaluation of conditions and trends, together with underlying root causes and facts, with a view to their policy implication. And the third is the vision of a framework to develop international environmental agreements and support their forum for the exploration and discussion of newly emerging issues. These three themes have always been part and parcel of UNEP's mandate. Focusing on the three main themes – synthesizing, deriving policy implications and supporting international negotiations – calls for a strong knowledge-based organization capable of mobilizing international environmental science and technology, and its related social, political, economic and legal components.

While universal participation underpins the decision-making process of the United Nations, it is also been achieved on UNEP's governing structures, like the Environment Assembly, which has replaced the Governing Council, where universal participation, on the biennial sessions of the Environment Assembly, has been emphasized, which is highly desirable, particularly considering the universal nature of environmental problems. At the same time, UNEP has to become proactive rather than reactive. The designation of a smaller intercessional body is crucial in order to provide the Secretariat with timely and substantive policy guidance and present a powerful collective voice on emerging and evolving environmental issues.

The Malmo Declaration

The importance that the world attaches to this crucial issue of environmental protection and sustainable development was most clearly attested to by the presence of 131 ministers and heads of delegations at UNEP's inaugural Global Ministerial Environment Forum held in Malmo, Sweden, from 29 to 31 May 2000.[83] The Malmo Declaration – which was the principal output of this forum addressed to the Millennium Assembly – acknowledged that the central challenge is to work out how the global ambitions contained in the increasing number of international environmental agreements can be turned into concrete local action and implementation.[84] The Declaration emphasized the 'need for reinvigorated international cooperation based on common concerns and a spirit of international partnership and solidarity'. The Malmo Declaration underlined the need to strengthen the engagement of civil society organizations through freedom of access to environmental information and broad participation in environmental decision-making.

The decadal review of the Earth Summit in 2002 provided a unique opportunity to the global community to reinvigorate the spirit of Rio and to inject a new spirit of cooperation and urgency based on agreed actions in the common quest for sustainable development. Today, the governance of global environmental issues is mediated by an intricate web of agreements, organizations and interrelationships. Key players include the United Nations system, with its member governments, its organs and agencies, particularly UNEP and UNDP; the Bretton Woods Institutions, particularly the World Bank; multilateral regional organizations, including the regional development banks, environmental treaties, agreements and conventions, which preceded Rio, with their support bodies; more recently created institutions and mechanisms, including the global environmental conventions on ozone depletion, biodiversity and climate change; the United Nations Commission on Sustainable Development and the Global Environment Facility; and other entities and forces that can strongly influence events, including the international scientific community, the private sector, the NGO community and the media. Together, these form a loose structure of institutions and activities, and these players are regarded as key components in an emerging system of governance of global environmental issues.

Some far-reaching proposals have attempted to address these deficiencies. In 1989, Britain proposed either creating a Security Council for the environment or entrusting the existing Security Council with environmental matters. The same year President Gorbachev proposed creating a global environment emergency capability, and the Hague Declaration called for a new institutional authority to set and implement environmental standards. There have been proposals to review the United Nations trusteeship authority – entrusting it with global environmental concerns – and informal discussions on the merit of bringing the global environmental conventions under a single umbrella.

Genuine progress in managing the global environment will require a move towards an over-reaching, coherent international structure. Some have proposed the establishment of a Global Environmental Authority, with regulatory powers of its own, as the appropriate approach. Any new system must provide a clear and unifying organizing principle. It should integrate the essential elements necessary in order to monitor global environmental conditions, develop the appropriate international policies, promote optimal strategies for collective actions and leverage implementation, and ensure compliance and timely achievement of effective results. Implementing a decentralized yet rigorously coherent system presents demanding organizational and management challenges. One important component of global environment governance, UNEP, needs deep and far-reaching reforms.

Imaginative developments towards an integrated system of governance of global environmental affairs contribute fresh thinking and innovative new approaches to international cooperation in general and thus greatly enhance humanity's ability peacefully to manage life on small, blue planet. The multilateral environmental negotiations put environmental diplomacy on the global centre stage. Enhancing sustainability is the ultimate purpose of these negotiations, and some major challenges are confronting these negotiations.

Though Western civil society and public concerns have largely driven the global green agenda, governments have taken over the agenda in the post-Rio dynamics.

The civil society, which led governments to take action in Rio, has since followed the inaction of the governments. The 'governmentalization' of the environmental agenda is at times providing disastrous consequences, as it has become cause without a concern. The civil society of both the North and the South has to play a bold and courageous role to develop an ability to keep governments accountable and responsive. A major problem with the current environmental diplomacy is its very agenda. At Rio Conference, environmental problems had already been categorized as local and global. Dividing environmental problems into global and local has created a standing conflict between industrialized and developing countries. While developing countries have a greater interest in immediate and short-term environmental problems, like land degradation, groundwater depletion, water scarcity and deforestation, which are classified as local environmental problems, industrialized countries are taking a greater interest in long-term environmental issues, like climate change, classified as global environmental problems.

Unless all environmental problems are addressed within an integrated perspective that takes into account the local and the global, there will be little confidence within the developing world that their concerns are being taken into account in the global environmental agenda. Agenda 21 talked at length about the relationship between poverty and environment, a problem that is of deep concern to poor countries, but no worthwhile attempt has been made in the post-Rio period to address it. The emerging problem of global warming and the resulting climate instability will make life for these people even more difficult in the decades to come. In many ways, there is no one else to blame for the state of affairs other than the Southern governments themselves. They have rarely been proactive in the setting of the global environmental agenda. Hardly any attempt has been made to develop proactive and coordinated positions in advance. No pressure has ever been put on the Global Environmental Facility (GEF) to include poverty and environment on its agenda. The weakness of Southern interventions lies in the failure of its political leadership to articulate and develop a coherent vision of a greener and equal world. In a highly divided world, getting the nations together to deal with their environmental challenges means that rich nations will have to provide good leadership, which generates confidence not just within their own populations but also in the populations of developing nations. In this context, the role of UNEP will be of immense importance in the years to come, in particular the role of developing South Asian countries. The role of the civil society, acting both at the national and at the global level, becomes critical. Though restricted to environmental concerns, the system must be built on principles of 'good governance': democracy, social justice, equality and transparency. The civil society has a key role to ensure that the outcomes are in the best interest of the maximum numbers; the purpose is to promote global equity and confidence in the fairness of the global environmental governance architecture and ensure that the outcomes lead to global sustainability. The process of economic and ecological globalization is changing the nature of the state. Ecological globalization is necessary to deal with cross-border environmental threats and threats to global public goods, as well as economic globalization will force the nation-state to give greater space to global economic and environmental governance system – WTO and global environmental treaties and UNEP, for instance. On the other hand, economic growth and the resulting pollution and environmental damage will force the nation-state to handover greater power to local authorities and governance systems. Several environmentalists, especially in the developing world, have argued that the answer to 'globalization' is 'localization', because it is at the local level that the trade-offs between economy and ecology will be best appreciated.

As the nation-state has to give over its powers to global governance, on the one hand, and local governance, on the other, it will have to play a stronger role within the globalized world. Globalization puts new demands on the nation-state to set domestic policies and to intervene proactively in the setting of global policies and rules so that these reflect the priorities for its people.

State of the Environment Report

The analysis presented in UNEP's Global Environment Outlook underscores the need for the global community to embark on major structural reforms. Clearly, there is a need to upgrade UNEP's capacity for progress, to empower and equip it to become a force for the protection and management of environmental issues on the global stage, fully equipped to face the challenge: an independent voice, dedicated to one purpose, the protection of the global environment, uniquely positioned as the only body in the United Nations system, able to address environmental issues with respect to the needs and aspirations of all the world's people – rich and poor.[85]

This aspect was nowhere better recognized than during the 19th session of UNEP's Governing Council, which endorsed UNEP's role as the leading environmental authority that promotes the coherent implementation of the environmental dimension of sustainable development.

The UNEP state of the environment report GEO-2000[86] sets out very clearly the challenges the world faces. According to the report, at the top of the list comes climate change, then water scarcity, which are then linked to other challenges of depletion of the ozone layer, the rapid disappearance of biodiversity, continued deforestation across the globe and the overexploitation and degradation of marine resources. Further the report focused attention on two factors that are key to understanding and remedying the global environmental crisis: poverty and consumption, which are driving the forces of environmental degradation. While a significant proportion of the world's population struggles with poverty, the affluent minority uses an ever-greater share of the world's resources. These two issues are driving the forces of environmental degradation.

The United Nations Environment Programme published the Global Environment Outlook (GEO) report as its flagship environment assessment. Till 2020, there have been six editions of Global Environment Outlook, launched by UNEP. UNEP released GEO-6 in March 2019. The sixth Global Environment Outlook, or GEO-6,[87] is the world's most comprehensive environmental report, covering a range of topics, issues and potential solutions. At present, the world is not on track to meet the Sustainable Development Goals. UNEP's GEO-6 calls for policy interventions that address entire systems in order to ensure healthy lives for all people.

The theme of the report is 'Healthy planet, healthy people'.[88] The purposes of this report are explained as: assessment of the current state of the environment, the challenges that the world community is facing, the measures taken to deal with the environmental challenges with due consideration given to gender, indigenous knowledge and cultural dimensions and to provide assistance to member nations to implement policies towards a sustainable future.[89] The Global Environment Outlook (2019) pressed on the need for immediate actions to be taken for the achievement of Sustainable Development Goals and important environmental agreements including the Paris Agreement. Four major concerns highlighted in the report are given:[90]

i. Greenhouse gas emissions (GHGs) where the top 10 emitting countries emit 45 per cent of global GHGs in comparison to the bottom 50 per cent countries that emit only 13 per cent, which is a serious concern that needs immediate action.
ii. Pollution exposure is another area for action, where nine million lives are lost annually with indoor and outdoor exposure to air and water pollution in 2015. East and South Asia have the highest total number of deaths (6–7 million deaths) attributable to air pollution. Global chemical pollution is another challenge to the environment.
iii. Immediate action is also needed, where as per the report, 2.1 billion people in 2015 lacked access to clean drinking water and 4.5 billion people lacked access to safely managed sanitation facilities.
iv. Urban growth leads to habitat loss and land use change. Land use practices like burning, livestock grazing and so on put pressure on the land. Besides these areas of concern, the report further mentions five drivers that bring changes in natural systems and social systems: population, economic development, urbanization, technology and climate change.[91] Basically, GEO-6 emphasizes that a healthy environment is both a prerequisite and a foundation for economic prosperity, human health and well-being. It addresses the main challenge of the 2030 Agenda for Sustainable Development: that no one should be left behind, and that all should live healthy, fulfilling lives for the full benefit of all, for both present and future generations.[92] Unsustainable production and consumption patterns and trends and inequality, when combined with increases in the use of resources that are driven by population growth, put at risk the healthy planet needed to attain sustainable development. Furthermore, the world is not on track to achieve the environmental dimension of the Sustainable Development Goals or other internationally agreed environmental goals by 2030; nor is it on track to deliver long-term sustainability by 2050. Urgent action and strengthened international cooperation are needed to reverse those negative trends and

restore planetary and human health.[93] For a healthy planet and healthy people, there is a need for clean energy that encompasses circular economy, adaptation to climate change, sustainable food systems and liveable cities.[94]

Conclusion

Thus, on the one hand, the UNEP continues to have a priority role, as it remains the crucial international vehicle for propelling development in environment-friendly channels. First, it has the resources and partnership to compile and disseminate the scientific information that has to be the basis of all sustainable development decisions. Second, it has the cross-sectoral partnerships to turn that information into action. Third, because UNEP is the officially designed voice for the environment within and beyond the United Nations system, it is able to effectively promote the interlinkages between the environment and sustainable development at the highest levels. Finally, UNEP is at the forefront of a growing global environmental movement and governance. People the world over are becoming increasingly aware of the importance of taking the environment into account. That, ultimately, is UNEP's strength.

However, as the critics have pointed out the main challenges are: paucity of finances, internal management, its location in Nairobi, the nature of UNEP's relationship with other UN agencies, the challenges from the Bretton Woods system and the domination by the advanced countries of the North, incompatibility between its role as conceptualizer of environmental policy and its commitment to action, and effective leadership within UNEP. Despite these problems as the chapter noted, the UNEP has tried to confront the challenges of implementing the recommendations of not only the foundational conferences like Stockholm and Rio but also the later World Summit on Sustainable Development of 2002 and Sustainable Conference of 2012, also known as Rio+20, including the SDGs and MDGs, which have become significant goals to be achieved by 2030. It has tried to develop different mechanisms to deal with rapidly evolving environmental issues. A significant mechanism in this context is Global Environment Outlook (GEO), which UNEP publishes, as its flagship environment assessment. This chapter briefly discussed the sixth Global Environment Outlook, or GEO-6, which is regarded as the world's most comprehensive environmental report (published in 2019, by UNEP), covering a range of topics, issues and potential solutions and calls for policy interventions that address entire systems in order to ensure healthy lives for all people.

After discussing all these issues, it has been observed that these declarations of principles and the interventions on the part of UNEP have been prime movers of the international environmental movement and governance. However, all these conference recommendations and their implementation are major landmarks in this effort of an evolving process. The conferences and UNEP are parts of a larger global development, and implementation of many of the conference recommendations will require time measured by decades.

Notes

1. Barry, R. Stephen, 'What Happened at Stockholm: A Special Report', *Science and Public Affairs: Bulletin of the Atomic Scientists* (28 September 1972), pp. 16–56.
2. Resolution 2997 (XXVII), recommended by the Second Committee, A/8901, adopted by Assembly on 15 December 1972, meeting 2112, by 116 votes in favour with 10 abstentions. Yearbook for the United Nations, 1972 (New York: United Nations, 1975), pp. 331–333.
3. Heilbronn, Bernice, 'UNEP Mandates: Stockholm Recommendations and Governing Council Decisions', *Earth Law Journal* (1975), pp. 161–167.
4. UNEP, *From Vision to Action: UNEP since UNCED* (Nairobi, Kenya: UNEP Publication, 1997), p. 39.
5. Ibid., 3rd and 4th paragraphs.
6. Ibid., 7th paragraph.
7. Ibid., 8th paragraph.
8. The Reports of the Governing Council. UN Monthly Chronicle 10 July 1973. 83, 84. The First Session of the Governing Council: 'On from Stockholm', Vanya Walker-Leigh, *The World Today*, 29 December 1973, pp. 543–546.
9. Ibid., p. 545.
10. https://digitallibrary.un.org/record (UNEP: Governing Council) (2nd session: 1974, Nairobi). Annexes (pp. 81–209): Decisions of the GC of UNEP.
11. https://digitallibrary.un.org/record (Report of the Governing Council of the United Nations Environment Programme).
12. UNEP Governing Council, 'Introductory Statement by the Executive Director, Apr. 1975', UNEP/GC/L. 27, p. 10.
13. Ibid., p. 11.
14. unep.org/environmental assembly/unea1. Accessed internet on 7 July 2022.
15. Unep.org/environment assembly/about united nations environment assembly. Accessed internet on 8 July 2022.
16. unep.org/environmental assembly/proceedings-and reports-resolutions and decisions-unep-1.
17. http://www.bmub.bund.de file admin/daten_ bmu/download pdf/Europa_ International/unea _ministerial_ outcome_ document_ en bf.pdf.
18. Unep.org/environment assembly/unea2/proceedings-reports-resolutions and decisions. (Outcomes of the report: Delivering on the 2030 agenda for Sustainable Development, UNEP/EA.2/Res5).
19. Ministerial declaration of United Nations Environment Assembly at its 3rd session: Towards a pollution free planet (UNEP/EA.3/HLS.1).
20. unep.org/environment assembly/unea.3/proceedings-reports-ministerial declarations-resolutions and decisions.
21. UNEP/EA.5/HLS.1.Ministerial declaration of the UN Environment Assembly at its fifth session: Strengthening actions for Nature to achieve Sustainable Development.
22. Unep.org/environment assembly/unea-5.2/proceedings/report/ministerial declaration/resolutions and decisions/unea-5.2.
23. Unep.org/environment assembly/about unea-5.
24. UNEP/EA.SS.1/3. Proceedings of the UN Environment Assembly at its first Special Session: Commemoration of the 50th anniversary of the establishment of the UNEP.
25. Sandbrook, Richard, 'The UK's Overseas Environmental Policy', in *The Conservation and Development Programme for the UK: A Response to the World Conservation Strategy* (London: Kogan Page, 1983), p. 388.
26. United Nations Environment Programme, Report on the Present State of UNEP, prepared by the Permanent Representatives of the European Community Countries, 1980 (Unpublish.).
27. Mustafa Kamal Tolba was born in 1928 in Egypt. He held the chair of microbiology at the University of Cairo and served as a government minister and aide to Anwar Sadat before

heading the Egyptian delegation to Stockholm. He served as deputy to Maurice Strong at UNEP for three years before being appointed UNEP executive director in 1976.
28 United Nations Environment Programme; see note 11.
29 United Nations Environment Programme, Review of Major Achievements in the Implementation of the Action Plan for the Human Environment, UNEP, doc No. 82–006–1142 (Nairobi, 26 January 1982), p. 58.
30 Clarke, Robin and Lloyd Timberlake, *Stockholm Plus Ten* (London: Earth scan, 1982), p. 48.
31 Hold gate, Martin, 'UNEP: Some Personal Thoughts', *Malinger*, Mar. 1982, pp. 17–20.
32 Clarke and Timberlake, op. cit., p. 48.
33 Ibid., p. 52.
34 Ibid., p. 56.
35 United Nations Environment Programme, op. cit., p. 20.
36 United Nations Environment Programme, The Environment in 1982: Retrospect and Prospect, Paper prepared for the Session of a special character, Nairobi (UNEP DOC UNEP/GC (SSC)/29 Jan 1982, Nairobi), Clarke and Timberlake, op. cit., p. 21.
37 Holgate, op. cit., p. 18.
38 Clarke and Timberlake, op. cit., p. 70.
39 United Nations Environment Programme, op. cit., p. 74.
40 The principle of a regional approach to marine management was not new; the International Council for the Exploration of the Seas, for example, has been active in the North Atlantic and Baltic as early as 1902.
41 Clarke and Timberlake, op. cit., p. 43.
42 Ibid., p. 44.
43 Erik Eck Holm, op. cit., p. 5.
44 United Nations Environment Programme, op. cit., p. 22.
45 Imber, Mark, *Environment Security and UN Reform* (Great Britain: Macmillan, 1992).
46 Stein, Robert E. and Johnson Brian. *Banking on the Biosphere? Environmental Procedures and Practices of Nine Multilateral Development Agencies*, Lexington Books (Lexington MA) 1979. (203 pages).
47 Veena Jha holds a doctoral degree in economics from the University of London and is the project-coordinator of UNCTAD in New Delhi affiliated to the International Trade Division of UNCTAD, Geneva, Switzerland.
48 Nirmal Andrews, UNEP's regional director, UNEP/ROADP, Bangkok, Thailand.
49 Excerpts from Veena Jha's address to the Seminar 'UNEP/SACEP: South Asian Regional Forum on Environmental Cooperation between Government and Private Sector', held in New Delhi on 13 July 2000.
50 Ibid.
51 Sudarshan Rodrigues, consultant, Environment and GEF Division at the UNDP office at New Delhi.
52 Dr. Pradeep Monga, Excerpts from the Address to the Seminar on 'UNEP/SACEP: South Asia Regional Forum on Environmental Cooperation between Government and Private Sector' held in New Delhi on 13–14 July 2000.
53 Excerpts from his speech on 'GEF Operational Projects in India'.
54 Ibid.
55 Sharma, Anju, research coordinator (New Delhi: CSE, interviewed in 2000).
56 World Environment Handbook: A Directory of Government Natural Resources Management Agencies in 144 Countries (Johnson, Hope and Johnson Janice Marie, *New York: World Environmental Policies in Developing Countries* (Berlin: Erich Schmidt Velar, 1977)).
57 Richard J. Rodd wig, *Green Bans, The Birth of Australian Environmental Politics: A Study in Public Opinion and Participation* (Montclair, NJ: Allan Held, Osman, 1978); Bradley, J.M., 'Green Parties in West Germany and Environmentally Parties in Austria World Environmental Problems', *Ambo* 12(1) (1982), pp. 20–26.

58 Myers, Norman and Dorothy Myers, 'How the Global Community Can Respond to International Environmental Problems', *Ambio* 12(1) (1983), pp. 20–25.
59 Buckminster, Fuller, *Operating Manual for Spaceship Earth* (New York: Simon and Schuster, 1970), p. 492.
60 UNEP, *From Vision to Action: The Earth Summit* (Nairobi: UNEP Publication, July 2000), p. 1.
61 Sachs, Wolfgang, *A Professor and a Writer on Development Issues from Germany. Interview with Amit Mitra, Down to Earth*, Special issues, 'Preparing You to Change the Future' (New Delhi: SZE, 2001), pp. 66–67.
62 Sachs, Wolfgang, *This Race Leads to an Abyss*, Special issue of *Down to Earth* titled, 'Preparing You to Change the Future' (New Delhi: CSE, 2001), p. 66.
63 Ibid., p. 67.
64 UNEP, *From Vision to Action: Building Consensus* (Nairobi: UNEP Publication, July 2000), p. 15.
65 Ibid., p. 16.
66 The implementation of the Convention on Biological Diversity has now entered its most challenging phase, particularly, in its implementation in certain new and sensitive areas, particularly agro-biodiversity and the role of indigenous and local communities. The Convention recognizes that the origins of conservations, whether spiritual, economic or recreational, lie buried in ancient cultures scattered around the world.
67 UNEP's approach to desertification encourages design of programmes that increase people's participation in the process of managing dry land schemes. Another issue in dry lands management that has received attention is the neglect of women.
68 The concept of 'Activities Implemented Jointly' is a means of stimulating additional investment and promoting the co-development of advanced technologies that could reduce the rate build-up of greenhouse gases with advancing national economic development priorities. It is also a way of encouraging new partnership between north and south.
69 In 1997 the Montreal Protocol on the Substances that deplete the Ozone Layer celebrates its tenth anniversary. Since UNCED, the protocol has registered significant progress in its implementation. Industrialized countries phased out the production and consumption of halons by 1 January 1994, and chlorofluorocarbons (CFC) by the end of 1995. The Multilateral Fund, established in 1993, assists developing countries in phasing out the controlled substances.
70 *From Vision on Action: Building Consensus*, op. cit., p. 19.
71 Ibid., p. 123.
72 Ibid., p. 25.
73 Ibid., p. 26.
74 The integrated programme that has been developed will promote global freshwater assessment and develop tools and guidelines for sustainable management and use of freshwater and coastal resources with special reference to Caribbean and South Pacific small island states and in selected megacities of Latin America, Asia and Africa.
75 *From Vision on Action: Building Consensus*, op. cit., p. 9.
76 Agarwal, Anil and Nara Sunita, *Towards a Green World: Should Global Environmental Management Be Built on Legal Conventions or Human Rights?* (New Delhi: CSE, 1992), p. 346.
77 UNEP mission statement.
78 Agarwal and Narayan, op. cit., p. 347.
79 Visions for the Future, op. cit., p. 10.
80 Agarwal and Narayan, *Suggestions for Global Governance Institutions*, p. 348.
81 Our planet, vol.8, No.5, Nairobi: UNEP, 1997, p. 3.
82 Our planet, vol.11, No.2, Nairobi: UNEP, 2000, p. 3.
83 Ibid., p. 4.
84 Agarwal and Narayan, op. cit., p. 30.

85 UNEP, *Global Environment Outlook Report (GEO-1): Trends and Recommendations* (Nairobi: UNEP, 1999).
86 UNEP: Annual Report (1999). Facing the Challenges of the Twenty-First Century, United Kingdom: Oxford, Apr. 2000, p. 42.
87 The Global Environment Outlook (GEO) is often referred to as UNEP's flagship environmental assessment. The first publication in 1997 was originally requested by member states. It is a flagship report because it fulfils the core functions of the organization, which date back to the UN General Assembly resolution that established UNEP in 1972. (www.unep.org/resources/global-environment-outlook-6). Accessed internet on 10 July 2022.
88 UN Environment (ed.), *Global Environment Report GEO-6 Healthy Planet, Healthy People* (Cambridge: Cambridge University Press, 2019).
89 Introduction and Context. In UN Environment (ed.), *Global Environment Outlook GEO-6: Healthy Planet, Healthy People* (Cambridge: Cambridge University Press, 2019), pp. 2–19.
90 State of the Global Environment. In UN Environment (ed.), *Global Environment Outlook GEO-6: Healthy Planet, Healthy People* (Cambridge: Cambridge University Press, 2019), pp. 104–105.
91 Drivers of Environmental Change. In UN Environment (ed.), *Global Environment Outlook GEO-6: Healthy Planet, Healthy People* (Cambridge: Cambridge University Press, 2019), pp. 20–55.
92 www.unep.org/resources/assessment/geo-6-key-messages? Accessed internet on 11 July 2022.
93 Some key messages of the Report, www.unep.org/resources/assessment/geo-6-key-messages? Accessed internet 12 July 2022.
94 Some key messages of the Report. Accessed internet on 12 July 2022.

6
CONCLUSION

Introduction

Evaluating 50 years of the role of UNEP as a leader, catalyst and coordinator, the book has assessed how UNEP has played its role as a bridge between science and policy, governments and non-governmental activists and environment and development. The book has, further, tried to underline the determining role of UNEP, with the help of the national governments: how it was able to create an environmental governance architecture, by establishing/creating ministries, agencies and policies in developing countries, to help them to participate fully in the process of international environmental cooperation and governance by strengthening their capacities to deal with their own domestic environment and development-related issues.

The second focus of the book was region-specific, and it attempted an evaluation of the steps taken by the SACEP in collaboration with UNEP in the creation and management of environmental and sustainable development issues in South Asian countries. In this context, the study explored the role of national conservation strategies in creating not only environmental awareness but also an environmental governance architecture and implementation process in South Asian countries. A significant component of this analysis was the role played by South Asia Co-operative Environment Programme (SACEP), which constituted an important landmark in regional cooperation in South Asia.

The book, in its conclusion, has also examined the role of the United Nations Environment Programme as an advocate for greater environmental responsibility and as a vehicle for new approaches to deal with pollution and conservation of natural resources. The book attempted to study the various strategies and measures

DOI: 10.4324/9781003370246-6

that were advocated to halt and reverse the effects of environmental degradation in the context of increased national and international efforts to promote sustainable and environmentally sound development in all countries. Ever since its founding, it has emerged as the main vehicle of international action on environmental issues: raising awareness, providing policy guidance and serving as a focus, within and outside the United Nations system, for a coordinated and integrated approach to this central challenge of our time.

The book is structured into six main chapters dealing with specific aspects of the theme. The first chapter attempted to provide a conceptual basis to the study as well as a brief background of Environment on the UN Agenda. It began by examining the intrinsic relationship between environment and economic development. In the process it focused on the emergence of the concept of sustainable development, and this was studied with special reference to the Brundtland Report. The chapter then explored the way in which environmental issues had been pursued within the United Nations, with a brief survey of environmental issues on the United Nations platform. This section further discussed the Bretton Woods Institutions and the UN and concluded with an analysis of Millennium development goals and Sustainable development goals.

While analysing the impact of the Brundtland Report and Sustainable Development, the study observed that changes in attitudes and behaviour were needed in all areas and all aspects of our societies to promote and achieve 'sustainable development' and to build a future that will be more prosperous, more just, more equal and secure.

Besides that it was noted that the Brundtland Report, written after three years of intensive work throughout the world by a distinguished commission (the World Commission on Environment and Development) under the chairmanship of Gro Harlem Brundtland, the former prime minister of Norway, chairperson of the World Health Organization, and its report 'Our Common Future', made a strong and persuasive case for sustainable development, integrating the environmental and social with the economic dimensions of development as the only viable pathway to the human future. The chapter underlined the fact that 'Our Common Future' succeeded in highlighting poverty, hunger, debt and economic growth as inherent aspects of environmental issues.

The chapter further explored the interlinkages between environment and developmental agenda, the problems of poverty and environmental degradation, the role of economic growth and the concepts of sustainability and participation. Another issue related to the Brundtland Report and Sustainable Development highlighted in the study was the Brundtland Commission's contribution in establishing sustainable development as the standard for international development. The study further underlined the importance of 'Our Common Future', which cemented the conceptual and political foundation on which the Rio Conference was to be erected. In the process, the study examined the objectives and definitions of sustainable development.

194 Conclusion

While explaining the role of 'Bretton Woods Institutions and the UN', the chapter examined how the Bretton Woods Institutions, which were originally conceived by Keynes as contributors to a world welfare system, were later manoeuvred to counter the growing strength of the Third World forums which were pressing for a more equitable role in defining the direction of the world economy. It was observed that the United Nations can become effective only when the international community is prepared to make the political commitment to act together for the common good. The chapter concluded with the study 'Environment on the UN Agenda', which highlighted environmental problems as global problems related to the protection of the global common heritage and thereby impacting on the security of all states, and an examination of Millennium Development Goals and Sustainable Development Goals in the evolution of the environmental governance.

The second chapter analysed two crucial aspects, first, the stellar role of leading environmentalists and environmental movements in the evolution of the UNEP, and, second, the emergence of significant differences in the perspectives of the advanced developed countries of the North and the developing countries of the South. Underlining the role played by environmental movements in awakening global consciousness, it examined the linkages between British environmentalism, American environmentalism and new environmentalism in the creation and evolution of the United Nations Environment Programme. The chapter further identified various environmental issues raised by both developed and developing countries during the Stockholm and Rio Conference proceedings and highlighted the achievements and constraints. While dealing with environmental movements, it highlighted three developments of major importance namely (1) the influence of professional ecologists like Rachel Carson and her book 'Silent Spring', Barbara Ward and Barry Commoner and activists like Brice Lalonde; (2) the Concern for the preservation of environment and conservation of the wildlife that grew in many countries outside Europe and North America; and (3) the adoption of a broader conception of the environment which had a scientifically more sophisticated perception of the relationship between man and environment.

Two major events which influenced British environmentalism were explained:

1 The 'study of natural history' that epitomized the Arcadian view of nature, which advocated simplicity and humility in order to restore man to peaceful co-existence with nature;
2 The anti-slavery movement and the crusade against cruelty to animals.

Two seminal events that influenced American environmentalism were analysed:

1 Publication of the book *Man and Nature* by George Perkins Marsh, and the ideas highlighted by him in the book influenced in the establishment of a

National Forestry Commission and represented the beginning of land wisdom in the United States;
2   Act of Congress of 1864 transferring the Yosemite Valley and Mariposa Grove of Big Trees to the State of California on the condition that wilderness preservation was important for the preservation of civilization.

The chapter further explored the linkages and divisions between the preservationists and the conservationists, the two camps into which American environmentalism was divided.

The study further probed the factors that played an important role in the emergence of new environmentalism. It underlined the fact that some environmental issues like marine pollution, whaling, fisheries, acid pollution and so on would not be solved by individual governments acting alone; the obvious response was greater international cooperation. And Stockholm provided a compelling spur to internationalism.

McCormick credited 'Maurice Strong' for incorporating developing countries views to the agenda of the Stockholm Conference and emphasized the new role of developing countries on the environmental debate. A detailed study of the major issues raised by both developed and developing countries during the proceedings of Stockholm and Rio Conferences, which were important milestones on the road to the establishment of the UNEP, were elaborately discussed. The chapter traced the developments leading to the creation of the United Nations Environment Programme, with a focus on its major institutional and functional components, namely environmental assessment, environmental management and supporting measures. It examined at length the achievements, outcomes, successes and constraints of both the conferences in the evolution of UNEP (1972–1992), besides highlighting the evolving role of UNEP from 1992 to 2022 (UNEP at 50) in the updated section. While dealing with major environmental issues, such as biological diversity, treaty on climate change, restructuring of Global Environment Facility and the forestry agreement, from the perspective of North-South divide, critical viewpoints of leading environmental activists like Vandana Shiva and Ashish Kothari were highlighted.

As a whole, the chapter concluded by observing that strong political and financial commitments were required, as support for a strengthened UNEP in the pursuit of environmental governance, in spite of its limited mandate and lack of funding.

The third chapter, 'UNEP and India', as the title suggests, was India-specific. It began with a study of the impact of the cultural traditions and thoughts that had provided the ideological underpinning to the present environmental movements in India. This was essential to underline (1) the extant environmental consciousness with the Indian ethos and (2) to provide the background to existing environmental activism in India. While it was outside the scope of the chapter to deal with all the environmental movements, the main movements were studied in the context of the discourse. These included the 'Chipko Movement', which had demonstrated the workability of a powerful model of people's participation, a model that can be emulated and replicated; 'Silent Valley Movement' in Kerala, as one of the most important milestones in the shaping of public opinion as well as the formulation of official policy which had been halted on environmental grounds, which offered a unique foresight of peoples' participation; and 'Narmada Bachao Andolan', led by Baba Amte, Medha Patkar and others, which focused on two major dams under construction that submerged almost as much area as it was meant to irrigate. The movement intensified its resistance on central government by peaceful means; 'Tehri Project', led by Sunderlal Bahuguna, prevented the further development of the project. This was undertaken to emphasize the extant environmental impulses within India.

The second part of the chapter tried to link the efforts of the UNEP in the management of various environmental issues in India. It was seen that these efforts were being made in a variety of areas. These included desertification, deforestation, aquatic ecosystem, multipurpose valley projects, biological diversity, climate change, natural resource management, urbanization, human health and welfare, energy, industry and transportation. In the management of these environmental issues the tools and methodologies of environmental planning were primarily addressed through the training, strategies and guidelines of UNEP. It was further examined how UNEP was guiding Indian initiatives and programmes.

It was also noted that in pursuance of the guidelines of UNEP, the Government of India has formulated strategies, action plans and programmes for combating desertification, deforestation, management and conservation of the tropical forest. However, it was seen that far greater commitment was needed on the part of the government. It was also noted that changes in attitudes and behaviour were needed to effectively implement the environmental policies. While dealing with the multipurpose river valley projects, the UNEP had the specific objective to improve and apply environmentally and economically sound methods to minimize waterloss, salinity, waterlogging and other environmental hazards in planned irrigation projects. It was further observed that in dealing with air pollution, climate change, biodiversity preservation, urbanization and health problems, the Government of India in the light of UNEP guidelines had the ultimate responsibility for national policy development and implementation and enforcing national environmental legislation.

The fourth chapter dealt with the environmental governance and sustainable development challenges in four South Asian countries, namely, Bangladesh, Nepal, Pakistan and Sri Lanka while analysing the National Conservation Strategies covering the period from 2015 to 2030 (updated). A significant component of this analysis was the role played by South Asia Co-operative Environment Programme (SACEP), which constituted an important landmark in regional cooperation in South Asia.

The chapter sought to present an environmental perspective of the South Asian region, where the study underlined the interlinkages between economic growth, population, natural disasters, forests, biodiversity, renewable freshwater resources, oil pollution, air pollution, land and food issues. It has been observed that effective environmental strategies have exerted a strong influence on development if they are an integral part of development planning and policy-making. It is examined that, on the one hand, the national conservation strategies of Bangladesh and Sri Lanka were developed in parallel to development planning, without any tangible links to the development planning systems of either country. As a result, their success has been limited. But, on the other hand, the strategies of Nepal and Pakistan are one step closer to achieving integration with development planning. These strategies have included environmental sustainability components in development plans, and even corporate strategies and programmes in each country have established some form of link with the development planning process. This has given them credibility within key economic and planning agencies; as a result, they have succeeded in influencing major policies and plans. The National Conservation Strategy and many of the state's strategies undertaken were planned in close collaboration with the economic planning unit in the Prime Minister's Office and the economic planning units in the states. The Nepal National Conservation Strategy was developed and implemented under the umbrella organization of the National Planning Commission. Pakistan has developed a formal link with the commissions, in the form of an environment unit, to help implement the NCS. Nepal has achieved the greatest degree of internalization within government and a solid working relationship between the National Planning Commission and sectoral ministries has been built up. The Nepal NCS has made significant contributions in the fields of environmental impact assessment, pollution control, environmental law, environmental planning, heritage conservation and environmental education. With top-level government and political support, the NCS has successfully been adopted in Pakistan as the environmental agenda. All the countries of South Asia have felt the importance of agencies such as the World Bank, United Nations Commission on Environment and Development, UNEP and Agenda 21 must fit in with all these initiatives. A special focus of the chapter was the role of National Conservation Strategies (updated) in creating not only environmental awareness but also environmental governance architecture to achieve sustainable development and challenges in their implementation in all four South Asian countries. This chapter underlined the importance of regional and sub-regional cooperative efforts in the management of environmental issues that culminated in the establishment of a South Asian Co-operative Environment Programme in the South Asian region. It was observed that the SACEP affords a unique opportunity for a collaborative approach towards the protection and management of the regional environment. This is the first occasion on which eight countries, namely Afghanistan, Bangladesh, Bhutan, India, Maldives, Nepal, Pakistan and Sri Lanka, are brought together solely on the subject of environment.

The fifth chapter attempted a critical appraisal of the United Nations Environment Programme covering a period of 50 years from 1972 to 2022. It tried to

198 Conclusion

detail the challenges and constraints faced by UNEP in the implementation of its policies, while taking stock of its achievements. 'An appraisal of UNEP' underlined the 'North-South' debate in which environment and development appeared to represent conflicting values. While development-related priorities played a dominating role for developing countries of the South, environment-related issues played a significant role for developed countries of the North. The priority areas put forward by the developing countries of the South were human settlements, land and water resources, desertification, trade, the transfer of technology, oceans, conservation of nature, wildlife, genetic resources and energy. Whereas the significant environmental issues addressed by the developed countries of the North were global warming, conservation of biodiversity, ozone depletion and deforestation. Further, the chapter, focused on the first three sessions of the Governing Council and the first three sessions of Environmental Assembly (which replaced GC in 2012) which were formative and set the course of the UNEP to the present date.

It was observed that UNEP faced four major handicaps in the management of local, regional and international environmental matters, which were elaborated and explained as financial, managerial, political and constitutional. Despite these handicaps, the Regional Seas Programme was widely regarded as UNEP's most effective undertaking in its first decade of operation, which was regional rather than global in nature. A special focus of this chapter was the views and opinions expressed by leading environmental activists as well as the officials conducting UNEP programmes on the evolving role of UNEP. Veena Jha, while critically analysing the role of UNEP, blamed the developed countries for their lack of commitment to implement the agendas on environmental issues. According to her, the causes responsible for the erosion of UNEP's credibility were several. Briefly, these included the creation of Commission on Sustainable Development, problem of financial resources, polarization of political issues and proliferation of other institutions like Asian Development Bank, UNDP, UNIDO, ESCAP. It was noted that for real and substantial change and progress on environmental issues, deeper involvement, understanding, commitment and strengthening the institutional capacities were required. Further the study detailed the views and opinions expressed by UNEP officials Mr. Nirmal Andrews, Sudarshan Rodrigues and Dr. Pradeep Monga on the role and functions of UNEP as they represent the major UN functionaries at the implementation level.

Mr. Nirmal Andrews, regional director of UNEP based at Bangkok, Thailand, emphasized on some important priority areas for successful environmental cooperation in South Asia. They include the implementation of regional action programmes, reviewing and monitoring by the ministries of environment, regional round table conferences, participation by international financial institutions and donor countries, regional review meetings and reviewing the government documents.

Sudarshan Rodriguez, who is the consultant, Environment and GEF division, New Delhi, was critical about UNEP's role and functions in the South Asian region for which he put the blame on the lack of presence of a regional office in South Asia. Dr. Pradeep Monga, assistant resident representative, United Nations Office, New Delhi, was critical about the UN agencies in general and UNEP in particular in tackling environmental issues of local people and emphasized on the need for a holistic approach involving the civil society. He was optimistic about resolving environmental problems with the involvement of good local, regional, national and international institutions and NGOs which, according to him, will always lead to good results.

This chapter also exclusively dealt with the challenges and revealed the constraints faced by UNEP in the implementation of environmental agendas. The major challenges and constraints faced by UNEP were underlined as finances, internal management, its location in Nairobi, the nature of UNEP's relationship with other UN agencies, incompatibility between its role as conceptualizer of environmental policy and its commitment to action and effective leadership within UNEP. The evaluation at the same time noted the significant role played by UNEP in planning a new way forward to ensure the survival of life on earth and to correct the flaws that created environmental degradation and protect the very systems that support life on earth.

In this context, UNEP's efforts to support the implementation of the Convention on Biological Diversity, the Convention to Combat Desertification, the Convention on Climate Change, the Montreal Protocol on the Substances that Deplete the Ozone Layer, the Convention on Forest Principles and assessment of their implications on global environment were underlined. It was observed that the prime challenge before UNEP was not only to promote and implement its environmental agenda but also to integrate it strategically with the goals of economic development and social well-being.

It was also observed that a strong regional architecture was central to UNEP's ability to manage and advance its global environmental agenda. On the whole, the analysis presented a detailed study of UNEP's role as the leading environmental authority that promoted the coherent implementation of the environmental dimension of sustainable development. The analysis further observed the need for the global community to embark on major structural reforms.

The chapter finally underlined the major environmental challenges faced by the new millennium. These included the destruction of the ozone layer, the rapid disappearance of biodiversity, continued deforestation, overexploitation and degradation of marine resources. These were intrinsically linked to the two major issues of poverty and consumption patterns that were driving the forces of environmental degradation. The chapter concluded with an examination of Global Environment Outlook-6 report (Healthy Planet, Healthy People, Cambridge University Press, 2019), which aims to build on sound scientific knowledge to provide governments, local authorities, businesses and individual citizens with the information needed to guide societies to a truly sustainable world by 2050.

UNEP, as an activator of the Stockholm and Rio Action Plans, had given the international environmental movement a universality, legitimacy and acceptability in the developing countries which under the circumstances could hardly have been obtained. The following findings have been drawn regarding the role of UNEP that represented environmental governance with reference to South Asia:

Findings:

- UNEP suffers from paucity of funding, lack of political and financial support from governments and from geographical isolation within the UN system. In spite of these constraints and challenges, UNEP has been active and partially successful in carrying out its mandate and contributing to the pursuit of environmental security.
- Major differences exist between the developed and developing countries' perspectives.
- Pressures emanating from the existing Bretton Woods systems allow the advanced countries of the North to dominate; hence structural changes are needed in the global system in favour of the developing countries of the South.
- In the South Asian region, with the creation of South Asia Co-operative Environment Programme to manage environmental issues, UNEP provides a forum for antagonistic nations to meet on neutral ground and come up with common solutions.
- The Stockholm and Rio Conference declarations and action plans set out a clear focus for UNEP's work, and the practical implementation of this will be a major mark of UNEP's progress in becoming a powerful global player in the twenty-first century. However, the work has only begun. Much needs to be accomplished and the progress is slow.
- A bold and creative vision for the UNEP of the future would be redesigning its programme, funding and governance. A global agenda for environmental innovation, tapping the significant potential of the private sector and civil society at large, will shape the twenty-first century in positive ways.
- To deliver on the promise of Stockholm, UNEP must evolve into an organization that acts clearly and unambiguously, as the world's environmental agency, providing principal environmental inputs into the sustainable development agenda and shaping global environment governance. It is a vision maker and forecaster, a consortium builder, monitor and auditor of progress and change. On balance, it is proactive rather than reactive. And through consistently encouraging excellence, the community of nations should adopt the best of approaches rather than the minimum negotiated.
- There has never been a paucity of ideas on enhancing and strengthening the role of the United Nations, especially since the end of the cold war, which

brought renewed optimism that the United Nations could at last begin to realize its potential.
- But all the ideas in the world amount to nothing unless there is political support for them on capitals around the world. The United Nation's great strength is its near-universal membership. But there must be broad consensus on the directions for change and a willingness to work to find a common basis for action.

BIBLIOGRAPHY

Primary Sources

United Nations Documents

Agenda 21 (www.un.org/esa/sustdev/documents/agenda21/english/agenda21toc).
Annual Reports of UNEP/UNDP/FAO/IBRD/UNESCO/WHO.
Conference Documents (Stockholm Conference to Paris to Glasgow Conference).
Everyone's United Nations (1992).
GEO-1 for Life on Earth (www.unep.org/resources/report/global-environment-outlook-1-life-earth).
GEO-3 (www.unep.org/resources/global-environment-outlook-3).
(www.unep.org/resources/global-environment-outlook-2000).
GEO-4 Environment for Development (www.unep.org/resources/global-environment-outlook-4).
GEO-5 for Local Government/ (www.unep.org/resources/geo-5-local-government).
GEO-6 for a Healthy Planet for Healthy People (www.unep.org/resources/global-environment-outlook-6) (2019 Cambridge University Press).
Global Environment Outlook 1–6 (1997–2019)
Keeping Track of Our Changing Environment: From Rio to Rio+20 (1992–2012); United Nations Environment Programme, Nairobi. Published October 2011 (www.unep.org/geo/pdfs/Keeping_Track.pdf).
Medium Term Strategy for People and Planet: The UNEP Strategy for 2022–2025.
Official Records of Economic and Social Council (1972–92).
Official Records of General Assembly.
Reports of the Executive Director of UNEP (1972–2022).
Reports of the Secretary-General (1971–2022).
Report of the UN Conference on Environment and Development (1992).
Report of the UN Conference on Human Environment (1972).
Speeches of Executive Directors of UNEP (1972–2022).
UN Environment Year Books (2003–2014, 10th Edition).

UNEP (2013) UNEP Year Book 2013 – Emerging Issues in Our Global Environment (www.unep.org/pdf/uyb_2013.pdf).
UNEP Programme of Work and Budget 2022–2023. Annex 1 of For People and Planet.
World Commission on Environment and Development (Brundtland Commission), Our Common Future, Oxford, Oxford University Press (1987).
Year Books of the United Nations (1972–2022).

UN General Assembly Resolutions and Reports/Stockholm, Rio, WSSD Conference Documents/UNEP Annual and Governing Council Reports

Development and International Economic Cooperation: Report of the World Commission on Environment and Development, UN Doc. A/42/42T, 4 August 1987.
Environment Programme. Review of Areas: Environment and Development and Environmental Management, Nairobi: UNEP, 1978.
The Global Environmental Monitoring System: Report of the Executive Director, UNEPIGC/31/Add. 2, 25 February 1975.
Question of Convening a Second United Nations Conference on the Human Environment: Note by the Executive Director. UNEP/GC/43, 29 January 1975.
Review of the Institutional Arrangements for International Environmental Cooperation: Report of the Executive Director, UNEP/GC/75, 16 February 1976.
Review of Major Achievements in the Implementation of the Action Plan for the Human Environment: Evaluation of the Implementation of the 1982 Goals, UNEP/GC 10/6/Add.1, 10 December 1981.
UN General Assembly Articles Prepared by Working Group I. UNEP/Bio-Div/N6–1NC4/WG 1/L 2/Md 3 Nairobi, 10 February 1992.
UN General Assembly Conservation of Biodiversity. A/Conf. 151IPC/28, 5 February 1991.
UN General Assembly Conservation of Biological Diversity: Background and Issues. A/Conf. 151/PC/66, 25 July 1991.
UN General Assembly Convention on Biological Diversity, 1992.
UN General Assembly Environmentally Sound Management of Biotechnology. A/Conf. 151/PC/29, 31 January 1991.
UN General Assembly Report of the Ad hoc Working Group of Legal and Technical Experts on Biological Diversity on the Work of Its Second Session. UNEP/Bio. Div./WG. 2/2/5. Nairobi, 7 March 1991.
UN General Assembly Report of the Intergovernmental Negotiating Committee for a Framework Convention on Climate Change on the Work of Its First Session Held at Washington, DC from 4 to 14 February 1991. A/AC 237/66.
UN General Assembly Report of the Intergovernmental Negotiating Committee for a Framework Convention on Climate Change on the Work of Its Second Session Held at Geneva from 19 to 28 June 1991. A/AC. 237/9, 19 August 1991.
UN General Assembly Report of the Intergovernmental Negotiating Committee for a Framework Convention on Climate Change on the Work of Its Third Session held at Nairobi from 9 to 20 Sept. 1991. A/AC. 237112, 25 October 1991.
UN General Assembly Resolution 43/53, 6 December 1988.
UN General Assembly Resolution 44/228, United Nations Conference on Environment and Development, 22 December 1989.
UN General Assembly, Development and Environment. UN Doc. A/CONF. 48/10, 22 December 1971.

204 Bibliography

UN, Global Outlook 2000: An Economic Social and Environmental Perspective, New York, 1990.

UN, Report of the United Nations Conference on the Human Environment. A/Conf. 48/14/Rev. 1, 1972.

UN/ESCAP, "Marine Environmental Problems and Issues in the ESCAP Region", Proceedings of the Regional Technical Workshop on the Protection of the Marine Environment and Related Ecosystems, Asian Institute of Technology, Bangkok, 20–28 February 1984.

UN/ESCAP, ST/ESCAP 618: Coastal Environmental Management Plan for Bangladesh, Vol.2, Final Report, Bangkok, 1982.

UNEP, Action on Ozone, Nairobi: UNEP, 1989.

UNEP, Annual Report of the Executive Director, 1987, Nairobi: UNEP, 1988.

UNEP, Annual Report of the Executive Director, 1988, Nairobi: UNEP, 1989.

UNEP, Annual Report of the Executive Director, 1989, Nairobi: UNEP, 1990.

UNEP, Annual Report of the Executive Director, 1991, Nairobi UNEP, 1992.

UNEP, Asia-Pacific Environment Outlook, Bangkok: UNEP, 1997.

UNEP, Corporate Profile: To Meet the Challenges of the New Millennium, the Global Environmental Authority, Nairobi: UNEP, 1999.

UNEP, Corporate Profile: To Meet the Challenges of the New Millennium, the Global Environmental Authority, Nairobi: UNEP, 2000.

UNEP, The Costs of Developing Countries in Meeting the Terms of the Montreal Protocol, UNEP/OZ L Pro. WG II (2)/3, 29 January 1990.

UNEP, Final Report. UNEP/OZL. Pro. WG. 11(1) 3, 25 August 1989.

UNEP, Montreal Protocol on Substances that Deplete the Ozone Layer, Nairobi: UNEP, 1987.

UNEP, Proceedings of the Governing Council at Its Second Special Session. UNEP/GCSS. 11/3, 8 August 1990.

UNEP, Remaining Issues to be Addressed at the Fourth Meeting of the Working Group: Note by the Executive Director. UNEP/OZL. Pro. WG. IV/3, 23 May 1990.

UNEP, Report of the Ad hoc Working Group of Legal and Technical Experts on Biological Diversity on the Work of Its First Session. UNEP/Bio. Div./WG.2/l/4. Nairobi, 28 November 1990.

UNEP, Report of the Executive Director of the United Nations Programme, Secretariat of the Montreal protocol UNEP/OZL. Pro. 2/2 Add. 4/Rev. 1, 28 May 1990.

UNEP, Report of the First Session of the Third Meeting of the Open-Ended Working Group of: The Parties to the Montreal Protocol. UNEP/OZL. Pro. WG.III(l)/3, 14 March 1990.

UNEP, Report of the Parties to the Montreal Protocol on the Work of Their First Meeting UNEP/OZL Pro 115, 6 May 1989.

UNEP, Report of the Second Meeting of the Parties to the Montreal Protocol on Substances that Deplete the Ozone Layer. UNEP/OZL. Pro. 2/3, 29 June 1990.

UNEP, Revised Draft Convention on Biological Diversity. UNEP/Bio. Div./WG. 2/3/3. Nairobi, 30 April 1991.

UNEP/UNCTAD, Report of the Expert Group on the Impact of Resource Management Problems and Policies in Developed Countries on International Trade and Development Strategies, Nairobi: UNEP, 1984.

UNEP/Vienna. Convention for the Protection of the Ozone Layer, Nairobi: UNEP, 1985.

UNEP/WMO, Ad hoc Working Group of Government Representatives to Prepare for Negotiations on a Framework Convention on Climate Change. UNEP/WMO Prep./FCCC/L.1/Report, Geneva, 24–26 September 1990.

UNEP/WMO, IPCC Supplement: Scientific Assessment of Climate Change, 1992.

UNESCO, UNCED Working Group on Biological Diversity: Promoting Education and Training, 12 April 1991.
United Nations, Annual Report of the Secretary-General of the United Nations (Kofi A. Annan) on the Work of the Organization: Partnerships for Global Community New York UN Publication, September 1998.
United Nations Conference on Environment and Development, Rio de Janeiro, 3–14 June 1992, UN Doc A/Conf 15 4 (Part I, II, III, IV), 1992.
United Nations Economic and Social Commission for Asia and the Pacific, Marine Environmental Problems and Issues in the ESCAP Region, Proceedings of Regional Technical Workshop on the Protection of the Marine Environment and Related Ecosystems, Asian institute of Technology, Bangkok, 20–28 February 1984.
United Nations Environment Programme, Achievements and Planned Development of UNEP's Regional Seas Programme and Comparable Programs Sponsored by Other Bodies (UNEP Regional Seas Reports and Studies No.1), Nairobi: UNEP, 1982.
United Nations Environment Programme, Action for the Environment: The Role of the United Nations, Nairobi: UNEP, 1989.
United Nations Environment Programme, Earth Summit: Agenda 21, The United Nations Programme of Action from Rio, New York: United Nations, 1993.
United Nations Environment Programme, Earth Watch: An In-depth Review (UNEP Report No.1), Nairobi: UNEP, 1981.
United Nations Environment Programme, Fifth Session, 9–25 May 1977, Nairobi: UNEP, 1977.
United Nations Environment Programme, In Defense of Earth: The Basic Texts on Environment: Founex, Stockholm, Coco Yoke, Nairobi: UNEP, 1981.
United Nations Environment Programme, The International Workshop on Environmental Education: Belgrade, October 1975. Final Report 1977.
The United Nations Environment Programme (Level One, Two and Three) Report of the Executive Director, UNEP/GC/90, 15 March 1977.
United Nations Environment Programme, Report of the Environmental Coordination Board in Sixth Session, New York, 20–21 October 1976 (UNEP/GC/89), Nairobi: UNEP, 1976.
The United Nations Environment Programme, Report of the Executive Director, UNEP/GC-7/7, 14 February 1979.
The United Nations Environment Programme, Report of the Executive Directors, UNEP/GC-815, 30 January 1980.
United Nations Environment Programme, Report of the Governing Council of the United Nations Environment Programme, Fourth Session, 30 March–14 April 1976, Nairobi: UNEP, 1976.
United Nations Environment Programme, Review of the Implementation of the • Stockholm Action Plan, Nairobi: UNEP, 1982.
United Nations Environment Programme, Review of Major Achievements in the Implementation of the Action Plan for the Human Environment (UNEP Doe. No.82–0006 1142 C), Nairobi: UNEP, 1982.
United Nations Environment Programme, Review of the Montevideo Programme for Development and Periodic Review of Environment Law 1981–1991, Nairobi: UNEP, 1992.
United Nations Environment Programme, State of the Environment, "A Decade After Stockholm." Vol.1 & 2, Nairobi, 1981.
United Nations Environment Programme, Trends in Environmental Education, Paris, 1977.
United Nations Environment Programme, UNEP Profile, Nairobi, 1990.
United Nations, World Economy: A Global Challenge, New York: UN Publication, March 1990.

United Nations Yearbook of the United Nations 1972, New York: Office of Public Information, United Nations. 1972.
World Climate Programme Impact Studies. Developing Policies for Responding to Climate Change. WCIP-l, WMOIFD – No.22.5, April 1988.

Other Documents

Council on Environmental Quality. The Global 2000. Report to the President: Entering the Twenty First Century. Harmondsworth: Penguin, 1982.
Environmental Study Board Commission on Natural Resources, NAS, Washington, DC, 1976.
European Fluorocarbon Technical Committee. (E}CTC) Policy on Technology Transfer to Developing Countries, 24 April 1989.
Final Consensus Report: Global Initiative for the Security and Sustainable Use of Plant Genetic Resources. Oslo Plenary Sessions, The Keystone Center: Keystone, 1991.
The Fourth World Congress on National Parks and Protected Areas, 1992.
GER, Global Environment Review, Ten Publication, Vol.2, no.2, December 2000.
Ghali-Boutros, Boutros. An Agenda on Democratization, New York: United Nations Publications, 1996.
I.L.I. Environmental Protection Act: An Agenda for Implementation, Indian Law Institute, Bombay: N.M. Tripathy, 1987.
India, 1991.
International Union for Conservation of Nature and Natural Resources: A Conservation Programme for Sustainable Development, Gland: IUCN, 1980–83.
International Union for Conservation of Nature and Natural Resources. National Conservation Strategies: A. Framework for Sustainable Development. Gland: CDC/IUCN, 1984.
Joint Communiqué of the SAARC Ministers' of Environment, New Delhi, 8–9 April 1992.
Kuala Lumpur Declaration on Environment and Development, 29 April 1992.
Latin American and Caribbean Commission on Development and Environment, Our Own Agenda New York UNDP, 1990.
Manaus Declaration on the United Nations Conference on Environment and Development, 10 February 1992.
N.A.S. Institutional Arrangement for International Environmental Cooperation. A Report to the Dept. of State by the Committee for International Environmental Program, NAS, Washington, DC, 1972.
National Academy of Sciences, Implementation of the Global Environmental Monitoring System: International Environmental Programme Committee.
National Academy of Sciences, International Arrangements of International Environmental Co-operation, Washington, DC: NAS, 1972.
ODA, Biological Diversity and Developing Countries – Issues and Options, London: ODA, 1991.
OECD, The Polluter Pays Principle, Paris, OECD, 1975.
OECD, The State of Environment in OECD Countries, Paris, 1985.
Report of People's Commission on Environment and Development (PCED).
A Report: 'Public Hearing on Environmental. Problems and Development Strategies', Bombay, 16–17 November 1991.
Singapore Resolution on Environment and Development, 18 February 1992.

South Centre, Environment and Development – Towards a Common Strategy of the South in the UNCED Negotiations and Beyond, Geneva: South Centre, 1991.
TERI, Report on Global Warming and Associated Impacts, New Delhi: TERI, June 1990.
United Nations, Yearbook of United Nations 1972, New York: Office of Public Information, United Nations, 1972.
United States Department of State, Stockholm and Beyond, Department of State Publication, 8657, Washington, DC, 1972.
US Environmental Protection Agency, Summaries of Foreign Govts. Environment Reports. Washington, DC, 197.
WMO, Conference Proceedings, World Conference on the Changing Atmosphere: Implications for Global Security, Toronto: WMO, No.710, June 1988, 27–30.
World Bank, Environment and Development, Washington, DC: World Bank, 1979.
World Bank, Environmental Health and Human Ecological Consideration in Economic Development Projects. Washington, DC, 1973.
World Bank, Environmental Health and Human Ecology Consideration in Economic Development Projects. Washington, DC: World Bank, 1972.
World Resources Institute, World Resources 1990–91, Oxford: Oxford University Press, 1990.
WPC, Peace, Disarmament and Environment. Report of the Commission on Environment. World Peace Council. Helsinki, 1979.

Government Publications, Documents and Non-government Reports

Government of India Documents/Publications

Annual Action Plan 1985–90, Do En & Forest (1990).
Annual Report 1986–87. New Delhi: MOEF, 1987.
Annual Report 1987–88. New Delhi. MOEF, 1988.
Annual Report 1988–89. New Delhi. MOEF, 1989.
Annual Report 1989–90. New Delhi. MOEF, 1990.
Annual Report 1991–92. New Delhi: MOEF, 1992.
Annual Reports of Ministry of Environment, Forest and Wildlife, 1985–92.
Annual Reports of Wildlife Institute of India, New Delhi, Dehradun, 1988–92.
Chairman's Summary: Conference of Select Developing Countries on Global Environmental Issues. New Delhi: MOEF, 23–25 April 1990.
Conference of Select Developing Countries on Global Environmental Issues. New Delhi: MOEF, 23–25 April 1990.
Documents on National Wildlife Action Plan, Do En., Govt. of India, 1983.
Draft Eighth Five Year Plan, 1990.
DST. Report of the Committee for Recommending Legislative Measures and Administrative Machinery for Ensuring Environmental Protection. New Delhi: DST, 1980.
Environment and Development: Traditions, Concerns and Efforts in India. New Delhi: MOEF, June 1992.
Inter-Ministerial Meeting Regarding Matters Related to Beijing Ministerial Conference on Environment and Development. New Delhi: MOEF, 10 May 1991.
Ministry of Environment and Forest (MOEF). India – Country Study on the Environment. New Delhi: MOEF, October 1990.

National Conservation Strategy and Policy Statement on Environment and Development. New Delhi: MOEF, June 1992.
National Conservation Strategy and Sustainable Development: Report of the Core Committee. New Delhi: MOEF, April 1990.
National Parks and Sanctuaries: A Note of Consultative Committee of Members of Parliament, Wildlife Division. New Delhi: MOFF, 1987.
Policy Statement for Abatement of Pollution. New Delhi: MOEF, 1992. Annual Report 1992-93. New Delhi: MOEF, 1993.
Resolution on National Forest Policy. New Delhi: MOEF, 7 December 1993.
Second Meeting of the Expert Advisory Committee on Global Environmental Issues. New Delhi: MOEF, 24 May 1990.
Seventh Five Year Plan (1985-90) Vol.1 and 2~ Planning Commission. New Delhi, October 1985.
Sixth Five Year Plan (1980-85), Vol.1, 2 and 3. Planning Commission. New Delhi, 1980.
Souvenir, Center of Forest Education in India (1881-1981), Forest Research Institute, New Forest, Dehradun, 19 December 1981.
Statement by the Leader of the Indian Delegation. New Delhi: MOEF, 19 June 1991.

Government Publications from South Asia/SACEP Documents and Non-governmental Reports

Bangladesh Environment and Natural Resources~ Assessment, Final Report, prepared for the U.S. Department of Development and Environment, Washington, DC, 1992.
The Environment and Development in Bangladesh: An Overview and Strategy for the Future, Prepared by the Canadian International Development Agency for Bangladesh Programme, Asia Branch, October 1989.
Environmental Policy, Report of the Task Force on Bangladesh Development Strategies for the 1990's, Vol.4, Dhaka, 1991.
Government of Bangladesh, the Ganges Water Crisis in Bangladesh, Memorandum Circulated at the Organization of Islamic Conference, Foreign Ministers Conference in Istanbul, May 1976.
Government of Bangladesh, Study on the Causes and Consequences of Natural Disasters and Protection and Preservation of the Environment, Dhaka, December 1989.
Government of Pakistan: National Human Settlement Policy Study, Pakistan Environmental Planning and Architectural Consensus (PEPAC), Islamabad, Government, of Pakistan, 1983.
Government of Sri Lanka: Citizens' Report on Environment and Development to UNCED, Colombo, June 1992.
His Majesty's Government of Nepal (HMG), The Causes and Consequences of Natural Disasters and the Protection and ~Preservation of the Environment, Kathmandu, March 1990.
His Majesty's Government of Nepal (HMG/N), Population and Development: Nepal, Ministry of Population and Environment, Kathmandu, July 1999.
His Majesty's Government of Nepal (HMG/N), Sate of the Environment: Nepal, Ministry of Population and Environment, Kathmandu, June 2000.
International Institute for Environment and Development, IIED Annual Report 1981-82, London: IIED, 1982.
Managing the Development Process, Report of the Task Force on Bangladesh Development Strategies for the 1990's, Vol.ii, Dhaka: UPL, 1991.

National Commission on Agriculture (NCA), Islamabad, 1988.
Nepal, Ministry of Population and Environment, Kathmandu, July 1999.
Nepal, Ministry of Population and Environment, Kathmandu, June 2000.
Pakistan National Report, Submitted to the United Nations Conference on Environment, August 1991, Environment and Urban Affairs Division, Government of Pakistan, Islamabad, 1991.
SAARC, Country Wise Study Report for Sri Lanka on the Causes and the Consequences of Natural Disasters and the Protection and Preservation of the Environment, Colombo, March 1991.
SAARC, Regional Study on the Causes and Consequences of Natural Disasters and the Protection and Preservation of the Environment (SAARC Secretariat, Kathmandu, 1992).
Sadik, S. The State of the World Population 1990, New York: World Population Fund, 1990.
State of India's Environment: The Citizens' Fifth Report, Part I, II, New Delhi: CSE, 1999.
State of India's Environment: A Citizens' Report. Dying Wisdom: Rise, Fall and Potential of India's Traditional Water Harvesting Systems, New Delhi: CSE, 1997.
State of India's Environment: A Citizens' Report. 'Floods, Food, Plains and Environmental Myths', New Delhi: CSE, 1991.
State of India's Environment: 'The First Citizens' Report', New Delhi: CSE, 1982.
State of India's Environment: 'The Second Citizens' Report, 1984–85', New Delhi: CSE, 1985.
Task Force on Migration (TFM), Internal and International Migration in Nepal, Kathmandu: The National Commission on Population, 1983, pp. 51–73.
Towards Sustainable Development: The National Conservation Strategy of Bangladesh, First Draft, Ministry of Environment and Forest, Government of Bangladesh, Dhaka, 1991.
Towards Sustainable Development: National Conservation Strategy of Bangladesh, Ministry of Environment and Forest, Government of Bangladesh/IUCN, Dhaka, 1991.
The World Bank, Nepal: Politics of Improving Growth and Alleviating Poverty, Washington, DC: WB, 1989.
UNICEF: The State of the World's Children 1987, New York: UNICEF, 1987.
UNPF the United Nations Population Fund, the State of World Population 1992, New York: UNPF, 1992.
World Bank Development Report 1995, Oxford: Oxford University Press, 1995.
World Bank, World Development Report 1987, New York: Oxford University Press, 1987.
World Bank, World Development Report 1988, New York: Oxford University Press, 1988.
World Bank, World Development Report 1992, Oxford: Oxford University Press, 1992.
World Commission on Environment and Development, Our Common Future, the Brundtland Report, New York: Oxford University Press, 1987.

Encyclopaedias, Dictionaries and Year Books

Encyclopedia of International Environmental Laws (Vol.1–10). New Delhi: The Global Open University Press in Association with Indian Institute of Ecology and Environment, 1995.
Gaur, G., International Encyclopedia of Environmental Pollution and Its Management. New Delhi, 1997, Vols. I-V (Vol.1 – Air Pollution and Its Management; Vol-II-Water Pollution and Its Management; Vol-III-Soil and Solid Waste Pollution and Its Management; Vol-IV-Noise Pollution and Its Management; Vol-V-Nuclear Waste Pollution and Its Management).
International Union for Conservation of Nature and Natural Resources IUCN Year Book 1971 Morges IUCN, 1972.

210 Bibliography

International Union for Conservation of Nature and Natural Resources. IUCN Year Book 1972 Morges, Switzerland, IUCN, 1973.
International Union for Conservation of Nature and Natural Resources IIJCN Year Book 1973 Morges, Switzerland, IUCN, 1974.
International Union for Conservation of Nature and Natural Resources IUCN Year Book 1977–76 Morges, Switzerland, IUCN, 1976.
Kemp, David D. *The Environmental Dictionary*. London: Routledge, 1998.
Mayur, Rashmi (ed.). *Indian Environment Dictionary*. Bombay: Astra 1991.
Paehike, Robert (ed.). *Conservation and Environmentalism: An Encyclopedia*. London: Fitzroy Dearborn, 1995.
Swarup, R., S.N. Mishra, & V.P. Johari (eds.). *Encyclopedia of Ecology, Environment and Pollution Control*. New 1~élhi: Mittal, 1992 (5 Vols.).

Interviews

Interview with Anju Sharma, Research Co-ordinator and Environmentalist from CSE, New Delhi.
Interview with Dr Pradeep Monga, Assistant Resident Representative Environment and GEF Division UNDP, United Nations Office in New Delhi.
Interview with Mr. Sudarshan Rodriguez, Consultant, Environment and GEF Division, United Nations Office in New Delhi.
Interview with Nirmal Andrews, UNEP's Regional Director at Bangkok, Thailand.
Interviews with Officials Conducting UNEP Programs and Environmentalists.
Interview with Veena the, Project Co-ordinator UNCTAD, United Nations Office in New Delhi.

Secondary Sources

Books and Articles

Abel, G.J., B. Barakat, K.C. Sameer, and W. Lutz. Meeting the Sustainable Development Goals Leads to Lower World Population Growth. *Proceedings of the National Academy of Sciences of the United States of America*, Vol.113, no.50, 2016, pp. 14294–14299.
Adam, Roberts, and Kingsbury Benedict (eds.). *United Nations in a Divided World*. London: Clarendon Press, 1988.
Adams, W.M. *Green Development: Environment and Sustainability in the Third World*. London: Routledge, 1990.
Adams, W.M. *The Future of Sustainability: Re-thinking Environment and Development in the Twenty-first Century*. (Geneva: IUCN, The World Conservation Union), 2006 (Report of the IUCN Renowned Thinkers Meeting, 29–31 January 2006).
Adger, W.N. et al. Adaptation to Climate Change in the Developing World. *Progress in Development Studies*, Vol.3, no.3, 2003, pp. 179–195.
Adiseshiah, Malcom S. (ed.). Some Economic Issues in the Rural Energy. *The Asian Age*, 4 August 2000.
Adiseshiah, Malcolm S. (ed.). *Sustainable Development: Its Content, Scope and Prices*. New Delhi: Lancer International, 1990.
Aerts, J., and W. Botzen. Adaptation: Cities' Response to Climate Risks. *Nature Climate Change*, Vol.4, 2014, pp. 759–760.

Afshar, H.R. The Earth, Impending Physical Environment Round the Year 2000 and Problems of the Developing Countries. Tehran: Penguin, 1973–1975, Vol.1 and 2.

Agarwal, Anil (ed.). The Challenge of the Balance: Environmental Economics in India. In *Proceedings of the National Environment and Economics Meeting*. New Delhi: CSE, January 1994.

Agarwal, Anil. Green Trade Wars' Down to Earth. New Delhi: CSE, Vol.1, no.6, 15 August 1992.

Agarwal, Anil. The North-South Perspective: Alienation or Independence? *Ambio*, Vol.19, no.2, April 1990, pp. 94–96.

Agarwal, Anil. The North-South Perspective Issues in the Rural Energy. *The Asian Age*, 4 August 2000.

Agarwal, Anil (ed.). *Proceedings of a Seminar on the Economics of the Sustainable Use of Forest Resources*. New Delhi: CSE, 1992.

Agarwal, Anil. *State of the Environment Series: (I, II, III, IV): Slow Murder: The Deadly Story of Vehicular Pollution in India*. New Delhi: CSE, 1996.

Agarwal, Anil. Towards Global Environmental Movement. *Social Action*, Vol.42, no.2, April–June 1992, pp. 111–119.

Agarwal, Anil, and Narayan Sunita. Earth Needs a Summit: It Is Going to Be a Grand Affair Full of Little Glitches. *Down to Earth* (special issue named Preparing You to Change the Future, Vol.200), February 2001.

Agarwal, Anil, and Narayan Sunita. *Global Warming in an Unequal World: A Case of Environmental Colonialism*. New Delhi: CSE, 1991.

Agarwal, Anil, and Narayan Sunita. A New Morality. *The Illustrated Weekly of India*, 15, 24 December 1989, pp. 84–87.

Agarwal, Anil, and Narayan Sunita. *Towards a Green World: Should Global Environmental Management Be Built on Legal Conventions or Human Rights*. New Delhi: CSE, 1992.

Agrawal, K.C. *Environmental Biology*. Bikaner: Agro Botanical Publication, 1989.

Agrawal, S., and S. Yamamoto. Effect of Indoor Air Pollution from Biomass and Solid Fuel Combustion on Symptoms of Pre-eclampsia/Eclampsia in Indian Women. *Indoor Air*, Vol.25, 2015, pp. 341–352.

Agrawala, Shardul, and Samuel Fankhauser (eds.). *Economic Aspects of Adaptation to Climate Change: Costs, Benefits and Policy Instruments*. Paris: Organization for Economic Cooperation and Development, 2008.

Ahamad, Yusuf J. (ed.). *Managing the Environment: An Analytical Examination of Problems and Procedures*. Oxford: Pergamon, 1983.

Ahmad, Yusuf J. *Economics of Survival: The Role of Cost-Benefit Analysis in Environmental Decision-Making*. Oxford: Pergamon, 1984.

Ahmed, Iftikhar. *Biotechnology: A Hope or a Threat?* London: The Macmillan Press, 1992, pp. 1–14.

Ahmed, Iftikhar, and Doel Neman Jacobus A. *Beyond Rio: The Environmental Crisis and Sustainable Livelihoods in the Third World (ILO Studies Series)*. London: Macmillan Press, 1995.

Ahmed, M., and S. Suphachalasai. *Assessing the Costs of Climate Change and Adaptation in South Asia*. Manila: Asian Development Bank, 2014.

Ahmed, Mushir. Solar Energy Use Sees Major Growth. *Financial Express* (Dhaka), Vol.18, no.77, 30 January 2011. REGD NO DA 1589. Available at: www.thefinancialexpress-bd.com/more.php?news_ id=124464&date=2011–01–30.

Ahmed, S. End Poverty in All Its Forms Everywhere. In *India and Sustainable Development Goals*. New Delhi: Research and Information System for Developing Countries (RIS), 2016, pp. 1–11.

Bibliography

Aijaz, Rumi. Preventing Hunger and Malnutrition in India. *ORF Issue Brief (18) Observer Research Foundation*, June 2017. Available at: https://www.orfonline.org/wp-content/uploads/2017/06/ORF_IssueBrief_182_ Hunger.pdf.

Alabi, Michael. Environment. *Britannica Book of the Year*, 1986, pp. 242–248.

Alagh, Y.K., Mahesh Pathak, and D.T. Buch. *Narmada and Environment: An Assessment*. New Delhi: Har Anand, 1995.

Aldous, Tony. *Battle for the Environment*. London: Fontana Books, 1972.

Alexander Atos, N. *World-Agri Culture Towards 2(XX): An FAO Study*. London: Belhaven Press, 1988.

Alfred, J.R.B. Perspectives on Biodiversity: A Vision for Megadiverse Countries. In D.D. Verma, S. Arora, and R.K. Rai (eds.), *Ministry of Environment & Forests*. New Delhi: Government of India, 2006, pp. 272–293.

Alire, Rod. *The Reality Behind Biodegradable Plastic Packaging Material: The Science of Biodegradable Plastics*. Redwood City, CA: FP International. Available at: www.fpintl.com/resources/wp_ biodegradable_plastics.htm.

Allen, Robert. *How to Save the World: Strategy for World Conservation*. London: Kogan Page, 1980.

Alliston, Philip. International Regulation of Toxic Chemical. *Ecology Law Quarterly*, Vol.67, 1978, pp. 397–456.

Altieri, Miguel A. *Small Farms as a Planetary Ecological Asset: Five Key Reasons Why We Should Support the Revitalization of Small Farms in the Global South*. Oakland, CA: Food First/Institute for Food and Development Policy, 2008. Available at: www.foodfirst.org/en/node/2115. Alvarez, Benjamín, and others (1999).

Alvaredo, Facundo, Lucas Chancel, Thomas Piketty, Emmanuel Saez, and Gabriel Zucman. *World Inequality Report*, 2018. Available at: https://wir2018.wid. world/files/download/wir2018-full-report-english.pdf.

Amato, Anthony D. et al. Do We Owe a Duty to Future Generations to Preserve the Global Environmental Responsibility? *American Journal of International Law*, Vol.84, no.1, January 1980, pp. 10–91.

Anand, R.P. (eds.). Law, Science and Environment. Papers edited in a Conference on Law, Science and Environment, New Delhi, Lancer Books, 1985.

Anand, R.P. *South Asia: In Search of a Regional Identity*. New Delhi: Banyan Publications, 1991.

Andresen, Steinar, and Willy Ostreng. *International Resource Management: The Role of Science and Politics*. London: Belhaven Press, 1989.

Archibugi, D., and C. Pietrobelli. The Globalization of Technology and Its Implications for Developing Countries: Windows of Opportunity or Further Burden? *Technological Forecasting and Social Change*, Vol.70, no.9, 2003, pp. 861–883.

Arden-Clarke, Charles. South-North Terms of Trade: Environmental Protection and Sustainable Development. *International Environmental Affairs*, Vol.4, no.2, Spring 1992, pp. 122–138.

Arefin, Sadia. Climate Refugees and Position of Bangladesh. *The Independent*, 20 May 2017. Available at: www.theindependentbd.com/arc print/details/95322/2017–05–20.

Arif, T. *Asian Development Outlook 2017: Transcending the Middle-Income Challenge*. Manila: Asian Development Bank, 2017.

Arif, T. Bangladesh's Climate Change Response and Adaptation Efforts. *Department of Environment*, 2013. Available at: https://unfccc.int/files/adaptation/groups_committees/ldc_expert_group/application/pdf/bangladesh.pdf Asian Development Bank. *Addressing Climate Change and Migration in Asia and the Pacific*. Manila, 2012.

Arif, T. *A Region at Risk: The Human Dimensions of Climate Change in Asia and the Pacific*. Manila, 2017. Available at: www.adb.org/sites/default/files/publication/325251/region-risk-climate-change.pdf.

Ashish, Madhu. In the Area of Environment. *Seminar Annual*, Vol.269, January 1982, pp. 72–74.

Asthana, Vandana. *The Politics of Environment: A Profile*. New Delhi: Ashish Publications, 1992.

Asthana, Vandana et al. Regional Co-operation for the Protection of Coastal Environment in South Asia. *South Asia Journal*, Vol.3, nos.1–3, July–December 1989, pp. 109–120.

Atapattu, S. Climate Change in South Asia: Towards an Equitable Legal Response Within a Framework of Sustainable Development and Human Security. International Development Law Organization (IDLO), IDLO Sustainable Development Law on Climate Change Working Paper Series 1, Rome, 2011.

Atpatyev, A.M. *Problems of the Development, Transformation and Protection of the Environment*. Moscow: Gerome Izzat Pub., 1978.

Bahuguna, Sunder Lal, Vandana Shiva, and M.N. Buch. *Environment and Sustainable Development*. Dehradun: Natraj Pub., 1992.

Bai, Z.G. et al. *Global Assessment of Land Degradation and Improvement: 1. Identification by Remote Sensing. Report 2008/01*. Wageningen, Netherlands: ISRIC – World Soil Information, Food and Agriculture Organization of the United Nations, 2008.

Bajwa, G.S. Environmental Management: Problems and Prospects. In R.K. Sapru (ed.), *Environmental Management in India*. New Delhi: Ashish Publishing House, Vol.2, 1987, pp. 207–217.

Baker, Elaine et al. *Vital Waste Graphics*. Nairobi: Basel Convention Secretariat, UNEP Division of Environmental Conventions, Grid-Arendal and UNEP Division of Early Warning Assessment, Europe, 2004.

Bandara, J.S., and Y. Cai. The Impact of Climate Change on Food Crop Productivity, Food Prices and Food Security in South Asia. *Economic Analysis and Policy*, Vol.44, no.4, 2014, pp. 451–465.

Bandyopadhyay, J., and D. Gyawali. Himalaya's Water Resources: Ecological and Political Aspects of Management. *Mountain Research and Development*, Vol.14, no.1, 1994.

Bandyopadhyay, Jayanta, and Vandana Shiva. Development, Poverty and the Growth of the Green Movements in India. *The Ecologist*, Vol.19, no.3, May–June 1989, pp. 111–117.

Bandyopadhyay, T. et al. (eds.). *India's Environmental Crisis and Responses*. Dehradun: Natraj Publications, 1985.

Bangladesh Bureau of Statistics. *Population Density and Vulnerability: A Challenge for Sustainable Development of Bangladesh*. Dhaka: Bangladesh Bureau of Statistics, 2015.

Barnett, J., and W.N. Adger. Climate Change, Human Security and Violent Conflict. *Political Geography*, Vol.26, no.6, 2007, pp. 639–655.

Barret, Richard N. *International Dimensions of the Environmental Crisis*. London: Westview, 1982.

Barros, James (ed.). *The United Nations: Past, Present and Future*. New York: Free Press, 1972.

Barrow, C.T. *Land Degradation: Development and Breakdown of Terrestrial Environments*. Cambridge: Cambridge University Press, 1994.

Bartelmus, P. *Environment and Development*. London: Allen and Unwin, 1986.

Basiago, A.D. Economic, Social, and Environmental Sustainability in Development Theory and Urban Planning Practice. *The Environmentalist*, 19, 1999, pp. 145–161. Available at: www.amherst.edu/system/files/media/0972/fulltext.pdf

Bass, S. The National Conservation Strategies for Nepal. *Landscape Design*, Vol.2, 1989.

Basu, Kaushik. Examples of Gender Progress from Bangladesh and India, 10 March 2015. Available at: www.weforum.org/agenda/2015/03/examples-of-gender-progress-from-bangladesh-and-india/.

Basu, Rumki. *The Global Environment and the United Nations with Special Reference to India's Environmental Policy.* New Delhi: National Publishing House, 1998.

Basu, Rumki. Politics and Economics of the Global Environment Debate Between the Developed and Developing Countries. *Indian Journal of Political Science*, Vol.52, no.1, January–March 1991, pp. 74–84.

Baumert, Kevin A., Timothy Herzog, and Jonathan Pershing. *Navigating the Numbers: Greenhouse Gas Data and International Climate Policy.* Washington, DC: World Resources Institute, 2005.

Baviskar, Amita. Ecology and Development in India: A Field and Its Future. *Sociological Bulletin*, Vol.46, no.2, September 1997, pp. 193–209.

Baviskar, Amita. *In the Belly of the River: Tribal Conflicts Over Development in the Narmada Valley.* New Delhi: Oxford University Press, 1995.

Baviskar, Amita. Tribal Politics and Discourses of Environmentalism. *Contributions to Indian Sociology*, Vol.31, no.2, July–December 1997, pp. 195–225.

Baxi, Upendra. *Inconvenient Forum and Convenient Catastrophe: The Bhopal Case.* New Delhi: Indian Law Institute, 1986.

Baxi, Upendra, and Thomas Paul. *Mass-Disasters and Multinational Liability: The Bhopal Case.* New Delhi: Indian Law Institute, 1986.

Baxter, Craig, Yogendra K. Malik, Charles H. Kennedy, and Robert C. Oberst. *Government and Politics in South Asia.* London: Westview Press, 1987.

Beckerman, Wilfred. Global Warming and International Action: An Economic Perspective. In Andrew Hurrell and Benedict Kingsbury (eds.), *The International Politics of the Environment.* Oxford: Clarendon Press, 1992, pp. 253–289.

Beehacker, A.A. *Kaleidoscopic Circumspection of Development Planning with Contextual Reference to Nepal.* Rotterdam: Rotterdam: University Press, 1973.

Begum, Ferdousi Sultana. Gender Equality and Women's Empowerment: Suggested Strategies for the 7th Five Year Plan. Prepared for General Economics Division, Planning Commission, Government of Bangladesh, 2014.

Begum, Khurshida. *Tension Over the Farrakka Barrage: A Techno Political Tangle in South Asia.* Calcutta: K.P. Bag chi & Co., 1988.

Begum, Sharifa. *Population, Birth, Death and Growth Rates in Bangladesh: Census Estimates.* Dhaka: Bangladesh Institute of Development Studies, 1990.

Benedick, Richard Elliot. Environment in the Foreign Policy Agenda. *Department of State. Bulletin*, Vol.86, no.2111, June 1986, pp. 55–58.

Benedick, Richard Elliot. Protecting the Ozone Layer. *Department of State Bulletin*, Vol.85, no.2097, April 1985, pp. 63–64.

Benedick, Richard Elliot et al. *Greenhouse Warming: Negotiating a Global Regime.* Washington, DC: WRI, 1991.

Bernie, Patricia. The Role of International Law in Solving Certain Environmental Conflicts. In John E. Carroll (ed.), *International Environmental. Diplomacy.* Cambridge: Cambridge University Press, 1988, pp. 95–121.

Bertelsmann, Stiftung, and Sustainable Development Solutions Network. SDG Index and Dashboards Report 2018, Global Responsibilities, Implementing the Goals. Available at: www.sdgindex.org/assets/files/2018/01%20 SDGS%20GLOBAL%20EDITION%20 WEB%20V9%20180718.pdf.

Bibliography

Besley, Timothy, and Louise J. Cord (eds.). *Delivering on the Promise of Pro-Poor Growth: Insights and Lessons from Country Experiences*. Washington, DC: World Bank; Basingstoke: Palgrave Macmillan, 2007.

Bhagwati, Jagdish. Development Aid: Getting It Right. *OECD Observer*, no.249, May 2005.

Bhargava, Gopal. *Ecological Imbalances and Quality of Life*. Bombay: Commerce Publication. 1982.

Bhatia, Arti, H. Pathak, and P.K. Aggarwal. Inventory of Methane and Nitrous Oxide Emissions from Agricultural Soils of India and Their Global Warming Potential. *Current Science*, Vol.87, no.3, August 2004, pp. 317–324.

Bhatt, S. *Environmental Law and Water Resource Management*. New Delhi: Radiant, 1986.

Bhogal, Parminder S. India's Security Environment in the 1990: The South Asia Factor. *Strategic Analysis*, Vol.13, no.7, October 1989, pp. 167–177.

Bhushan, Vijay. *Environmental Planning and Its Implementation*. New Delhi: IPL, 1980.

Birkmann, Jörn, and Korinna von Teichman. Integrating Disaster Risk Reduction and Climate Change Adaptation: Key Challenges – Scales, Knowledge, and Norms. *Sustainability Science*, Vol.5, no.2, 2010, pp. 171–184.

Birkmann, Jörn et al. Adaptive Urban Governance: New Challenges for the Second Generation of Urban Adaptation Strategies to Climate Change. *Sustainability Science*, Vol.5, no.2, 2010b, pp. 185–206. Extreme Events and Disasters: A Window of Opportunity for Change? Analysis of Organizational, Institutional and Political Changes, Formal and Informal Responses After Mega-Disasters. *Natural Hazards*, Vol.55, no.3, December 2010a, pp. 637–655.

Biswas, Asit K., and Qugeping (eds.). *Environmental Impact Assessment (EIA) for Developing Countries*. London: Tycoolly Int., 1987.

Bizikova, L., and L. Pinter. *Indicator Preferences in National Reporting of Progress Toward the Sustainable Development Goals*. Winnipeg: International Institute for Sustainable Development, 2017.

Blake, R.O., B.J. Lewche et al. *Aiding the Environment: A Study of the Environmental Policies, Procedures and Performance of the US Agency for International Development*. Washington, DC: Natural Resources Defense Council, 1980.

Blondel, A. Climate Change Fuelling Resource-Based Conflicts in the Asia Pacific, Asia-Pacific Human Development Report Background Papers Series, 2012.

Boardman, Robert. *International Organizations and the Conservation of Nature*. London: Macmillan, 1981.

Bossone, Biagibn. Environment Protection: How Should We Pay for It? *International Journal of Social Sciences*, Vol.17, no.1, 1990, pp. 3–15.

Bothe, Michael (ed.). *Trends in Environmental Policy and Law*. Gland: IUCN, 1980.

Bowonder, B. Environmental Quality and Water Resources Management in India, Hyderabad Administrative Staff College, 1983.

Boyce, James. *Agrarian Impasses in Bengal: Institutional Constraints to Technological Change*. Oxford: Oxford University Press, 1987.

Brandt, Willy. *North-South. A Programme for Survival: The Report of the Independent Commission on International Development Issues*. London: Pan Books, 1980.

Brandt, Willy, et al. *North-South, a Programme for Survival. Report of the Independent Commission on International*. Cambridge: MIT Press, 1980.

Brecher, Michael. *The New States of Asia: A Political Analysis*. London: Oxford University Press, 1963.

Bregman, Jacob I. *Environmental Impact Statements*. Boca Raton: Lewis Publication, 1999.

Brookfield, Harold, and Yvonne Byron. *South-East Asia's Environment Future: The search for sustainability*. Tokyo: United Nations University Press, 1993.

Bibliography

Brown, Lester R. *The Twenty ninth Day: Accommodating Human Needs and Numbers to. Earth Resources.* New York: W.W. Norton, 1978.

Brown, Lester R. *Building a Sustainable Society.* New York: W.W. Norton, 1981.

Brown, Lester R. *The State of the World.* New York: World Watch Institute, W.W. Norton, 1984.

Brown, Lester R., Flann R. Christopher, and Postal Sandra. *Saving the Planet: How to Shape an Environmentally Sustainable Global Economy.* London: Earthscan, 1992.

Brown, Lester R., and Jodi L. Jacobson. The Future Urbanization: Facing the Ecological and Economic Constrains, World Watch Paper 77, May 1987.

Brown, Lester R. et al. (ed.). *State of the World.* New Delhi: World Watch Institute, 1991.

Brundtland, Gro Harlem. After: Unlearnt Lessons. *Hindu Survey of the Environment*, 2000.

Burney, A. Environmental Health, Islamabad National Conservation Strategy, 1989 (mimeographed).

Burhan, Wolfgang (ed.). *International Environmental Law. Multilateral Treaties.* Berlin: Erich Schmidt Verlag, 1985.

Butler, C.D. Food Security in the Asia-Pacific: Malthus, Limits and Environmental Challenges. *Asia Pacific Journal of Clinical Nutrition*, Vol.18, no.4, 2009, pp. 577–584.

Caldwell, Lyndon K. *Between Two Worlds: Science, the Environmental Movement and Policy Choice.* Cambridge: Cambridge University Press, 1990.

Caldwell, Lyndon K. *Environment.* London: Penguin, 1984.

Caldwell, Lyndon K. *In Defense of Earth: International Protection of the Biosphere.* Bloomington: Indiana University Press, 1972.

Caldwell, Lyndon K. *International Environmental Policy: Emergence and Dimensions.* Durham: Duke University Press, 1984.

Caldwell, Lyndon K. *International Environmental Policy: Emergence and Dimensions* (2nd ed.). London: Duke University Press, 1990.

Cambell, Leyton Keith. Co-operation and Conflict; International Response to 1 Environmental Issues. *Environment*, Vol.27, no.1, January–February 1985, pp. 10–14.

Carroll, John E. (ed.). *International Environmental Diplomacy: The Management and Resolution of Trans Frontier Environmental Problems.* Cambridge: Cambridge University Press, 1988.

Carson, Rachel. Silent Spring. Houghton Mifflin, 27 September 1962.

Castro, Joao Augusto de Araujo. Environment and Development: The Case of the Developing Countries. *International Organization*, Vol.26, 1972, pp. 401–416.

Chadha, S.K. (ed.). *Environmental Holocaust in Himalaya.* New Delhi: Ashish, 1989.

Chand, Attar. *Environmental Challenges. A Global Survey.* New Delhi: UDH Pub., 1985.

Chandler, David L. Parts of India, Pakistan and Bangladesh Could Become Too Hot for Humans by 2100, 10 August 2017. Available at: www.weforum. org/agenda/2017/08/pakistan-india-and-bangladesh-will-be-too-hot-for humans-this-century-research-find.

Chatterjee, Manas (ed.). *Partial Environmental Policies in the Developing Countries.* London: Prager, 1985.

Chishti, Sumitra. India's Foreign Economic Policy. In Bimal Prasad (ed.), *India's Foreign Policy – Studies in Continuity and Change.* New Delhi: Vikas Publishing House Private Limited, 1979, pp. 35–36.

Christopher, Penny. Greening the Security Council: Climate Change as an Emerging "Threat to International Peace and Security." *International Environmental Agreements*, Vol.7, 2007, pp. 35–71. Available at: www.springerlink. com/content/j4857081065920w1/fulltext.pdf.

Clarke, Peter B. Environmental Destruction. *Economic and Political Weekly*, Vol.29, no.17, 1976, pp. 17–19.

Clarke, Robin, and Lloyd Timberlake. *Stockholm Plus Ten*. London: Earthscan, 1982.

Climate Change Cell, Department of Environment, Government of the People's Republic of Bangladesh. Climate Change and Bangladesh. Published with Support from Comprehensive Disaster Management Program of the Government of Bangladesh and UNDP and DFID, 2007. Available at: www. climatechangecell-bd.org/publications/13ccbd.pdf.

Colby, Michael E. *Environmental Management in Development: The Evolution of Paradigms*. Washington, DC: IBRD, 1989.

Commoner, Barry. *The Closing Circle: Nature, Man and Technology*. New York: Knopf, 1971.

Cosbey, Aaron. Are There Downsides to a Green Economy? The Trade, Investment and Competitiveness Implications of Unilateral Green Economic Pursuit. In United Nations Conference on Environment and Development, The Road to Rio+20: The Green Economy, Trade and Sustainable Development, Geneva, UNCTAD, 2011b.

Cosbey, Aaron. Trade, Sustainable Development and a Green Economy: Benefits, Challenges and Risks. In United Nations, United Nations Environment Programme and United Nations Conference on Trade and Development, the Transition to a Green Economy: Benefits, Challenges and Risks from a Sustainable Development Perspective. Report by a Panel of Experts to the Second Preparatory Committee Meeting for the United Nations Conference on Sustainable Development, New York, 7 and 8 March 2011a, pp. 39–67.

Court, Thijs Dela. *Beyond Brundtland: Green Development in the 1990s*. London: Zed Books Ltd., 1990.

Cowen, C. The Rise of Co-diplomacy. *Technology Review*, May–June 1988, pp. 91–118.

Cowen, R.C. Environment Protection for 1990s. *Environment*, September 1990, pp. 12–15.

Cox, George W. *Conservation Ecology: Biosphere and Bio Survival*. Dubuque, Iowa: Wm C. Brown, 1993.

CSE, Statement of the South Asian NGO Summit, New Delhi: CSE, 17–19 February 1992, Reprinted in R.S. Digest, Vol.9, nos. 1 and 2, June 1992, pp. 32–42.

Daoud, Alden L. International. Environmental Developments: Perceptions of Developing and Developed Countries. *Natural Resources Journal*, Vol.12, October 1972, pp. 520–529.

Das, R.C., J.K. Baral, and N.C. Sahoo (eds.). *Dilemma of Developing Countries: The Environmental Divide*. Daryaganj, Delhi: APH Publishing House, 1998.

Dasgupta, S., Mainul Huq, Zahirul Huq Khan, Manjur Murshed Zahid Ahmed, Nandan Mukherjee, Malik Fida Khan, and Kiran Pandey. Vulnerability of Bangladesh to Cyclones in a Changing Climate, Potential Damages and Adaptation Cost, Policy Research Working Paper 5280, 2010.

Das Mann, R.F., J.P., Milton, and P.H. Freeman. *Ecological Principles for Economic Development*. London: Wiley, 1973.

Dassman, Raymond F. *Environmental Conservation, IUCN*. New York: John Willey and Sons, 1973.

Dassman, Raymond F. *The Lost Horizon*. New York: Collier Book, 1963.

Dassnian, Raymond F., J.P. Milton, and P.H. Freeman. *Ecological Principles for Economic Development*. London: John Wiley, 1973.

Dassman, Raymond F., F.R. Thibodeau, and H.H. Field (ed.). An Introduction to World Conservation in Sustaining Tomorrow, Hanover University 5 Press of New England, 1985.

De Hien, Haring. Agricultural Development and Environmental Protection: Some Key Issue of Potential Relevance to Pakistan. *The Pakistan Development Review*, Vol.32, no.4, Part-I, Winter 1993.

Demonte, Darryl. *Temples and Tombs? Industry vs. Environment*. New Delhi: CSE, 1985.

Department of Economic and Social Affairs. Climate Change: Technology Development and Technology Transfer. Background Paper prepared for the Beijing High-level

Conference on Climate Change: Technology Development and Technology Transfer, Beijing, 7 and 8 November 2008.

Department of Economic and Social Affairs. Climate Change: Technology Development and Technology Transfer. Background Paper prepared for the Delhi High-level Conference on Climate Change: Technology Development and Transfer, New Delhi, 22 and 23 October 2009. Available at: www.un.org/esa/dsd/dsd_aofw_cc/cc_pdfs/confl009/Background_paperDelhi_CCTT_12Oct09.pdf.

Department of Environment, Government of Assam. 2015. *Assam State Action Plan on Climate Change (2015–2020)*. Available at: www.moef.gov.in/sites/default/files/Final%20draft%20ASAPCC%20document.pdf.

Desai, Bharat H. Coming Out of Coma. *Down to Earth*, Vol.9, no.20, 15 March 2001, pp. 48–49.

Desh, Bandhu, and G. Barbaret (ed.). *Environmental Education for Conservation and Development Proceedings of the 2nd International Conference on Environmental Education*. New Delhi: Indian Environmental Society, 1987.

Desh, Bandhu, and N.L. Ramanathan (ed.). *Education for Environmental Planning and Conservation*. New Delhi: Indian Environmental Society, 1982.

Deva, Indra. Towards a More Meaningful Study of Ecology. *Society and Culture. Sociological Bulletin*, Vol.46, no.1, March 1997, pp. 1–21.

Dewdney, Daniel. The Case Against Linking Environmental Degradation and National Security. *Millennium*, Vol.19, no.3, Winter 1990, pp. 416–476.

Dewdney, Daniel. Rethinking the Link Between Environment and National Perspectives on War and Peace. *Centre for International Cooperation and Security Studies*. Madison: University of Wisconsin, Vol.7, no.2, Spring 1990, p. 12.

Di Liberto, Tom. India Heat Wave Kills Thousands. *National Oceanic and Atmospheric Administration (NOAA)*, 9 June 2015. Available at: https://www.climate.gov/news-features/event-tracker/india-heat-wavekillsthousands.

Dixon, Thomas F. Homer. Environmental Change and Economic Decline in Developing Countries. *International Studies Notes*, Vol.16, no.1, Winter 1991.

Dodd's, Felix. *Earth Summit 2002: A New Deal*. London: Earthscan Pub., 2000.

Dubey, Much Kund. India's Foreign Policy in the Evolving Global Order. *International Studies*, Vol.30, no.2, April–June 1993, pp. 117–129.

Dugarova, Esuna, and Tom Lavers. *Social Inclusion and the Post-2015 Sustainable Development Agenda*. Geneva: United Nations Research Institute for Social Development, 2014. Available at: www.unrisd.org/unitar-socialinclusion.

Durmng, A. Poverty and the Environment Reversing the Downward Spiral, World Watch Paper 92, World Watch Institute, Washington, DC, 1989.

Dutta, Ritwick, Sunita Dubey, Cohn Gonsalves, and Aparna Bhatta. *The Environmental Activists*. New Delhi: Paul's Press, Vol.11, 2000.

Dwivedi, O.P. *India's Environmental Policies, Programs and Stewardship*. London: Macmillan, 1997.

Dwivedi, O.P., and R.B. Jam. *India's Administrative State*. New Delhi: Gitanjali Publishing House, 1985.

Dwivedi, O.P., and B. Kishore. Protecting the Environment from Pollution. *Asian Survey*. September 1982, pp. 12–35.

Dwivedi, O.P., and B.N. Tiwari. *Environmental Crisis and the Hindu Religion*. New Delhi: Geetanjali Publications, 1987.

Dwivedi, O.P., and D.K. Vajpayee. *Environmental Policies in the Third World: A Comparative Analysis*. Westport, CT: Mansell Pub. Limited, 1995.

Dwivedi, Ranjit. Parks, People and Protest. The Mediating Role of Environmental Action Groups. *Sociological Bulletin*, Vol.46, no.2, September 1997, pp. 209–245.

Dworkin, Danier (ed.). *Environment and Development*. Bloomington, IN: Indiana University Press, 1974.

Eck Holm, Erik. *Down to Earth*. New York: W.W. Norton and Co., 1982.

The Economic Times. India Ranks 116 Out of 157 Nations on SDG Index. 13 July 2017. Available at: https://economictimes.indiatimes.com/news/economy/indicators/india-ranks-116-out-of-157-nations-on-sdg-index/articleshow/59582186.cms.

Ehrlich, P.R., A.H. Ehrlich, and J.P. Hoyden. *Eco Science: Population, Resource and Environment*. San Francisco: W.H. Freeman, 1977.

Ehrlich, P.R., A.H. Ehrlich, and J.P. Hoidren. *The Global Ecology*. Harper: Jovan Ovich, 1977.

Ehrlich, P.R., and R.L. Harriman. *How to Be a Survivor: A Plan to Save Spaceship Earth*. New York: Ballantine Books, 1971.

Ehrlich, P.R., and J.P. Holdren. Impact of Population Growth. New Series. *Science*, Vol.171, no.3977, 1971, pp. 1212–1217. World Bank.

El-Hinnavi, Essam (ed.). *Global Environmental Issues: United Nations Environment Programme*. Dublin: Tycooly International, 1982, pp. 20–25.

El-Hinnawi, E. *The Environmental impact of Production and Use of Energy: An Assessment Prepared by the UNEP*. Dublin: Tycooly, Int., 1981.

El-Hinnawi, E., and M.H. Hashmi (eds.). *Global Environmental Issues: UNEP*. Dublin: Tycooly Int., 1982.

Elliott, Jennifer A. *An Introduction to Sustainable Development: The Developing World*. New York: Routledge, 1994.

Environment India, *India International Centre Quarterly*, Vol.9, no.3, 4 December 1982, pp. 215–388.

Ewald, Willian R. Jr. (ed.). *Environment and Policy: The Next Fifty Years*. Bloomington, IN: Indiana University Press, 1969.

Falk, Richard A. *This Endangered Planet: Prospects and Proposals for Human Survival*. New York: Vintage Books, 1972.

Falk, Richard A. Environmental Policy as a World Order Problem. *Natural Resources Journal*, Vol.12, no.2, April 1972, pp. 161–171.

Falk, Richard A. *A Study of Future Worlds*. New York: Free Press, 1975.

Farmer, B.H. Perspectives on the Green Revolution in South Asia. *Modern Asian Studies*, Vol.20, 1986, pp. 175–199.

Farvar, Taghi M., and Milton John P. (eds.). *The Careless Technology: Ecology and International Development*. New York: Natural History Press, 1972.

Finger, Mathias. New Horizons for Peace Research: The Global Environment. In Kakonen Jyrki (ed.), *Perspectives on Environmental Conflict and International Politics*. London: Pintid Publishers, 1992, pp. 5–30.

Flavin, Christopher. *Slowing Global Warming. A Worldwide Strategy*. Washington, DC: World Watch Institute, 1989.

Food and Agriculture Organization of the United Nations. Global Forest Resources Assessment 2005: Progress Towards Sustainable Forest Management, Rome, 2006.

Food and Agriculture Organization of the United Nations. Investing in Food Security, Rome, November 2009.

Food and Agriculture Organization of the United Nations. *Land Degradation in South Asia: Its Severity, Causes and Effects Upon the People*. Rome: FAO, United Nations Development Programme and United Nations Environment Programme, 1994.

Food and Agriculture Organization of the United Nations. The State of Food Insecurity in the World 2009: Economic Crises – Impacts and Lessons Learned, Rome, 2009.

Food and Agriculture Organization of the United Nations. World Agriculture: Towards 2015/2030 Prospects for Food Nutrition, Agriculture and Major Commodity Groups – Interim Report, Rome, 2003.

Foresight. *The Future of Food and Farming: Challenges and Choices for Global Sustainability*. London: Government Office for Science, 2011.

Fox, Stephen. *John Muir and His Legacy: The American Conservation Movement*. Boston: Little Brown and Co., 1981.

Franda, Marcus. *Bangladesh. The First Decade*. New Delhi: South Asian Publishers, 1981.

Freedman, Andrew. Global Warming Is Sharply Raising the Risk of 'Unprecedented' Events. *Mashable India*, 25 April 2017. Available at: https://mashable.com/2017/04/24/global-warming-risk-unprecedented-heat-/#8LGflIVNLiqg.

Freeman, Peter H. *Large Dams and The Environment: Recommendations for Development Planning: A Report for the UN Water Conference*. Argentina: Mar Del Plats, March 1977.

Friedmann, John. *Empowerment: The Politics of Alternative Development*. Cambridge: Blackwell Publishers, 1992.

Gadgil, Madhav. Biodiversity: Time for Bold Steps. *The Hindu Survey of the Environment*, 1992, pp. 21–23.

Gadgil, Madhav. Conservation: Taking Care of All Life. *Hindu Survey of the Environment*, 2000.

Gadgil, Madhav. Conserving India's Bio-diversity: The Human Context. In T.N. Kho shoo and Manju Sharma (eds.), *Sustainable Management of Natural Resources*. New Delhi: Malhotra Publishing House, 1992, pp. 243–255.

Gadgil, Madhav. Conserving India's Biodiversity: The Societal Context. *Evolutionary Trends in Plants*, Vol.5, no.1, 1991, pp. 3–8.

Gadgil, Madhav. *Deforestation: Problems and Prospects*. New Delhi: Society for Promotion of Wastelands Development, 1989.

Gadgil, Madhav, and Ramachandra Guha. Towards a Perspective on Environmental Movements in India. *The Indian Journal of Social Work*, Vol.59, no.1, January 1998, pp. 450–472.

Galbraith, J.K. *The Affluent Society*. New York: Houghton Mifflin Company, 1958 (First edition).

Gallagher, Kelly Sims et al. Harnessing Energy: Technology Innovation in Developing Countries to Achieve Sustainable Prosperity. Background Paper prepared for World Economic and Social Survey, 2011.

Galtung, Johan. International Development. In Roger A. Coates and Jerrell A. Rosati (eds.), *The Power of Human Needs in World Society*. London: Lynne Reiner Publishers, 1988.

Gandhi, Indira. Man, and Environment (Plenary, Session of UNCHE, 14 June 1972). In *DOE, Indira Gandhi on Environment*. New Delhi: DOE, 1984, pp. 20–29.

Gandhi, Indira. *Safeguarding Environment*. New Delhi: Wiley Eastern Int., 1992.

Gandhi, Indira. Speech at the Stockholm Conference, Man and Environment. Plenary Session of United Nations Conference on Human Environment, 14 June 1972. Stockholm, Sweden.

Gann, Narottam. *Environment and National Security. The Case of South Asia*. New Delhi: South Asian Publishers, 1999–2000.

Gareth, Porter. Environmental Security as a National Security Issue. *Current History*, May 1995.

Garner, Robert. *Environmental Politic*. London: Prentice Hall, 1996.

Gayawali, Dipak. Water in Nepal: An Interdisciplinary Look at Resource Uncertainties Evolving Problems and Future Prospects. East West Centre. Environment and Policy Institute, Occasional Paper no.8, 1989.

Geetha, Krishnan. Sustainable Development in Operation. In Malcolm S. Adiseshiah (ed.). *Sustainable Development – Its Content, Scope and Prices*. New Delhi: Lancer International in association with India International Centre, 1990, pp. 7–21.

General Economic Division, Bangladesh Planning Commission, Government of the People's Republic of Bangladesh. 2015. *Millennium Development Goals Bangladesh Progress Report 2015*. Bangladesh. Available at: www.bd. undp.org/content/Bangladesh/en/home/library/mdg/mdg-progress report-2015.html.

George, A. *Ecology as Politics*. London: Pluto Press, 1980.

George, Jose. Development and Environmental Hazards. *Law and Environment*, 1984, pp. 224–278.

George, P. Castille, and Gilbert Kushner (eds.). *Persistent Peoples Cultural Enclaves in Perspective*. Tuson: University of Arizona Press, 1981.

Ghai, Dharam, and Jessica M. Vivian (eds.). *Grassroots Environmental Action: People's Participation in Sustainable Development*. London: Routledge, 1992.

Ghose, Sisir Kumar. *Mediation on Matricides: Some Ecological Essays*. New Delhi: Ajanta Publications, 1986.

Ghosh, Partha S. *Cooperation and Conflict in South Asia*. New Delhi: Manohar, 1989.

Gland Islam, M.R., and M. Hasan. Climate-Induced Human Displacement: A Case Study of Cyclone Aila in the South-West Coastal Region of Bangladesh. *Natural Hazards*, Vol.81, no.2, 2016, pp. 1051–1071.

Glasbergen, Pieter, and Andrew Blowers. *Environmental Policy in an International Context: Perspectives on Environmental Problems*. London: Arnold, 1995.

The Global Environment Facility. Cover Story. *Our Planet*, Vol.3, no.3, 1991, pp. 10–13.

Goldsmith, Edward (eds.). *A Blueprint for Survival*. Boston: Houghton Mifflin, 1972.

Goldsmith, Edward, and N. Hildyard. *The Social and Environmental Effects of Large Dams, An Overview*. Cornwall: Wadebridge Ecological Centre, 1984.

Goodland, Robert, and Herman Daly. Environmental Sustainability: Universal and Non-Negotiable. *Ecological Applications*, Vol.6, no.4, 1995, pp. 1002–1017.

Gopinathan, S. *The Struggle Against Pollution*. Singapore: Federas, 1975.

Gore, Albert. Strategic Environmental Initiative (SEI). *SAIS Review*, Vol.10, 1990, pp. 59–71.

Gormley, W. Paul, *Human Right: And Environment. The Need for International Co-operation*. Netherland: Lydon, 1976.

Gosovic, Branislav. *The Quest for World Environmental Cooperation: The Case of the UN Global Environment Monitoring System*. London: Routledge, 1992.

Goudie, A.S. *Environmental Changes*. Oxford: Clarendon Press, 1977.

Goyal, A.K., and Sujata Arora. India's Fourth National Report to the Convention on Biological Diversity, 18–19. Ministry of Environment and Forests Government of India, 2009. Available at: www.moef.nic.in/sites/default/files/India_Fourth_National_Report-FINAL_2.pdf.

Graham, Frank. *Since Silent Spring*. Boston: Houghton Mifflin, 1970.

Greenpeace. Blue Alert, Climate Migrants in South Asia: Estimates and Solutions – A Report by Greenpeace, 2008. Executive Summary. Available at: www.greenpeace.at/uploads/media/blue-alert- report_web.pdf.

Greenpeace International. 2008. *Annual Report 2008*. Amsterdam: Greenpeace International.

Griffith-Jones, Stephany, and Krishnan Sharma. Climate Change 2007: Impacts, Adaptation and Vulnerability – Contribution of Working Group II to the Fourth Assessment Report of the Intergovernmental Panel on Climate Change, M.L. Parry and others, eds. Cambridge, United Kingdom, Cambridge University Press, 2007.

Griffith-Jones, Stephany, and Krishnan Sharma. Climate Change 2007: Synthesis Report. Geneva, 2007.

222 Bibliography

Griffith-Jones, Stephany, and Krishnan Sharma. GDP-Indexed Bonds: Making It Happen. DESA Working Paper, No.21, April 2006, New York, Department of Economic and Social Affairs of the United Nations Intergovernmental Panel on Climate Change (2001). Climate Change 2007: Mitigation – Contribution of Working Group III to the Fourth Assessment Report of the Intergovernmental Panel on Climate Change, B. Metz and others, eds. Cambridge, United Kingdom, Cambridge University Press. Available at: www.grida.no/publications/other/ipcc_tar/?src=/climate/ipcc_tar/.

Griffith-Jones, Stephany, and Krishnan Sharma. Managing the Risks of Extreme Events and Disasters to Advance Climate Change Adaptation. Scoping Paper – IPCC Special Report, Submitted to IPCC at Its Thirtieth Session, Antalya, Turkey, 21–23 April 2009. Available at: www.ipcc.ch/meetings/session30/doc14.pdf.

Grubb, Michael. *Green House Effect: Negotiating Targets*. London: Royal Institute of International Affairs, 1989.

Grubb, Michael, Koch Matthias, Koy Thomson, Munson Abby, and Francis Sullivan. *The 'Earth Summit' Agreements: A Guide and Assessments. An Analysis of the Rio 92 UN Conference on Environment and Development*. London: Routledge, 1993.

Guha, Ramachandra. Ecological Roots of Development Crisis. *Economic and Political Weekly*, Vol.21, no.15, 12 April 1986, pp. 623–626.

Guha, Ramachandra. *The Unquiet Woods: Ecological. Change and Peasant Resistance in the Himalaya*. New Delhi: Oxford University Press, 1989.

Gundlach, Erich R. Oil Tanker Disasters. *Environment*, Vol.19, 9 December 1977, pp. 16–27.

Güney, Taner. Governance and Sustainable Development: How Effective Is Governance? *The Journal of International Trade & Economic Development*, Vol.26, no.3, 2017, pp. 316–335.

Gupta, Avijit. *Ecology and Development in the Third World*. London: Routledge, 1988.

Gupta, Joyeeta. *The Climate Change Convention in Developing Countries: From Conflict to Consensus* (Environment & Policy Volumes). London: Kluwer Academic Publishers, 1997.

Gupta, S.P. *Planning and Development in India: A Critique*. New Delhi: Allied Publication, 1989.

Haider, A., M. Ud Din, and E. Ghani. Consequences of Political Instability, Governance and Bureaucratic Corruption on Inflation and Growth: The Case of Pakistan. *The Pakistan Development Review*, Vol.50, no.4, 2011, pp. 773–807. Available at: www.jstor.org/stable/23617732.

Haier, Marten A. *The Politics of Environmental Discourse. Ecological Modernization and the Policy Process*. Oxford: Clarendon Press, 1995.

Haley, Mary Jean (ed.). *Open Options: A Guide to Stockholm's Alternative Environmental Conferences*. London: Nicholas Brealey Publishing, 1972.

Halim, M.A. Sustainable Development and Bangladesh. *Bangladesh Economic Association*. Available at: http://bea-bd.org/site/images/pdf/044. pdf.

Hallegate, Stephane, Colin Green, Robert J. Nicholls, and Jan Corfee-Morlot. Future Flood Losses in Major Coastal Cities. *Nature Climate Change*, Vol.3, 2013, pp. 802–806.

Hambro, E. The Human Environment: Stockholm and After. *Yearbook of World Affairs*, 1974, pp. 200–210.

Hamilton, D. *Technology, Man and Environment*. London: Faber and Faber, 1973.

Hamilton, Lawrence S. What Are the Impacts of Himalayan Deforestation on the Ganges-Brahmaputra Low Lands and Deltas? Assumption and Facts. *Mountain Research and Development* (Kathmandu), Vol.7, no.3, August 1987.

Hansen, Roger D. *Beyond the North-South Stalemate*. London: McGraw Hill Book Company, 1979.

Hanumanth Rao, C.H. *Agricultural Growth, Rural Poverty and Environmental Degradation in India.* New Delhi: Oxford University Press, 1994.

Haqq, Mahbub U. *The Poverty Curtain: Choices for the Third World.* New York: Columbia University Press, 1976.

Haque, Masroora, and S. Huq. Bangladesh and the Global Climate Debate. *Current History*, 2015, pp. 144–148. Available at: www.icccad.net/wp-content/uploads/2015/12/Current-History-Bangladesh-and-the-GlobalClimate-Debate-Haque-Huq.pdf.

Haque, M. Shamsul. The Fate of Sustainable Development Under Neo-Liberal Regimes in Developing Countries. *International Political Science Review*, Vol.20, no.2, 1999, pp. 197–218.

Hardin, Charles M., and Stuart S. Nagel (ed.). Observations on Environmental Politics. In *Environmental Politics.* London: Prager, 1974.

Hardy, Michael. The United Nations Environment Programme. *Natural Resources Journal*, 13 April 1973, pp. 235–255.

Hare, F. Kenneth. Climate, Drought and Desertification. *Nature and Resources*, Vol.20, no.1, January–March 1984.

Hargrove, John Lawrence (ed.). *Law, Institutions and the Global Environment.* New York: Dubs Ferry, 1972.

Harrison, Paul. *The Third Revolution: Environment, Population and a Sustainable World.* New York: Tauris and Co. Ltd., 1992.

Harvey, Brian, and John D. Hallett. *Environment and Society. An Introductory Analysis.* London: The Macmillan Press, 1977.

Hassan, Shaukat. Problems of Internal Stability in South Asia, PSIS Occasional Paper no.1, 1988, Geneva, The Graduate Institute of International. Studies, June 1988.

Hassani-Mahmooei, B., and B.W. Parris. Climate Change and Internal Migration Patterns in Bangladesh: An Agent-Based Model. *Environment and Development Economics*, Vol.17, no.6, 2012, pp. 763–780.

Hays, Samuel P. *Beauty, Health, and Permanence: Environmental Politics in the United States, 1955–85.* Cambridge: Cambridge University Press, 1987.

Hazarika, Sanjoy. Bangladesh and Assam: Land Pressures, Migration and Ethnic Conflict, Occasional Paper Series of the Project on Environmental Change and Acute Conflict, a Joint Project of the University of Toronto and the American Academy of Arts and Sciences, no.3, March 1993.

Hazarika, Sanjoy. *Bhopal: The Lessons of a Tragedy.* New Delhi: Penguin Books (India) Pvt. Ltd., 1987.

Hazarika, Sanjoy. India Admits Failure to Cut Bangladesh Influx. *New York Times*, 16 December 1992.

Herely, Michael. *United Nations Environment Programme.* Toronto: Dubious Bros. Inc., 1975.

Hillary, Edmund (ed.). *Ecology 2C(X).: The Changing Faces of the World.* London: Michael Joseph Pub., 1984.

Hinnawi, H.L.E., and H. Manzurul Haq (eds.). *Global Environmental Issues and UNEP.* Dublin: Tycooly Int., 1982.

Hoekstra, A.Y., and T.O. Wiedemann. 2014. Humanity's Unsustainable Environmental Footprint. *Science*, Vol.344, no.6188, pp. 1114–1117.

Holdden, Susan O. Statues and Standards for Pollution Control in India. *EPW*, Vol.XXII, no.16, 18 April 1987.

Holdgate, Martin W., M. Kassas, and G.F. White (eds.). *The Issues of Environmental Protection.* Dublin: Tycooly Int., 1985.

Holdgate, Martin W., M. Kassas, and G.F. White (eds.). *The World Environment 1972–82: A Report by UNEP.* Dublin: Tycooly Int., 1982.

Hossain, Hamida. *From Crisis to Development: Coping with Disasters in Bangladesh.* Bangladesh: University Press Limited, 1992.

Howell, John M. An Approach to the Development of International Jurisdiction to Deal with Environmental Problems. *Asian Perspectives*, Vol.5, no.2, Fall–Winter 2018, 218–246.

Hubrich, Harold W., Jr. *The Environmental Crisis.* New Haven: Yale University Press, 1990.

Huda, M.N., and J.U. Choudhury. Floods and Erosion. Paper presented at a Regional Conference on Floods and Erosion, Dhaka, 7–10 September 1989.

Hurrell, Andrew, and Benedict Kingsbury (eds.). *The International Politics of the Environment.* Oxford: Clarendon Press, 1992.

Hurries, A. Radical Economics and Natural Resources. *International Journal of Environmental Studies*, Vol.21, 1983, pp. 43–53.

Hussain, Akmal. *Strategic Issues in Pakistan's Economic Policy, Introductory Essay: Is Pakistan Growth Plan Sustainable?* Lahore: Progressive Publishers, 1988.

Hussain, Akmal. *Resource Depletion and Sustainable Development in South Asia.* Asia Panel Meetings of IUCN. Perth-Australia: The World Conservation Union, 26–27 November 1990.

Imber, Mark. *Environmental Security and UN Reform.* Great Britain: Macmillan, 1992.

Indian Green Building Council. *About us- Green Building Movement in India.* 2015. Available at: https://igbc.in/igbc/redirectHtml.htm?redVal=showAboutusnosign&id.

Inglehart, Ronald. *The Silent Revolution: Changing Values and Political Styles Among Western Publics.* Princeton: Princeton University Press, 1977.

Intergovernmental Panel on Climate Change (IPCC). *Climate Change 2014: Impacts, Adaptation, and Vulnerability.* Contribution of Working Group II to the Fifth Assessment Report of the Intergovernmental Panel on Climate Change. Cambridge and New York: Cambridge University Press, 2014.

Intergovernmental Panel on Climate Change (IPCC). *Climate Change 2014: Synthesis Report.* Contribution of Working Groups I, II and III to the Fifth Assessment Report of the Intergovernmental Panel on Climate Change (IPCC), eds. Core Writing Team, R.K. Pachauri, and L.A. Meyer, 2. Geneva: IPCC, 2015.

International Conference on Global Warming and Climate Change: African Perspectives. *The Nairobi Declaration on Climatic Change.* Nairobi: UNEP, 1990.

Islam, A., R. Shaw, and F. Mallick. 2013. National Adaptation Program of Action. In R. Shaw, F. Mallick, and A. Islam (eds.), *Climate Change Adaptation Actions in Bangladesh.* Tokyo, Heidelberg, New York, Dordrecht and London: Springer.

Islam, Muinul. Natural Calamities and Environment Refugees in Bangladesh. *Refugee*, Vol.12, no.1, June 1992.

IUCN, Species Survival Commission. 2004 Red List of Threatened Species: A Global Species Assessment, J.E.M. Baillie and others. Gland, Switzerland, 2004.

Jackson, Tim. Philosophical and Social Transformations Necessary for the Green Economy. Background Paper prepared for World Economic and Social Survey 2011, 2010.

Jackson, Tim. *Prosperity Without Growth: Economics for a Finite Planet.* London: Earthscan, 2009.

Jackson, Tim. *Prosperity Without Growth? The Transition to a Sustainable Economy.* London: Sustainable Development Commission, 2009.

Jain, R.B., and Renu Khatun. *Bureaucracy, Citizen Interface: Conflict and Consensus.* New Delhi: B.R. Publishing, 1999.

Jam, Renu. The Road from Rio. *Development Alternatives*, Vol.3, no.1, 1993, pp. 1, 3.

Jamal, Amir H. The Socio-Economic Impact of New Biotechnologies in the Third World. In Carry Fowler et al. (eds.), *The Laws of Life: Another Development and the New Biotechnologies, Development Dialogue*, no.1–2, 1988, pp. 5–8.

Jha, Veena, Hewson Grant, and Underhill Maree. *Trade Environment and Sustainable Development: A South Asian Perspective*. London: Macmillan Press, 1997.

Johnson, R.J. *Environmental Problems: Nature, Economy and State*. London: Belhaven Press, 1989.

Johnson, Stanley. *The Politics of Environment: The British Experience*. London: Tom Stacey, 1973.

Johnston, R.J. *Environmental Problems: Nature, Economy and State*. New York: Belhaven Press, 1989.

Kabraji, Aman Marker, and Anil Agarwal. *Beyond Shifting Sands: The Environment in India and Pakistan*. New Delhi: IUCN and CSE, May 1994.

Kahn, M. Concepts, Definitions, and Key Issues in Sustainable Development: The Outlook for the Future. Proceedings of the 1995 International Sustainable Development Research Conference, Manchester, 27–28 March 1995, Keynote Paper, 2–13.

Kattamuri, R. *Sustainable Development: The Goals and the Challenges Ahead*. London: London School of Economics, 2015. Available at: http://blogs.lse.ac.uk/southasia/2015/09/25/sustainabledevelopmentthegoalsandthechallenges.

Kay, David A., and Harold K. Jacobson (ed.). *Environmental Protection: The International Dimension*. Totowa: Allanheld, Osmun & Co, 1983.

Kemal, A.R. et al. *Sustainable Development in Pakistan with a Focus on Environmental Issues*. Islamabad: Pakistan Institute of Developmental Economics, 1990.

Khan, Humidor Rahman. Water Resources Development in Bangladesh: Problems and Prospects. Paper presented at a Regional Seminar on Water Resources Policy in Agro-Socio-Economic Development, Dacca, 4–5 August 1985.

Khan, Rahmatullah, and Bharat Desai. *Environmental Law Series, Environmental Laws of India*. New Delhi: Lancers, 1994.

Khanna, G.N. *Environmental Problems and the United Nations*. New Delhi: Ashish, 1990.

Khanna, G.N. Environmental Protection Act: A Critical Analysis. *Indian CI Journal of Social Work*, Vol.LI, no.1, 1990, pp. 181–194.

Khanna, G.N. Problems of Human Environment: People's Participation and Legal Solution. *Social Action*, Vol.39, July–September 1989, pp. 275–297.

Khanna, G.N. United Nations Environment Programme: Assessing Its Implementation with Special Reference to India. JNU Ph.D. Thesis, 1992.

Khanna, K.P., and Ashok Modi. Environment and Development. *EPWP*, Vol.18, 1983, pp. 638–642.

Khatun, Renu. Organizational Response to the Environmental Crisis in India. *Indian Journal of Political Science*, Vol.49, no.1, January–March 1989, pp. 20–30.

Khor, Martin. Challenges of the Green Economy Concept and Policies in the Context of Sustainable Development, Poverty and Equity. In United Nations, United Nations Environment Programme and United Nations Conference on Trade and Development, The Transition to a Green Economy: Benefits, Challenges and Risks from a Sustainable Development Perspective. Report by a Panel of Experts to the Second Preparatory Committee Meeting for the United Nations Conference on Sustainable Development, New York, 7 and 8 March 2011b, pp. 68–96.

Khor, Martin. The Climate and Trade Relation: Some Issues. South Centre Research Paper, No.29, May 2010, Geneva, South Centre.

226 Bibliography

Khor, Martin. Global Debate on Green Economy. Star online (Peta ling Jaya, Malaysia). 24 January 2011a. Available at: http://thestar.com.my/columnists/story.asp?col=globaltrends&file=/2011/1/24/columnists/global trends/7856802&sec=Global%20Trends.

Khoshoo, T.N. *Environmental Concerns and Strategies*. New Delhi: Ashish Publishing House, 1988.

Khoshoo, T.N. *Environmental Concerns and Strategies*. New Delhi: Times Press, 1984.

Khushoo, T.N. *Environmental Protection in India and Sustainable Development*. New Delhi: Ashish, 1986.

Kiessling, Kerstin, and Nathan Key Fitz. The Population Debate: Urgency of the Problem. Paper presented at the Population Summit of the World's Scientific Academies Held in New Delhi from 24–27 October 1993.

Kirdar, Uner (ed.). *Ecological Change: Environment, Development and Poverty Linkages*. New York: United Nations, 1992.

Kneeze, Allen V. *Economics and the Environment*. Harmondsworth: Penguin Books, 1997.

Kothari, Ashish. India's Forests: Can Money Save Them? *Hindu Survey of the Environment*, 2000.

Kothari, Ashish. Politics of Biodiversity Convention. *Economic and Political Weekly*, Vol.XXVII, Nos.15–16, 11–18 April 1992, pp. 749–755.

Kothari, Ashish, Nena Singh, and Salone Sun (eds.). *People and Protected Areas: Towards Participatory Conservation in India*. New Delhi: Sage Publications, 1996.

Kothari, Rajni. *Footsteps into the Future: Diagnosis of the Present World and a Design for the Alternative*. New York: The Free Press, 1974.

Kothari, Rajni. *Transformation and Survival: In Search of Human World Order*. London: Aspect Publications, 1990.

Kothari, Smith. Ecology Versus Development: The Struggle for Survival. *Social Action*, Vol.35, 4 October–December 1985, pp. 146–153.

Kroll, C. Poor Implementation: Rich Countries Risk Achievement of the Global Goals, SDG-Index. *Bertelsmann Stiftung*, 2018. Available at: www. bertelsmann-stiftung.de/en/topics/aktuelle-meldungen/2018/juli/poorimplementation-rich-countries-risk-achievement-of-the-global-goals.

Kroll, C., and R. Schwarz. Ahead of G20 Summit: 'My Country First' Approach Threatens Achievement of Global Goals. *Bertelsmann Stiftung*, 6 July 2017. Available at: www.google.co.in/search?q=rising+trend+of+national ism+and+protectionism+impede+the+implementation+of+the+goals+SDG+Index&aqs=chrome.69i57.7584j1j 7&sourceid.

Kulz, Helmut. Future Water Disputes Between India and Pakistan. *International and Comparative Law Quarterly*, Vol.18, July 1969, pp. 718–738.

Kumar, R. *Environmental Pollution and Health Hazards in India*. New Delhi: Ashish, 1987.

Lake, Laura M. (ed.). *Environmental Mediation: The Search for Consensus*. London: West View Press, 1980.

Lake, Laura M. *The Environmental Regulations: The Political Effects of Implementation*. New York: Prager, 1982.

Landsberg, Hans H. *Reflections on the Stockholm Conference*. Unpublished Paper. Washington, DC, 1972.

Lashoff, Daniel A., and Dennis A. Terpak (eds.). *Policy Options for Stabilizing Global Climate*. London: Hemisphere Publishing Corporation, 1990.

Laurent, Jaffrey, and Layman Francisco. *One Earth, Many Nations, the International System and the Problem of the Global Environment*. The Sierra Club: UNAUS, 1989.

Lavasa, Ashok. Foreword by the Secretary, Ministry of Environment, Forest and Climate Change. Government of India in India- First Biennial Update Report to the United

Nations Framework Convention on Climate Change, Ministry of Environment, Forest and Climate Change, Government of India (6–7), 2015.

Lee, James A. *Environment, Public Health and Human Ecology: Consideration for the Ecodevelopment.* Baltimore: John Hopkins University Press, 1985.

Leggett, Jeremy (ed.). *Global Warming: The Greenpeace Report.* Oxford: Oxford University Press, 1990.

Leggett, Jeremy. Global Warming: A Greenpeace View. In Jeremy Leggett (ed.), *Global Warming: The Greenpeace Report.* Oxford: Oxford University Press, 1990, pp. 457–480.

Lele, S.M. Sustainable Development: A Critical Review. *World Development*, Vol.19, no.6, 1991, pp. 607–621.

Levin, A.L. *Protecting the Human Environment: Procedures and Principles for Prevention and Resolving International Controversies.* New York: UNITAR, 1977.

London School of Hygiene & Tropical Medicine. Good Health at Low Cost. 25 Years on What Makes an Effective Health System? 2011. Available at: http://blogs.lshtm.ac.uk/ghlc/files/2011/12/Policy-Briefing-No1-Bangladesh.pdf.

Lowe, Philip, and Jones Goyder. *Environmental Groups in Politics.* London: George Allen and Unwin, 1983.

Madhukar, Uday. Battle Royal. *India Today*, 31 January 1991, pp. 66–68.

Mahtab, F.U. *Effects of Climate Change and Sea Level Rise in Bangladesh.* London: Commonwealth Secretariat, 1989.

Malhotra, Inder. *Indira Gandhi: A Personal and Political Biography.* London: Hodder and Stoughton, 1989.

Malviya, R.A. *Environmental Pollution and Its Control Under International Law.* Allahabad: Chough Publication, 1987.

Manzar, L.E. The CFC-Ozone Issue: Progress on the Development of Alternatives to CFCs. *Science*, Vol.249, 6 July 1990, pp. 31–35.

Marcus, Alfred A. *Promise and Performance Choosing and Implementing Environmental Policy.* West-Port Country: Greenwood Press, 1980.

Mathur, Deepa. Some thoughts on Women, Environment and Development. *Man, and Development*, Vol.XIX, no.4, December 1997, pp. 79–83.

Mazur, Ann (ed.). *Beyond the Numbers: A Reader on Population, Consumption and the Environment.* Washington, DC: Island Press, 1994.

McCombie, L. *The Quality of the Environment.* New York: Free Press, 1972.

McConnell, Grant. The Conservation Movement: Past and Present. *Western Political Quarterly*, Vol.7, 3 September 1954, pp. 463–478.

McCormick, John. *Acid Earth: The Global Threat of Acid Pollution.* London: Earthscan, 1985.

McCormick, John. *International Environmental Movements: Reclaiming Paradise.* London: Belhaven, 1989.

McNeely, J., and D. Pitt (eds.). *Culture and Conservation: The Human Dimensions in the Environmental Planning.* London: Croom Helm, 1985.

McPherson, Poppy. Dhaka: The City Where Climate Refugees Are Already a Reality. *The Guardian*, 1 December 2015. Available at: www.theguardian.com/cities/2015/dec/01/Dhaka-city-climate-refugees-reality.

Meadows, D.H. et al. *The Limits to Growth.* London: Earth Island Ltd., 1972.

Mehta, M.C. *Environmental Cases: What the Judiciary Can Do.* The Hindu Survey of the Environment. The Hindu. Chennai: Kasturi & Co, 1992, pp. 161–163.

Mehta, Simi et al. Lessons in Sustainable Development from Bangladesh and India, Comparative Studies of Sustainable Development in Asia, 2018. https://doi.org/10.1007/978-3-319-95483-7.

Mehta, Simi, Vikash Kumar, and Rattan Lal. Climate Change and Food Security in South Asia. In Sara Hsu (ed.), *Routledge Handbook of Sustainable Development in Asia*. New York: Routledge, 2018.

Meleshkin, T.M. *The Economy and the Environment: Interaction and Management*. Moscow: Economika Publications, 1979.

Middleton, Nick. *The Global Casino: An Introduction to Environmental Issues*. London: Edward Arnold, 1995.

Milisup, William. *Applied Social Sciences for Environmental Planning*. London: West View Press, 1987.

Miller, Marion A.L. *The Third World in Global Environmental Politics*. Buckingham: Open University Press, 1995.

Ministry of Environment, Forest and Climate Change, Government of India. India: First Biennial Update Report to the United Nations Framework Convention on Climate Change, 2015.

Ministry of Statistics and Program Implementation. Millennium Development Goals- Final Country Report of India. New Delhi, 2017. Available at: www. mospi.gov.in/sites/default/files/publication_reports/MDG_Final_Country_ report_of_India_27nov17.pdf.

Mishra, R.P. *Environmental Ethics: A Dialogue of Culture*. New Delhi: Sustainable Development Foundation and Gandhi Bhavan, Delhi University, 1992.

Mitra, A.P. (ed.). *Greenhouse Gas Emissions in India. A Preliminary Report*. New Delhi: CSIR, 1991.

Mitra, A.P. Status and Policy Implications of Global Change: The Indian Scene. In S. Gupta and R.K. Pachauri (eds.), *Proceedings of the International Conference on Global Warming and Climate Change*. New Delhi: Tata Energy Research Institute, 1989.

Mohan, I. (ed.). Environmental pollution and Management. *New World Environment Series*, Vol.1 to 4. New Delhi: Ashish, 1989.

Mohan, I. *The Fragile Environment*. New Delhi: Ashish, 1991.

Monga, Pradeep, and P. Venkata Ramana (ed.). *Energy, Environment and Sustainable Development in the Himalayas*. New Delhi: Indus Publ. Company, 1992.

Montiel, Lenni. Social Inclusion in the Age of Sustainable Development: Who Is Left Behind? *UN DESA News*, 5 October 2016. Available at: www. un.org/development/desa/en/news/social/who-is-left-behind.html.

Mortimer, Robert A. *The Third World Coalition in International Politics*. New York: Praeger Publishers, 1980.

Mundanthra, Balakrishnan. *Environmental Problems and Prospects in India*. New Delhi: Oxford University Press, 1993.

Murthy, C.S.R. 'Reforming the UN' in the World Focus, 'United Nations at Fifty: Needs Restructuring'. New Delhi, Vol.16, no.9, September 1995.

Murti, C.R. Krishna. Environmental Challenges in Developing Society. *Gandhi Marg*, Vol.1, 1979, pp. 142–148.

Myers, Norman. The Environmental Dimension to Security Issues. *The Environmentalist*, Vol.6, no.4, Winter 1986.

Myers, Norman. Environment and Security. *Foreign Policy*, no.74, Spring 1989, pp. 23–41.

Myers, Norman. *Ultimate Security: The Environmental Basis of Political Stability*. New York: W.W. Norton, 1993.

Myrdal, G. *Asian Drama: An Enquiry into the Poverty of Nations*, 3 vols. New York: Pantheon, 1980.

Nag Choudhury, B.D., and S. Bhatt. *The Global Environmental Movement: A New Hope for Mankind*. New Delhi: Sterling, 1987.

Naser, M.M. Climate Change, Environmental Degradation, and Migration: A Complex Nexus. *William & Mary Environmental Law and Policy Review*, Vol.36, no.3, 2011, pp. 712–768.

Nath, Anita. India's Progress Toward Achieving the Millennium Development Goals. *Indian Journal of Community Medicine*, Vol.36, no.2, 2011, pp. 85–92. Available at: www.ncbi.nlm.nih.gov/pmc/articles/PMC3180952/.

Nath, D.K. Sustainable Development Goals: Challenges for Bangladesh. *Daily Sun*, 14 September 2015. Available at: www.daily-sun.com/post/75877/Sustainable-Development-Goals: -Challenges-for-Bangladesh.

National Sustainable Development Strategy (NSDS). Planning Commission. Government of Bangladesh. 2013. Available at: www.plancomm.gov.bd/wpcontent/uploads/2013/09/National-Sustainable development strategy.pdf.

Negroponte, John D. Protecting the Ozone Layer. *Department of State Bulletin*, Washington, Vol.87, no.2123, June 1987, pp. 58–60.

Nicholas Stern: Cost of Global Warming 'Is Worse Than I Feared'. Interviewed by Robin McKie. *The Guardian*, November 6. Available at: The Guardian, 6 November 2016. Available at: www.theguardian.com/environment/2016/nov/06/nicholas-stern climate-change-review-10-years-on-interview-decisive-years humanity.

Nicholson, Max. *The Environmental Revolution: A Guide for the New Masters of the World*. London: Holder and Stoughton, 1970.

Nicholson, Max. *The New Environmental Age*. Cambridge: Cambridge University Press, 1987.

Ocampo, José Antonio. The Macroeconomics of the Green Economy. In United Nations, United Nations Environment Programme and United Nations Conference on Trade and Development, the Transition to a Green Economy: Benefits, Challenges and Risks from a Sustainable Development Perspective. Report by a Panel of Experts to the Second Preparatory Committee Meeting for the United Nations Conference on Sustainable Development, New York, 7 and 8 March 2011, pp. 14–38.

Ocampo, José Antonio. Summary of Background Papers. In United Nations, United Nations Environment Programme and United Nations Conference on Trade and Development, the Transition to a Green Economy: Benefits, Challenges and Risks from a Sustainable Development Perspective. Report by a Panel of Experts to the Second Preparatory Committee Meeting for the United Nations Conference on Sustainable Development, New York, 7 and 8 March 2011, pp. 1–14.

Oldenburg, Philip (ed.). *India Briefing*. Boulder: Westview Press, 1993.

Oliver, S. Owen. *Natural Resource Conservation: An Ecological Approach*. New York: Macmillan, 1971.

Osborn, D., A. Cutter, and F. Ullah. Universal Sustainable Development Goals, Understanding the Transformational Challenge for Developed Countries. Report of a Study by Stakeholder Forum. 2015. Available at: https://sustainabledevelopment.un.org/content/documents/1684SF_-_SDG_Universality_ Report_-_May_2015.pdf.

Oxfam India. Richest 1 Percent Bagged 73 Percent of Wealth Created Last Year – Poorest Half of India Got 1 Percent Says. *Oxfam India*, 22 January 2018. Available at: www.oxfamindia.org/pressrelease/2093.

Oxley, N. *5 Challenges for Least Developed Countries in the Post-2015 Era*. 2016. Available at: http://stepscentre.org/2016/blog/5challengesforleastdevelope dcountriesinthepost2015era/

Panchmukhi, V.R., and Nagesh Kumar (eds.). *Biotechnology Revolution and the Third World*. New Delhi: Research and Information System for the Non-Aligned and Other Developing Countries, 1988.

Pandaval Guna Nath, and Dinesh L. Shrestha. On the Choice of the Chispani Project: A Case of Inadequate Planning. Kathmandu Meeting, Cooperative Development of Himalayan Water Resources, Kathmandu, 27–28 February 1993.

Papadakis, Elim. The Green Party in Contemporary West German Politics. *Political Quarterly*, Vol.54, no.3, July–September 1983, pp. 302–307.

Parikh, Jyoti, and Kirti Parikh. Role of Unsustainable Consumption Patterns and Population in Global Environmental Stress. *Sustainable Development*, Vol.1, no.1, October 1991, pp. 108–118.

Parsons, Hayward L. (ed.). *Marx and Engel's on Ecology*. West-Port Country: Greenwood Press, 1977.

Passmore, John. *Man's Responsibility for Nature: Ecological Problems and Western Traditions* (2nd ed.) London: Hackworth, 1980.

Patel, H.M. *Policy for the National Conservation*. New Delhi: S. Chand and Co., 1980.

Pearce, David. Anil Markandeya and Edward Barbier. *Blueprint for a Green Economy*. London: Earthscan, 1989.

Pearce, D.W., E. Barbier, and A. Markandya. *Sustainable Development: Economics and Environment in the Third World*. Aldershot: Elgar, 1990.

Pearce, D.W., A. Markandya, and A. Barbier. *Blueprint for a Green Economy*. London: Earthscan, 1990.

Pearce, D.W., and M. Redclift (eds.). Sustainable Development. *Futures*, 20 Special Issue, 1988.

Pearce, Fred. War Over Water. *Down to Earth*, Vol.2, no.11, 31 October 1993, pp. 25–28.

Pearson, Charles. *Environment, North and South: An Economic Interpretation*. New York: John Wiley, 1978.

Peduzzi, P., H. Dao, C. Herold, and F. Mouton. Assessing Global Exposure and Vulnerability Towards Natural Hazards: The Disaster Risk Index. *Natural Hazards and Earth System Sciences*, Vol.9, 2009, pp. 1149–1159.

Pepper, D. *The Roots of the Modern Environmentalism*. London: Crown Helm, 1984.

Permitta, John C. Impacts of Climate Change and Sea Level Rise on Small Island States. *Global Environmental Change*, Vol.2, no.1992, pp. 19–31.

Perrings, Charles. *Economy and Environment: A Theoretical Essay on the Interdependence of Economic and Environmental Systems*. Cambridge: Cambridge University Press, 1987.

Peters, R., and T. Lovejoy (ed.). *Global Warming and Biological Diversity*. New Haven: Yale University Press, 1992.

Petulla, Joseph M. *American Environmentalism: Values, Tactics, Priorities*. New York: Texas A&M University Press, 1980.

Phadnis, Urmila, S.D. Muni, and Kalim Bahadur (eds.). *Domestic Conflicts in South Asia: Political Dimensions*. New Delhi: South Asia Publishers, 1986.

Pimental, D. et al. Conserving Biological Diversity in Agricultural/Forestry Systems. *Bioscience*, Vol.42, no.5, 1992, pp. 354–363.

Pinchot, Gifford. *The Fight for Conservation*. New York: Doubleday Page & Co., 1910.

Plant, Glen. Institutional and Legal Responses to Global Environmental Change. In Ian H. Rowlands and Malroy Greene (eds.), *Global Environmental Change and International Relations*. London: Macmillan Academic and Professional Ltd., 1992, pp. 122–144.

Polyani, Karl. *The Great Transformation: The Political and Economic Origins of Our Times*. Boston, MA: Beacon Press, 1944.

Porritt, Jonathon. *Seeing Green: The Politics of Ecology Explained*. Oxford: Basil Blackwell, 1984.

Porter, Gareth, and Tanet Welsh Brown. *Global Environmental Politics*. Boulder: Westview Press, 1991.

Postell, Sandra, and Ryan John C. Reforming Forestry. In *State of the World 1991, a World Watch Institute Report on Progress Towards a Sustainable Society*. New York: W.W. Norton and Company Inc., 1991.

Prakash, Ishwar (ed.). *Desert-Ecology: Proceedings of National Symposium on Desert Ecology* (Organized by the University of Rajasthan, Jaipur and National Academy of Sciences, Allahabad). Jodhpur: Scientific Publishers, 1988.

Pretty, J.N. et al. Resource-Conserving Agriculture Increases Yields in Developing Countries. *Environmental Science and Technology*, Vol.40, no.4, 2006, pp. 1114–1119.

Pursell, Carroll (ed.). *From Conservation to Ecology: The Development of Environmental Concern*. New York: Thomas Y. Crowell Co., 1973.

Quarrie, Joyce (ed.). *United Nations Conference on Environment and Development Held at Rio*. London: Regency Press, 1992.

Quraishi, G.S. *Climate Change and Sea Level Rise in the South Asia Seas Region*. Nairobi: UNEP, 1988.

Rahman, I.A. South Asian Perspective on Human Rights and Environment. In V.A. Pal Panandiker and Navnita Chadha Behera (eds.), *Perspectives on South Asia*. New Delhi: Centre for Policy Research, Konark Publishers, 2000, pp. 417–423.

Ramakrishna, Kilaparti. The Emergence of Environmental Law in the Developing Countries: A Case Study of India. *Ecology Law Quarterly*, Vol.12, no.4, 1985, pp. 907–935.

Ramakrishna, Kilaparti. Third World Countries in the Policy Response to Global Environmental Change. In Jeremy Leggett (ed.), *Global Warming: The Greenpeace Report*. Oxford: Oxford University Press, 1990, pp. 421–437.

Ramakrishnan, P.S. Tropical Forests: Exploitation, Conservation and Management. *Environment and Development*, no.166, 1993.

Ramanathan, N.L., and Desh Bandhu. *Declaration and Recommendations of the International Conference on Environmental Education*. New Delhi: Indian Environment Society (IES), 1982.

Rana, Ratana. *Notes for a Design, Environment and Development Planning: Mountain, Environment and Development*. Kathmandu: Tribhuvan University, 1976.

Rangarajan, Mahesh. Ecology: Dilemmas and Problems. *EPW*, Vol.XXXIII, no.1, 2 January 1998, pp. 23–25.

Rao, J. Mohan. Economic Reform and Ecological Refurbishment: A Strategy for India. *EPW*, Vol.XXX, no.28, 15 July 1975.

Rao, K.L. *India's Water-Wealth*. New Delhi: Orient Longman, 1975.

Rao, P.K. *The Economics of Global Climatic Change*. Armonk: M.E. Sharpe, 2000.

Rao, R. Rama. *Environment: Problems of Developed and Developing Countries*. New Delhi: Economic and Scientific Research Foundation, 1976.

Rapoport, Anatole. *Conflict in a Manmade Environment*. Baltimore: Penguin, 1974.

Rapaport, Anatole. *Development and Environmental Crisis: Red or Green Alternative*. London: Methuen, 1984.

Rawat, Ajay S. (ed.). *Indian Forestry: A Perspective*. New Delhi: Indus Publishing Co., 1993.

Razzaque, Jona. 2002. Human Rights and the Environment: The National Experience in South Asia and Africa. Joint UNEP-OHCHR Expert Seminar on Human Rights and the Environment, 14–16 January 2002. Geneva: Background Paper No.4. Available at: http://eprints.uwe.ac.uk/18403/1/Joint%20UNEP%20-%20Razzaque.pdf.

Redcliff, Michael. Sustainable Development and Global Environmental S Change. *Global Environmental Change*, Vol.2, no.1, March 1992, pp. 32–42.

232 Bibliography

Reddiff, Michael. *Sustainable Development: Exploiting the Contradictions.* London: Methuen, 1987.

Renner, Michael. *National Security: The Economic and Environmental Dimensions.* World Watch paper 89, Washington, DC: World Watch Institute, 1989.

Rhodes, Steven L. Climate Change Management Strategies. *Global Environmental Change,* Vol.2, no.3, 1992, pp. 205–214.

Rich, Bruce. *Mortgaging the Earth: The World Bank Environmental Impoverishment and Crisis of Development.* London: Earthscan, 1994.

Riordan, Timothy. *Environmentalism.* London: Pion Ltd., 1981.

Roger, John R. *River Bank Erosion: Flood and Population Displacement in Bangladesh.* Dacca: Jahangirnagar University, 1990.

Rohrlich, George F. (ed.). *Environmental Management Economic and Social Dimensions.* Cambridge: Ballenger Pub. CO., 1978.

Roosevelt, Theodore. *An Autobiography.* New York: Macmillan, 1913.

Rosa, Fiona De Preservationism and the Place of People in the Environmental Movement *Social Alternatives.* Vol.17, no.1, January 1998, pp. 21–25.

Rose, Leo E., and John T. Scholz. *Nepal: Profile of a Himalayan Kingdom.* Boulder, CO: Westview Press, 1980.

Rosenbaum, Walter A. *Environmental Politics and Policy.* Washington, DC: CQ Press, 1985.

Rothstein, Robert L. Epitaph for a Monument to a Failed Protest? A North-South Retrospective. *International Organization,* Vol.42, no.4, Autumn 1988, pp. 725–748.

Rowlands, Ian H. The International Politics of Environment and Development: The Post-UNCED Agenda. In *The International Library of Politics and Comparative Government; The United Nations,* Vol.11. Ashgate: Dartmouth, 2000, pp. 209–224.

Rowlands, Ian H., and Malory Greene (eds.). *Global Environmental Change and International Relations.* London: Macmillan Academic and Professional Ltd., 1992.

Rudig, Wolfgang, and Philip Lowe. The Withered Greening of British Politics: A Study of the Ecology Party. *Political Studies,* Vol.34, no.2, June 1986, pp. 262–284.

Runte, Alfred. *National Parks: The American Experience.* Lincoln: University of Nebraska Press, 1979.

Runte, Alfred. *The Politics of Environmental Concern.* New York: Praeger Publications, 1973.

Ryan, Stephen. *The United Nations and International Politics. Studies in Contemporary History.* London: Routledge, 1998.

Sand Back, Francis, *Environment: Ideology & Policy.* Oxford: Blackwell, 1980.

Saward, M. Green Theory. *Environmental Politics,* Vol.2, no.3, 1993, pp. 509–512.

Saxena, K.D. *Environmental Planning, Policies and Programs in India.* New Delhi: Shipra, 1993, pp. 34–48.

Saxena, K.D. *Environmental Planning Policies and Programs in India, Annexure II.* New Delhi: Shipra Publications, 1988, pp. 201–209.

Saxena, K.P. Period of Tribulations. *World Focus,* New Delhi, Vol.16, no.9, September 1995, pp. 8–11, 11–13.

Saxena, K.P. *United Nations and Co-operation in Development. The International Library of Politics and Comparative Government; The United Nations,* Vol.11. Ashgate: Dartmouth, 2000.

Scheffran, J. Climate Change and Security in South Asia and the Himalaya Region: Challenges of Conflict and Cooperation. In S. Aneel, U.T. Haroon, and I. Niazi (eds.), *Sustainable Development in South Asia: Shaping the Future.* Islamabad: Sustainable Development Policy Institute and Sang-e Meel Publishers, 2014, pp. 439–458.

Secretariat of the Convention of Biodiversity. Communique on India Offers to Host the Eleventh Meeting of the Conference of the Parties to the Convention on Biological Diversity in 2012. United Nations Environmental Program, Canada, 2009.

Sen, Amartya. How Is India Doing? In Iqbal Khan (ed.), *Fresh Perspectives on India and Pakistan*. Oxford: Bougainvillea Books, 1985, pp. 86–96.

Sen, Gupta R. et al. State of the Marine Environment in the South Asia Seas Region. *UNEP Regional Seas Reports and Studies*, no.123, UNEP, 1990.

Serageldin, I. *Developmental Partners: Aid and Cooperation in the 1990's*. Stockholm: Swedish International Development Cooperation Agency, 1993.

Seth, Pravin. *Environmentalism: Politics, Ecology and Development*. Jaipur: Rawat Publications, 1997, pp. 55–68.

Shah, T. Climate Change and Groundwater: India's Opportunities for Mitigation and Adaptation. *Environmental Research Letters*, Vol.4, no.3, 2009, p. 035005.

Shea, Kevin P. A Celebration of Silent Spring. *Environment*, Vol.15, no.1, January–February 1973, pp. 4–5.

Shelton, David. Aborigines, Environment and Waste: A Post-Colonial Perspective. *Social Alternatives*, Vol.17, no.1, January 1998, pp. 7–10.

Sheth, Pravin. Gandhi: Eco-World View and Its Reinterpretation. *The Fourth World*, no.5, April 1995, pp. 57–72.

Sheth, Pravin. *Narmada Project: Politics of Eco-Development*. New Delhi: Har-Anand, 1994, pp. 22–26.

Shiva, Vandana. *Biodiversity: A Third World Perspective*. Pinang: Third World Network, 1991.

Shiva, Vandana. *Biotechnology and the Environment*. Pinang: Third World Network, 1991.

Shiva, Vandana. *Towards Hope: An Ecological Approach to Future*. New Delhi: Indian National Trust for Art and Cultural Heritage, 1992.

Shiva, Vandana. *The Violence of Green Revolution: Third World Agriculture, Ecology and Politics*. Penang, Malaysia: Third World Network, 1991.

Shiva, Vandana, Y.M. Meher -Hornjo, and N.D. Tayal. *Forest Resources: Crisis and Management*. Dehradun: Natraj Publishers, 1992.

Shiva, Vandana et al. *The Future of Progress: Reflections on Environment and Development*. Dehradun: Natraj Pub., 1994.

Shukla, S.P., and Nand Eshwar. *Sustainable Development Strategy: Indian Context*. New Delhi: Mittal Publications, 1996.

Sills, David L. The Environmental Movement and Its Critics. *Human Ecology*, Vol.3, no.1, 1975, pp. 1–41.

Simon, Julian L., and Herman Kohn (ed.). *The Resourceful Earth; A Response to Global 2000*. Oxford: Basil Blackwell, 1984.

Singh, Abhay. Our Ecosystem in the 21st Century. *Vision*, Vol.XVI, no.3–4, January–June 97, pp. 1–8.

Singh, L.R. *Environmental Management: Some Issues*. Allahabad: G.~B. Pant Social Science Institute, 1982.

Singh, Manmohan. *Environment and the New Economic Policies*. New Delhi: Society for Promotion of Wastelands Development, 1992.

Singh, Narinder. *Economics and the Crisis of Ecology*. New Delhi: Oxford University Press, 1976.

Singh, R.P. (ed.). *Neem and Environment*. New Delhi: Oxford University Press, 1996.

Singh, Shekhar (ed.). *Environmental Policy in India*. New Delhi: Indian Institute of Public Administration, 1984.

Smith, D., and A. Blowers. Passing the Buck: Hazardous Waste Disposal as an International Problem. *Talking Politics*, Vol.4, no.1, 1991, pp. 4–9.

Smith, Keith. *Environmental Hazards: Assessing Risk and Reducing Disaster* (2nd ed.). London: Routledge, 1996.

Smith, Tony. Changing Configurations of Power in North-South Relations Since 1945. *International Organization*, Vol.32, no.1, 1977, pp. 1–27.

Sneha, Rishikesh. Politics of Water Power in Nepal. Kathmandu Meeting on Co-operative Development of Himalayan Water Resources, Kathmandu, February 27–28, 1993.

Social Statistics Division, Central Statistics Office, Ministry of Statistics and Program Implementation, Government of India. 2015. Millennium Development Goals, India Country Report 2015. Available at: http://mospi.nic. in/sites/default/files/publication reports/mdg_2july15_1.pdf.

Soheim, Erik. UN Sees 'Worrying' Gap Between Paris Climate Pledges and Emissions Cuts Needed. *UN News Centre*, 2017. Available at: www.un.org/apps/news/story.asp? News ID=57999#.

Sohn, Lows B. The Stockholm Declaration on the Human Environment. *Harvard International Law Journal*, Vol.14, 1973, pp. 423–515.

South Asia Co-operative Environment Program. *Bangladesh: Country's Environmental Profile*. 2018. Available at: www.sacep.org/?page_id=15.

Speth, James Gustave. A Post-Rio Compact. *Foreign Policy*, no.88, Fall 1992, pp. 145–161.

Sprout, Harold, and Sprout Margaret. *The Ecological Perspective in Human Affairs with Special Reference to International Politics*. Princeton, NJ: Princeton University Press, 1965.

Stern, N. *Climate Change in South Asia: A Conversation with Sir Nicholas Stern*. 2016. Available at: www.etourisminsight.com/index2.php?option=com_eti& view=news &doc id=1588&region=&print=1.

Sterner, Thomas. *The Market and the Environment: The Effectiveness of Market-Based Policy Instruments for Environmental Reform*. Cheltenham: Edward Elgar, 1999.

Strong, Maurice F. The International Community and the Environment. *Environmental Conservation*, Vol.4, no.3, Autumn 1977, pp. 165–172.

Strong, Maurice F. *One Year After the Stockholm: An Ecological Approach to Management*. London: Earthscan, 1973.

Strong, Maurice F. The Way Ahead. *Our Planet*, Vol.8, no.5, 1997, pp. 6–8.

Swaminathan, M.S. Biodiversity: Equity in Benefit Sharing. *Hindu Survey of the Environment*, 2000.

Swaminathan, M.S. Biodiversity: Promoting Efficiency in Conservation and Equity in Utilization. RGICS (Rajiv Gandhi Institute for Contemporary Studies) Paper no.21, 1995.

Talbot, Lee M. *The World Conservation Strategy in Sustaining Tomorrow: A Strategy for World Conservation and Development*. Hanover: The University Press of New England, 1984.

Taylor, Paul, and A.J.R. Groom (eds.). *Global Issues in the United Nations Framework*. London: Macmillan, 1989.

Thakur, Kailash. *Environmental Protection Law and Policy in India*. New Delhi: Deep and Deep Publications, 1997.

Thomas, C. Beyond UNCED: An Introduction. *Environmental Politics*, Vol.2, no.4, 1993, pp. 1–27.

Thomas, C. *The Environment in International Relations*. London: The Royal Institute of International Affairs, 1992.

Tolba, M.K. Development Without Destruction: Evolving Environmental Perceptions. *Natural Resources Environment Series*, Vol.12, Dublin: Tycooly International Publishing Ltd., 1982.

Tolba, M.K. *Development Without Destruction: Evolving an Environmental Perspective*. Nairobi: UNEP, 1982.

Tolba, M.K. *Earth Matters: Environmental Challenges for 1980's*. Nairobi, UNEP, 1983.

Tolba, M.K. (ed.). *Evolving Environmental Perceptions: From Stockholm to Nairobi*. London: Butterworths, 1989.

Tolba, M.K. Redefining UNEP. *Our Planet*, Vol.8, no.5, 1997, pp. 9–11.

Trivedi, P.R., and K. Sudarshan. *Cherry. Global Environmental Issues.* New Delhi: Commonwealth Publication, 1995.

Trzyna, Thaddeus C., and Julia K. Osborn (eds.). *A Sustainable World: Defining and Measuring Sustainable Development.* London: Earthscan, 1995.

Turner, R. Kerry (ed.). *Sustainable Environmental Economies and Management: Principles and Practice.* London: Belhaven Press, 1993.

Udall, Stewart L. *The Quiet Crisis.* New York: Holt, Rinehart and Winston, 1963.

UN. *Back to Our Common Future Sustainable Development in the 21st Century (SD21) Project Summary for Policymakers.* New York: United Nations, 2012.

UN Chronicle. Education as the Pathway Towards Gender Equality, Vol.50, no.3, 2013. Available at: https://unchronicle.un.org/article/education-pathway-towards gender-equality.

UN DESA. Monterrey Consensus on Financing for Development. Final Text of Agreements and Commitments Adopted at the International Conference on Financing for Development, Monterrey, Mexico, March 18–22, 2002, 2003, Paragraph1. Available at: www.un.org/esa/ffd/monterrey/MonterreyConsensus. Pdf.

UN Development Program. *Human Development Report 2007/8, Fighting Climate Change: Human Solidarity in a Divided World.* London: Palgrave Macmillan, 2008.

UN Development Program. *Human Development Report 2016, Human Development for Everyone,* UNDP: New York, 2016.

UN Economic and Social Commission for Asia and the Pacific. *Achieving the Sustainable Development Goals in South Asia: Key Policy Priorities and Implementation Challenges.* New Delhi: UN-ESCAP Publication, 2017.

UNESCAP. India and the MDGs- Towards a Sustainable Future for All. *UN India,* 2015. Available at: www.unescap.org/sites/default/files/India_ and_the_MDGs_0.pdf.

UNESCAP. *The Millennium Development Goals Report.* New York, 2015. Available at: www.un.org/millenniumgoals/2015_MDG_Report/pdf/MDG%20 2015%20rev%20(July%201).pdf.

UN General Assembly (UNGA). Objective and Themes of the United Nations Conference on Sustainable Development Report of the Secretary-General (A/CONF.216/PC/7). Preparatory Committee for the United Nations Conference on Sustainable Development Second Session, 7–8 March 2011. Available at: www.un-documents. net/aconf216pc7.pdf.

UN General Assembly (UNGA). Rio Declaration on Environment and Development. Report of the United Nations Conference on Environment and Development, Rio de Janeiro, June 3–14, 1992. Available at: www.un.org/documents/ga/conf151/aconf15126–1annex1.htm.

UN General Assembly (UNGA). UN Millennium Declaration (A/55/L.2). Adopted by the UN GA, 2000. Available at: www.un.org/millennium/declaration/ares552e.html.

UN Women Watch. Women, Gender Equality and Climate Change. *Fact Sheet,* 2011. Available at: www.un.org/womenwatch/feature/climate_change/.

United Nations. Energy for a Sustainable Future: Summary Report and Recommendations of the Secretary-General's Advisory Group on Energy and Climate Change (AGECC), 2010a. 28 April. Available at: www. un.org/wcm/web day/site/climate change/shared/Documents/AGECC%20 summary%20report%5B1%5D.pdf.

United Nations. MDG Gap Task Force Report 2010: The Global Partnership for Development at a Critical Juncture. Sales No. E.10.I.12, 2010b.

United Nations. Report of the United Nations Conference on Environment and Development, Rio de Janeiro, 3–4 June 1992, Vol.I, Resolutions Adopted by the Conference. Sales No. E.93.I.8 and corrigendum. Resolution 1, annex I (Rio Declaration on Environment and Development). Resolution 1, annex II (Agenda 21), 1993.

United Nations. World Economic Situation and Prospects 2011. Sales No. E.11. II. C.2., 2011.

United Nations. World Economic and Social Survey 2010: Retooling Global Development. Sales No. E.10. II. C.1, 2010c.

United Nations Economic Commission for Africa. *Institutional and Strategic Frameworks for Sustainable Development in Africa.* Ethiopia: Addis Ababa, 2012.

United Nations Environment Programme. *The Emissions Gap Report 2017 a UN Environment Synthesis Report.* Nairobi: United Nations Environment Programme (UNEP), 2017. Available at: www.unenvironment.org/resources/emissions-gap-report.

United Nations Environment Programme. *Global Environment Outlook 3: Past, Present and Future Perspectives.* London: Earthscan, 2002.

United Nations Environment Programme. *Green Economy: Developing Country Success Stories.* Geneva: Division of Technology, Industry and Economics, 2010. Available at: www.unep.org/pdf/GreenEconomy_SuccessStories.pdf.

United Nations Environment Programme. Institutional Framework for Sustainable Development. Eighteenth Meeting of the Forum of Ministers of Environment of Latin America and the Caribbean, Quito, Ecuador, 31 January–3 February 2012, UNEP/LAC-IG.XVIII/4, Regional Office for Latin America and the Caribbean, 2011. Available at: www.pnuma.org/forodeministros/18- Ecuador/Reunion%20Expertos/Marco%20Institucional%20para%20el%20Desarrollo%20Sostenible/ENGLISH%20Marco%20Institucional%20para%20el%20Desa%2016%20DEC%202011.pdf.

United Nations Environment Programme. *Overview of the Republic of Korea's National Strategy for Green Growth.* Prepared by the Programme as Part of Its Green Economy Initiative. Geneva: Division of Technology, Industry and Economics, Economics and Trade Branch, April 2010.

United Nations Environment Programme. *Towards a Green Economy: Pathways to Sustainable Development and Poverty Eradication – a Synthesis for Policy Makers.* Nairobi: United Nations Environment Programme, 2011.

United Nations Environment Programme. Two Decades of Achievement and Challenge. *Our Planet,* Vol.4, no.5, 1992.

United Nations Environment Programme. *UNEP Background Paper on Green Jobs.* Nairobi, 2008. Available at: www.unep.org/labour_environment/pdfs/green-jobs-background-paper-18-01-08.pdf.

United Nations Framework Convention on Climate Change. *Investment and Financial Flows to Address Climate Change.* Bonn, 2007. Available at: http://unfccc.int/resource/docs/publications/financial_flows.pdf.

United Nations Framework Convention on Climate Change. Report of the Conference of the Parties on Its Sixteenth Session, Held in Cancun from 29 November to 10 December 2010: Addendum. Part Two: Action Taken by the Conference of the Parties at Its Sixteenth Session. FCCC/CP/2010/7/Add.1, 2011. Available at: http://unfccc.int/resource/docs/2010/cop16/Eng./07a01.pdf#page=4.

United Nations Framework Convention on Climate Change. *Technologies for Adaptation to Climate Change.* Bonn: Adaptation, Technology and Science Programme of the UNFCCC Secretariat, 2006. Available at: http://unfccc.int/resource/docs/publications/tech_for_adaptation_06.pdf.

United Nations, General Assembly. *Climate Change and Its Possible Security Implications. Report of the Secretary-General.* A/64/350. New York: United Nations [Online: Web], 2009. Available at: www.un.org/ga/search/view_doc.asp?symbol=A/64/350.

United Nations, General Assembly. Five-Year Review of the Mauritius Strategy for the Further Implementation of the Programme of Action for the Sustainable Development of Small Island Developing States. Report of the Secretary-General. A/65/115, 2010.

United Nations, General Assembly. Official Records of the General Assembly, Forty-Fourth Session, Supplement No.25. A/44/25. Annex I, Decision 15/3. (2009). Progress Report of the Secretary-General on Innovative Sources of Development Finance. 29 July. A/64/189 and Corr.1, 1989.

United Nations, General Assembly. Progress to date and remaining gaps in the implementation of the outcomes of the major summits in the area of sustainable development, as well as an analysis of the themes of the Conference, 2010. Report of the Secretary-General prepared for the first session of the Preparatory Committee for the United Nations Conference on Sustainable Development, 17–19 May 2010. A/CONF.216/PC/2.1 April.

Ursul, A.J. (ed.). *Philosophy and Ecological Problems of Civilization* (Translated by H. Campbell Creighton). Moscow: Progress Pub., 1983.

Vaidya, Andrew P. *War in Ecological Perspectives*. New York: Plenum, 1976.

Vajpayee, Atal Behari. India's Foreign Policy: Today. In Bimal Prasad (ed.), *India's Foreign Policy – Studies in Continuity and Change*. Vikas: Original from the University of California, 1979.

Val Diya, K.S. *Environmental Geology: An Indian Context*. New Delhi: Tata McGraw Hills Pub. Ltd., 1987.

Van den Bergh, Jeroen C.J.M. et al. *Evolutionary Economics and Environmental Policy: Survival of the Greenest*. Cheltenham: Edward Elgar Publishing, 2007.

Van Vuuren, D.P., and Keywan Riahi. Do Recent Emission Trends Imply Higher Emissions Forever? *Climatic Change*, Vol.91, no.3, 2008, pp. 237–248.

Van Vuuren, D.P. et al. Stabilizing Greenhouse Gas Concentrations at Low Levels: An Assessment of Reduction Strategies and Costs. *Climatic Change*, Vol.81, no.2, 2007, pp. 119–159.

Varshney, C.K. and D.R. SarDesai. *Environmental Challenges*. New Delhi: Wiley Eastern Limited, 1993.

Vasilyev, V.S. *Ecology and International Relations: Environmental Politics in the World Politics and Economics*. Moscow: M.O. Publication, 1978.

Vig, Norman J., and Michael-Craft. Environmental Policy from the Seventies to the Eighties. In *Environmental Policy in the 1980s: Reagans New Agenda*. Washington, DC: CQ Press, 1984.

Vincent, A. The Character of Ecology. *Environmental Politics*, Vol.2, no.2, 1993, pp. 248–76.

Visvanathan, Shiv. Health: Semiotics of Waste. *Hindu Survey of the Environment*, 2000.

Vivekananda, J. *Practice Note: Conflict-Sensitive Responses to Climate Change in South Asia*. London: Initiative for Peacebuilding, 2011.

Vivekananda, J., J. Schilling, S. Mitra, and N. Pandey. On Shrimp, Salt and Security: Livelihood Risks and Responses in South Bangladesh and East India. *Environment, Development and Sustainability*, Vol.16, no.6, 2014, pp. 1141–1161.

Vogler, John, and Mark F. Imber (ed.). *The Environment and International Relations Global Environmental Change Programme*. New ~York: Routledge, 1996.

Vohra, B.B. *Land and Water Towards a Policy for the Life Support Systems*. New Delhi: INTACH, 1985.

Waid, S. Thinking Global, Acting Local? British Local Authorities and Their Environmental Plans. *Environmental Politics*, Vol.2, no.3, 1993, pp. 453–478.

Walsham, M. *Assessing the Evidence: Environment, Climate Change and Migration in Bangladesh*. Dhaka: International Organization for Migration, 2010.

Walter, Ingo (ed.). *Studies in International Environmental Economics*. New York: Willey Inter Sciences, 1976.

Ward, Barbara. *The Home of Man*. Harmondsworth: Penguin Books, 1976.

Ward, Barbara. Speech on the UN Conference on the Human Environment, June 1992. Unpublished.

Ward, Barbara, and Rene Dubos. *Only One Earth*. Harmondsworth: Penguin, 1972.

Washington, DC: Department for International Development. *Growth – Building Jobs and Prosperity in Developing Countries*. Great Britain, 2008.

Weal, A., and A. Williams. Between Economy and Ecology? The Single Market and the Integration of Environmental Policy. *Environmental Politics*, Vol.1, no.4, 1992, pp. 45–64.

Weber, Madeline. *Q&A with Mark Robinson: The Role of Good Governance in Sustainable Development*. Washington, DC: World Resources Institute, 18 February 2015. Available at: www.wri.org/blog/2015/02/qa-markrobinson-role-good-governance-sustainable-development.

Weiner, Jonathan. *The Next One Hundred Years: Shaping the Fate of Our Living*. New York: Publisher Bantam Books, 1990.

Weiss, Edith Brown. *In Fairness to Future Generations: International Law Common Patrimony, and intergenerational Equity*. Tokyo: United Nations University; New York: Transnational Publishers, 1989.

Westing, A.H. The Atmosphere as a Common Heritage of Humankind: Its Role in Environmental Security. *Scientific World* (London), Vol.34, no.4, 1990.

Westing, A.H. (ed.). *Global Resources and International Conflict: Environmental Factors in Strategic Policy and Action*. Oxford: Oxford University Press, 1986.

Wheeling, Kate. How Climate Change Contributed to Massive Floods in South Asia. *Pacific Standard*, 2017, August 30. Available at: https://psmag.com/environment/how-climate-change-contributed-to-massive-floods-in-south-asia.

White, Rodney R. *North, South and the Environmental Crisis*. London: University of Toronto Press, 1991.

Wickremasinghe, Anoja. Environmental Deterioration in the Hill Country of Sri Lanka. *Malaysian Journal of Tropical Geography*, Vol.19, June 1989.

Williams, Marc. re-Articulating the Third World Coalition: The Role of the Environmental Agenda. *Third World Quarterly*, Vol.14, no.1, 1993, pp. 7–29.

Williams, R. *Towards 2000*. London: Penguin, 1984.

Willye, A. Which Countries Are Achieving the UN Sustainable Development Goals Fastest? *World Economic Forum*, 2017. Available at: www.weforum.org/agenda/2017/03/countries-achieving-un-sustainable-development goals-fastest/.

Winpenny, James T. (ed.). *Development Research: The Environmental Challenge*. London: ODI, 1991.

Wisner, Ben Maureen Fordham, Ilan Kelman, Barbara Rose Johnston, David Simon, Allan Lavell, Hans Günter Brauch, Ursula Oswald Spring, Gustavo Wilches-Chaux, Marcus Moench, and Daniel Weiner. *Climate Change and Human Security*. 2007. Available at: www.disasterdiplomacy.org/cchswisneretal.pdf.

Wood, William B. Tropic Deforestation: Balancing Regional Development Demands and Global Environmental Concerns. *Global Environmental Change*, Vol.1, no.1, December 1990, pp. 23–41.

Worland, Justin. Climate Change Will Make Parts of South Asia Unlivable by 2100, Study Says. *Time*, 2 August 2017. Available at: http://time.com/4884648/climate-change-India-temperatures.

World Bank. *Economics of Adaptation to Climate Change: Country Study Bangladesh, Main Report*, Vol.1. Washington, DC: World Bank, 2010.

World Bank. *Environmental Aspects of Bank Work. The World Bank Operations Manual Statements, OMS 2.36*. Washington, DC: World Bank, 1986.

World Bank. *GDP Growth Rate- Bangladesh*. Washington, DC: World Bank, 2018.

World Bank. *Global Financial Inclusion*. 2018a. Available at: http://databank.worldbank.org/data/reports.aspx? source=1228

World Bank. *South Asia Economic Focus, Spring 2018: Jobless Growth?* Washington, DC: World Bank, 2018c. Available at: https://openknowledge.worldbank.org/handle/10986/29650.

World Bank. Warming Climate to Hit Bangladesh Hard with Sea Level Rise, More Floods and Cyclones. World Bank Report Says. Press Release, 19 June 2013. Available at: www.worldbank.org/en/news/press-release/2013/06/19/warmingclimate-to-hit-bangladesh-hard-with-sea-level-rise-more-floods-and-cyclones world-bank-report-says.

World Bank. *World Bank Draft Report on Climate Change in South Asia*. 2009. Available at: http://siteresources.worldbank.org/SOUTHASIAEXT/Resources/Publications/448813-1231439344179/5726136–1232505590830/1SARC CS January19, 2009.pdf.

World Bank. The World Bank in Bangladesh. Overview, 2018b. Available at: http://www.worldbank.org/en/country/bangladesh/overview.

World Bank. *World Development Indicators: Featuring the Sustainable Development Goals*. Washington, DC: World Bank Group, 2016.

World Bank. *World Development Report 2003: Sustainable Development in a Dynamic World – Transforming Institutions, Growth, and Quality of Life*. Washington, DC: World Bank; New York: Oxford University Press, 2003.

World Bank Conference on Development Economics Proceedings, edited by Boris Pleskovic and Joseph E. Stiglitz. Washington, DC: World Bank, 1998.

World Business Council for Sustainable Development, Innovating for Green Growth: Drivers of Private Sector RD&D. Geneva, 2011.

World Commission on Environment and Development. *Our Common Future*. Oxford: Oxford University Press, 1987.

World Economic Forum. The Global Gender Gap Report 2017. Geneva, 2017. Available at: http://www3.weforum.org/docs/WEF_GGGR_2017.pdf.

World Energy Council and Food and Agriculture Organization of the United Nations, *The Challenge of Rural Energy Poverty in Developing Countries*. London: World Energy Council, 1999.

World Environment Center. *The World Environment Handbook*. New York: WEC, 1983.

World Health Organization. Climate Change and Health, 1 February 2018. Available at: www.who.int/news-room/fact-sheets/detail/climate change-and-health.

World Health Organization. Gender, Climate Change and Health. Draft Discussion Paper. Geneva, 2009. Available at: www.who.int/globalchange/publications/reports/final_who_gender.pdf.

World Resources. *Word Resources Institute: International Institute for Environment and Development*. New York: Basic Books, 1986.

Worster, Donald (ed.). *Ends of the Earth: Perspectives on Modern Environmental History*. Cambridge: Cambridge University Press, 1988.

Yap, Namita. NGOs and Sustainable Development. *International Journal*, Vol.XLV, no.1, Winter 1989–90, pp. 75–105.

Young, Oran. Global Environmental Change and International Governance. In Ian H. Rowlands and Malory Greene (eds.), *Global Environmental Change and International Relations*. London: Macmillan Academic and Professional Ltd., 1992, pp. 6–18.

Young, Stephen C. The Different Dimensions of Green Politics. *Environmental Politics*, Vol.1, no.1, 1992, pp. 9–44.

Yuri, Izael. *Ecological and Environmental Control*. Moscow: Peoples Publication, 1982.

Zinger, Clem. *Richard Dalsemer, and Helen MA gargle, Environmental Volunteers in America*. Washington, DC: EPA, 1973.

Articles, Periodicals, Journals and Papers

Acharya, Keya. Climate change and Indian case studies. (*South Asian Journal* no.28; Apr–Jun 2010: pp.68–73) (ACH)

Ackerman, Frank & Stanton, Elizabeth A. Did the stern review underestimate US and global climate damages? (*Energy Policy* Vol.37, no.7; Jul 2009: pp.2717–2721) – ACK)

Agarwal, Anil. *Green Trade Wars' Down to Earth*. New Delhi: CSE, Aug 15, 1992, Vol.1, no.6.

Agarwal, Anil. The North-South perspective: Alienation or independence? (*Ambioo* Vol.19, no.2; Apr 1990: pp.94–96).

Agarwal, Anil. The North-South perspective issues in the rural energy. *The Asian Age*, Aug 4, 2000.

Agarwal, Anil. Towards global environmental movement. (*Social Action* Vol.42, no.2; Apr–Jun 1992: pp.111–119).

Agarwal, Anil & Narayan, Sunita. Earth needs a summit: It is going to be a grand affair full of little glitches. Down to Earth special issue named

Agarwal, Anil & Narayan, Sunita. A new morality. (*The Illustrated Weekly of India* Vol.24; Dec 15, 1989: pp.84–87).

Agrawal, Subhash. Small green steps for India. (*Far Eastern Economic Review* Vol.172, no.9; Nov 2009: pp.35–37) (AGR)

Ahmed, Iftekhar. *Biotechnology: A Hope or a Threat?* London: The Macmillan Press, 1992, pp.1–14.

Aliston, Philip. International regulation of toxic chemical. (*Ecology Law Quarterly* Vol.67; 1978: pp.397–456).

Allaby, Michael. Environment. (*Britannica Book of the Year*, 1986: pp.242–248).

Allaire, Julien. Impact of urban development in China on global warming. (*China Perspectives* no.1; 2007: pp.51–61) -(ALL)

Amato, Anthoni D. & others. Do we owe a duty to future generations to preserve the global environmental responsibility? (*American Journal of international Law* Vol.84, no.1; Jan 1980: pp.10–91).

Arden-Clarke, Charles. South-North terms of trade: Environmental protection and sustainable development. (*International Environmental Affairs* Vol.4, no.2; Spring 1992: pp.122–138).

Arpit, Claude. Strategic aspects of climate change. (*Indian Defense Review* Vol.25, no.3; Jul–Sep 2010: pp.113–121) – (ARP)

Ashish, Madhu. In the area of environment. (Seminar Annual Vol.269; Jan 1982: pp.72–74).

Asthana, Vandana and others. Regional Co-operation for the protection of coastal environment in South Asia. (*South Asia Journal* Vol.3, Nos. 1–3; July–Dec 1989: pp.109–120).

Avdeeva, T. 2009 Copenhagen summit. (*International Affairs* (Moscow) Vol.56, no.2; 2010: pp.130–145) – (AVD)

Baer, Paul, Fieldman, Glenn, Athanasios, Tom & Kartha, Sivan. Greenhouse development rights. (*Cambridge Review of International Affairs* Vol.21, no.4; Dec 2008: pp.649–669) – (BA)

Bajwa, G.S. Environmental management: Problems and prospects. In R.K. Sapru (ed.), *Environmental Management in India*. New Delhi: Ashish Publishing House, 1987, Vol.2, pp.207–217.

Balachandra, P., Ravindranath, Darshini & Ravindranath, N.H. Energy efficiency in India. (*Energy Policy* Vol.38, no.11; Nov 2010: pp.6428–6438) – (BA)

Bales, Carter F. & Duke, Richard D. Containing climate change. (*Foreign Affairs* Vol.87, no.5; Sep–Oct 2008: pp.78–89) – (BAL)

Bandyopadhyay, J. & Gyawali, D. Himalaya's water resources: Ecological and political aspects of management. (*Mountain Research and Development* Vol.14, no.1; 1994).

Bandyopadhyay, J. & Shiva, Vandana. Development, poverty and the growth of the green movements in India. (*The Ecologist* Vol.19, no.3; May/Jun 1989: pp.111–117).

Bang, Gury. Energy security and climate change concerns. (*Energy Policy* Vol.38, no.4; Apr 2010: pp.1645–1653) – (BAN)

Barnett, Jon. Titanic states? Impacts and responses to climate change in the Pacific Islands. (*Journal of International Affairs* Vol.59, no.1; Fall/Winter 2005: pp.203–219) 2005. – (BAR)

Bass, S. The national conservation strategies for Nepal. (*Landscape Design* Vol.2; 1989).

Basu, Rumki. Politics and economics of the global environment debate between the developed and developing countries. (*Indian Journal of S. Political Science* Vol.52, no.1; Jan–Mar 1991: pp.74–84).

Battie, Michele B. & Bern Auer, Thomas. National institutions and global public goods. (*International Organization* Vol.63, no.2; Spring 2009: pp.281–308) – (BAT)

Baviskar, Amita. Ecology and development in India a field and its future. (*Sociological Bulletin* Vol.46, no.2; Sept 1997: pp.193–209).

Baviskar, Amita. Tribal politics and discourses of environmentalism: Contributions to Indian sociology (Vol.31, no.2; Jul–Dec 1997: pp.195–22. New Delhi/Thousand Oaks/London: Sage Publications, 1997).

Bayliss-Smith, T., & Owens, S. Environmental challenge. In Gregory D. Martin, G.R. Smith (eds.), *Human Geography*. London: Palgrave, 1991, pp.25–28.

Beckerman, Wilfred. Global warming and international action: An economic perspective. In Andrew Hurrell & Benedict Kingsbury (eds.), *The International Politics of the Environment*. Oxford: Clarendon Press, 1992, pp.253–289.

Beckett, Margaret. Case for climate security. (*Ruisi Journal* Vol.152, no.3; Jun 2007: pp.54–59) – (BEC)

Bell, Ruth Greenspan. What to do about climate change. (*Foreign Affairs* Vol.85, no.3; May–Jun 2006: pp.105–113) – (BEL)

Below, Amy. U.S. presidential decisions on ozone depletion and climate change. (*Foreign Policy Analysis* Vol.4, no.1; Jan 2008: pp.1–20) – (BEL)

Bencala, Karin R. & Dabelko, Geoffrey D. Water wars. (*Journal of International Affairs* Vol.61, no.2; Spring/Summer 2008: pp.21–34) – (BEN)

Bennell, Richard. Linking as leverage. (*Cambridge Review of International Affairs* Vol.21, no.4; Dec 2008: pp.545–562) – (BE)

Bernie, Patricia. The role of international law in solving certain environmental conflicts. In John E. Carroll (ed.), *International Environmental. Diplomacy*. Cambridge: Cambridge University Press, 1988, pp.95–121.

Bertram, Christine. Ocean iron fertilization in the context of the Kyoto protocol and the post-Kyoto process. (*Energy Policy* Vol.38, no.2; Feb 2010: pp.1130–1139) – (BER)

Betsill, Michele M. & Bulkeley, Harriet. Cities and the multilevel governance of global climate change. (*Global Governance* Vol.12, no.2; Apr–Jun 2006: pp.141–159) – (BET)

Bhandary, Rishikesh. South Asian front. (*Himal* Vol.22, no.10–11; Oct–Nov 2009: pp.55–57) – (BHA)

Bhogall, Parminder S. India's security environment in the 1990: The South Asia factor. (*Strategic Analysis* Vol.13, no.7; Oct 1989: pp.167–177).

Bhuyan, Dasarathi. International bodies to combat global warming. (*World Affairs* Vol.14, no.1; Spring 2010: pp.12–24) – (BHU)

Bisht, Medha. India-Bhutan relations. (*Strategic Analysis* Vol.34, no.3; May 2010: pp.350–353) – (BI)

Blackwell, P.J. East Africa's pastoralist emergency. (*Third World Quarterly* Vol.31, no.8; 2010: pp.1321–1338) – (BLA)

Blunden, Margaret. New problem of arctic stability. (*Survival: The IISS Quarterly* Vol.51, no.5; Oct–Nov 2009: pp.121–142) – (BLU)
Bolen, Johannes, Hers, Sebastian & Zwaan, Bob van der. Integrated assessment of climate change, air pollution, and energy security policy. (*Energy Policy* Vol.38, no.8; Aug 2010: pp.4021–4030) – (BO)
Bossone, Biagibn. Environment Protection: How should we pay for it? (*International Journal of Social Sciences* Vol.17, no.1; 1990: pp.3–15).
Brigham, Lawson W. Arctic. (*Foreign Policy* no.181; Sep–Oct 2010: pp.70–74) – (BRI)
Brigham, Lawson W. Navigating the new maritime Arctic. (*US Naval Institute Proceedings* Vol.135, no.5; May 2009: pp.42–47) – (BRI)
Brown, Lester R. & Jacobson, Jodi L. The future urbanization: Facing the ecological and economic constrains. *World Watch Paper* Vol.77; May 1987.
Brown, Oli & Crawford, Alec. Climate change. (*African Security Review* Vol.17, no.3; Sep 2008: pp.39–57) – (BRO)
Brown, Oli, Hammill, Anne & McLean, Robert. Climate change as the 'new' security threat. (*International Affairs* Vol.83, no.6; Nov 2007: pp.1141–1154) – (BRO)
Brundtland, Gro Harlem. *Unlearnt Lessons*. Hindu Survey of the Environment. The Hindu. Chennai: Kasturi & Co, 2000.
Brzezinski, Zbigniew. Major foreign policy challenges for the next US President. (*International Affairs* Vol.85, no.1; Jan 2009: pp.53–60) – (BRZ)
Burch, Sarah. In pursuit of resilient, low carbon communities. (*Energy Policy* Vol.38, no.12; Dec 2010: pp.7575–7585) – (BUR)
Burt raw, Dallas, Palmer, Karen & Kahn, Danny. Symmetric safety valve. (*Energy Policy* Vol.38, no.9; Sep 2010: pp.4921–4932) – (BU)
Busby, Joshua W. Who cares about the weather? (*Security Studies* Vol.17, no.3; Jul–Sep 2008: pp.468–504) – (BUS)
Buskirk, Robert Van. Analysis of long-range clean energy investment scenarios for Eritrea, East Africa. (*Energy Policy* Vol.34, no.14; Sep 2006: pp.1807–1817) – (BUS)
Cam bell, Layton Keith. Co-operation and conflict; International response to environmental issues. (*Environment* Vol.27, no.1; Jan–Feb 1985: pp.10–14).
Carter, Neil. Vote Blue, go Green? Cameron's conservatives and the environment. (*Political Quarterly* Vol.80, no.2; Apr–Jun 2009: pp.233–242) – (CAR)
Castro, Joao Augusto de Araujo. Environment and development: The case of the developing countries. (*International Organization* Vol.26; 1972: pp.40l–416).
Chasek, Pamela S. Creating space for consensus. (*International Negotiation* Vol.16, no.1; 2011: pp.87–108) – (CHA)
Chawla, S.K. Climate change: Impact on Indian Ocean region. (*Journal of Indian Ocean Studies* Vol.16, no.1–2; Apr–Aug 2008: pp.69–75) – (CHA)
Chellaney, Brahma. Climate change and security in Southern Asia. (*Ruisi Journal* Vol.152, no.2; Apr 2007: pp.62–69) – (CHE)
Chen, Gang. China's diplomacy on climate change. (*Journal of East Asian Affairs* Vol.22, no.1; Spring/Summer 2008: pp.145–174) – (CHE)
Chishti, Sumitra. India's foreign economic policy. In Bimal Prasad (ed.), *India's Foreign Policy – Studies in Continuity and Change*. New Delhi: Vikas Publishing House Private Limited, 1979, pp.35–36.
Christofferson, Gaye. US-China energy relations and energy institution building in the Asia-Pacific. (*Journal of Contemporary China* Vol.19, no.67; Dec 2010: pp.871–889) – (CHR)
Churkin, V. UN, a matchless player on the international field. (*International Affairs* (Moscow) Vol.56, no.6; 2010: pp.113–118) – (CH)

Clarke, Peter B. Environmental destruction. (*Economic and Political Weekly* Vol.29, no.17; 1976: pp.17–19).

Compston, Hugh. Politics of climate policy. (*Political Quarterly* Vol.81, no.1; Jan–Mar 2010: pp.107–115) – (COM)

Costa, Oriol. Is climate change changing the EU? The second image reversed in climate politics. (*Cambridge Review of International Affairs* Vol.21, no.4; Dec 2008: pp.527–544) – (CO)

Cowen, R.C. Environment protection for 1990s. (*Environment*, Sept 1990: pp.12–15).

Cowen, R.C. The rise of eco-diplomacy. (*Technology Review*, May–Jun 1988: pp.91–118).

CSE, Statement of the South Asian NGO Summit, New Delhi: CSE, Feb 17–19, 1992, reprinted in (*R.S. Digest* Vol.9, Nos. 1 and 2; Jun 1992: pp.32–42).

Dadwal, Shebonti Ray. Is energy security the main driver for the West's debate on climate change. (*Strategic Analysis* Vol.33, no.6; Nov 2009: pp.836–848) – (DAD)

Dagoumas, A.S., Papagiannis, G.K. & Dokopoulos, P.S: An economic assessment of the Kyoto protocol application. (*Energy Policy* Vol.34, no.1; Jan 2006: pp.26–39) – (DAG)

Danes, Byron W. & Sussman, Glen. Greenless response to global warming. (*Current History* Vol.104, no.686; Dec 2005: pp.438–443) 2005. – (DAY)

Daoud, Alden L. International environment and development. (*Perceptions of Developing and Developed Countries Natural Resources Journal*); Oct 12, 1972: pp.520–529.

Das, Nikhilesh. Environmental management. (*Asian Studies* Vol.27, no.1; Jan–Jun 2009: pp.9–15) – (DAS)

Davies, Zoe G. & Armsworth, Paul R. Making an impact. (*Energy Policy* Vol.38, no.12; Dec 2010: pp.7634–7638) – (DAV)

D'Costa, Bina. Bangladesh in 2010. (*Asian Survey* Vol.51, no.1; Jan–Feb 2011: pp.138–147) – (D')

De Haen, Harting. Agricultural development and environmental protection: Some key issues of potential relevance to Pakistan. (*The Pakistan Development Review* Vol.32, no.4, Part-I; Winter 1993).

De, Prabir Kumar. Environmental agenda in SAARC countries. (*Asian Studies* Vol.27, no.1; Jan–Jun 2009: pp.16–25) – (DE)

Deere-Birkbeck, Carolyn. Global governance in the context of climate change. (*International Affairs* Vol.85, no.6; Nov 2009: pp.1173–1194) – (DEE)

Demonte, Darryl. Skepticism chic. (*Himal* Vol.22, no.10–11; Oct–Nov 2009: pp.72–73) – (DMO)

Desai, Bharat H. Changing the climate for climate change. (*World Focus* Vol.30, no.9; Sep 2009: pp.358–361) – (DES)

Desai, Bharat H. Coming out of coma. (*Down to Earth* Vol.9, no.20; Mar 15, 2001: pp.48–49).

Desai, Nitin: New race. (*India Quarterly* Vol.64, no.1; Jan–Mar 2011: pp.106–115) – (DES)

Detraz, Nicole & Betsill, Michele M. Climate change and environmental security. (*International Studies Perspectives* Vol.10, no.3; Aug 2009: pp.303–320) – (DET)

Deudney, Daniel. The case against linking environmental degradation and national security. (*Millennium* Vol.19, no.3; Winter 1990: pp.416–476).

Deutsch, John. Good news about gas. (*Foreign Affairs* Vol.90, no.1; Jan–Feb 2011: pp.82–93) – (DE)

Deva, Indra. Towards a more meaningful study of ecology, society and culture. *Sociological Bulletin* Vol.46, no.1; Mar 1997: pp.1–21).

Dewdney, Daniel. *Rethinking the Link between Environment and National Perspectives on War and Peace, Centre for International Cooperation and Security Studies*. Madison: University of Wisconsin, Spring 1990, Vol.7, no.2, pp.12.

Dixit, Kunda. Charting change. (*Himal* Vol.22, no.10–11; Oct–Nov 2009: pp.24–29) – (DIX)

Dixon Thomas F. Homer. Environmental change and economic decline in developing countries. (*International Studies Notes* Vol.16, no.1; Winter 1991).

Dobson, Andrew. Globalization, cosmopolitanism and the environment. (*International Relations* Vol.19, no.3; Sep 2005: pp.259–273) 2005. – (DOB)

Dolata-Kreutz Kamp, Petra. Canada-Germany-EU. (*International Journal* Vol.63, no.3; Summer 2008: pp.665–681) – (DOL)

Doyle, Timothy & Chaturvedi, Sanjay. Climate territories. (*Geopolitics* Vol.15, no.3; 2010: pp.516–535) – (DO)

Dupont, Alan. Strategic implications of climate change. (*Survival: The IISS Quarterly* Vol.50, no.3; May–Jun 2008: pp.29–54) – (DUP)

Durmng, A. *Poverty and the Environment: Reversing the Downward Spiral*. World Watch Paper 92. Washington, DC: World Watch Institute, 1989.

Dwivedi, O.P. & Kishore, B. Protecting the environment from pollution. (*Asian Survey* Sept 1982: pp.12–35).

Dwivedi, Ranjit. Parks, people and protest: The mediating role of environmental action groups. (*Sociological Bulletin* Vol.46, no.2; Sept 1997: pp.209–245).

Easterbrook, Gregg. Global warming. (*Atlantic* Vol.299, no.3; Apr 2007: pp.52–66) – (EAS)

Ebinger, Charles K. & Zambetakis, Evie. Geopolitics of Arctic melt. (*International Affairs* Vol.85, no.6; Nov 2009: pp.1215–1232) – (EBI)

Eckersley, Robyn. Ambushed. (*International Politics* Vol.44, no.2–3; Mar–May 2007: pp.306–324) – (ECK)

El-Hinnavi, Essam (ed.), *Global Environmental Issues: United Nations Environment Programme*. Dublin: Tycooly International, 1982: pp.20–25.

Elizabeth, L. C. Knowledge in sheep's clothing. (*Diplomacy and Statecraft* Vol.19, no.1; Mar 2008: pp.1–19) – (CHA)

Environment India. *India International Centre Quarterly* Vol.9, no.3:4 Dec 1982: pp.215–388.

Esty, Daniel C. Revitalizing global environmental governance for climate change. (*Global Governance* Vol.15, no.4; Oct–Dec 2009: pp.427–434) – (EST)

Evans, Alex. Hunger pains. (*Jane's Intelligence Review* Vol.22, no.4; Apr 2010: pp.26–29) – (EVA)

Falk, Richard A. Environmental policy as a world order problem. *Natural Resources Journal* Vol.12, no.2; Apr 1972: pp.161–171.

Faris, Stephan. Real roots of Darfur. (*Atlantic* Vol.299, no.3; Apr 2007: pp.67–69) – (FAR)

Farmer, B.H. Perspectives on the green revolution in South Asia. *Modern Asian Studies* 20; 1986, pp.175–199.

Finger, Mathias. New horizons for peace research: The global environment. In Kakonen Jyrki (ed.), *Perspectives on Environmental Conflict and International Politics*. London: Pinter Publishers, 1992, pp. 5–30.

Floyd, Rita. Environmental security debate and its significance for climate change. (*International Spectator* Vol.43, no.3; Sep 2008: pp.51–65) – (FLO)

Fortier, Francois. Taking a climate chance. (*Asia Pacific Viewpoint* Vol.51, no.3; Dec 2010: pp.229–247) – (FO)

Francisco, Herminia A. Adaptation to climate change. (*ASEAN Economic Bulletin* Vol.25, no.1; Apr 2008: pp.7–19) – (FRA)

Froggatt, Antony & Levi, Michael A. Climate and energy security policies and measures. (*International Affairs* Vol.85, no.6; Nov 2009: pp.1129–1141) – (FRO)

Gaan, Narottam. Transcendental values and sustainable development. (*India Quarterly* Vol.61, no.4; Oct–Dec 2005: pp.228–254) – (GAA)

Gadgil, Madhav. *Biodiversity: Time for Bold Steps*. The Hindu Survey of the Environment, The Hindu, Chennai: Kasturi & Co, 1992, pp. 21–23.

Gadgil, Madhav. *Conservation: Taking Care of all Life*. Hindu Survey of the Environment The Hindu, Chennai: Kasturi & Co, 2000.

Gadgil, Madhav. Conserving India's bio-diversity: The human context. In T.N. Khoshoo & Manju Sharma (eds.), *Sustainable Management of Natural Resources*. New Delhi: Malhotra Publishing House, 1992, pp.243–255.

Gadgil, Madhav. Conserving India's biodiversity: The societal context. (*Evolutionary Trends in Plants* Vol.5, no.1; 1991: pp.3–8).

Gadgil, Madhav & Guha, Ramachandra. Towards a perspective on environmental movements in India. *The Indian Journal of Social Work* Vol.59, no.1; Jan 1998: pp.450–472.

Gadihoke, Neil. Climate change implications for the Indian navy. (*Maritime Affairs* Vol.6, no.1; Sum 2010: pp.116–131) – (GAD)

Gallagher, Kelly Sims. China needs help with climate change. (*Current History* Vol.106, no.703; Nov 2007: pp.389–394) – (GAL)

Gandhi, Indira. Man, and environment (*Plenary, Session of UNCHE*, Jun 14, 1972), in DOE, Indira Gandhi on Environment'. New Delhi: DOE, 1984, pp.20–29.

Gang, He. Chinese society and climate change. (*China Perspectives* no.1; 2007: pp.77–82) – (GAN)

Garcia, Denise. Climate security divide. (*African Security Review* Vol.17, no.3; Sep 2008: pp.2–17) – (GAR)

Garcia, Denise. Warming to a redefinition of international security. (*International Relations* Vol.24, no.3: Sep 2010: pp.271–292) – (GAR)

Gareth, Porter. Environmental security as a national security issue. *Current History*, May 1995.

Gautam, P.K. Changing geographical factors in planning and conduct of Indian military operations. (*Strategic Analysis* Vol.32, no.2; Mar 2008: pp.245–258) – (GAU)

Gautam, P.K. Climate change. (*Journal of Indian Ocean Studies* Vol.16, no.3; Dec 2008: pp.185–193) – (GAU)

Gautam, P.K. Climate change and environmental degradation in Tibet. (*Strategic Analysis* Vol.34, no.5; Sep 2010: pp.744–755) – (GAU)

Gautam, P.K. Environmental OODA loop. (*USI Journal* Vol.137, no.567; Jan–Mar 2007: pp.95–101) – (GAU)

Gayawali, Dipak. *Water in Nepal: An Interdisciplinary Look at Resource Uncertainties Evolving Problems and Future Prospects*. East West Centre. Environment and Policy Institute, Occasional Paper no.8, 1989.

Geetha, Krishnan. Sustainable development in operation. In Malcolm S. Adiseshiah (ed.), *Sustainable Development: Its Content, Scope and Prices*. New Delhi: Lancer International in Association with India International Centre, 1990, pp.7–21.

George, Jose. Development and environmental hazards. *Law and Environment*; 1984: pp.224–278.

Gerst, Michael D., Howarth, Richard B. & Borsuk, Mark E. Accounting for the risk of extreme outcomes in an integrated assessment of climate change. (*Energy Policy* Vol.38, no.8; Aug 2010: pp.4540–4548) – (GE)

Gibson, Steyn D. Future roles of the UK intelligence system. (*Review of International Studies* Vol.35, no.4; Oct 2009: pp.917–928) – (GIB)

Gimenez, Eduardo L. & Rodriguez, Miguel. Reevaluating the first and the second dividends of environmental tax reforms. (*Energy Policy* Vol.38, no.11; Nov 2010: pp.6654–6661) – (GI)

Giovanni, Emily & Richards, Kenneth R. Determinants of the costs of carbon capture and sequestration for expanding electricity generation capacity. (*Energy Policy* Vol.38, no.10; Oct 2010: pp.6026–6035) – (GIO)

The Global Environment Facility. Cover story. (*Our Planet* Vol.3, no.3; 1991: pp.10–13).

Glover, David & Onn, Lef Poh. Environment, climate change and natural resources in Southeast Asia. (*ASEAN Economic Bulletin* Vol.25, no.1; Apr 2008: pp.1–6) – (GLO)
Godement, Francois. United States and Asia in 2009. (*Asian Survey* Vol.50, no.1; Jan–Feb 2010: pp.8–24) – (GOD)
Gore, Albert. Strategic environmental initiative (SEI). (*SAIS Review* Vol.10; 1990: pp.59–71).
Greising, David. Carbon frontier. (*Bulletin of the Atomic Scientists* Vol.64, no.3; Jul–Aug 2008: pp.32–37) – (GRE)
Grove, Kevin J. Insuring "our common future?" Dangerous climate change and the biopolitics of environmental security. (*Geopolitics* Vol.15, no.3; 2010: pp.536–563) – (GR)
Guha, Ramachandra. Ecological roots of development crisis. (*Economic and Political Weekly* Vol.21, no.15; Apr 12, 1986: pp.623–626).
Gundlach, Erich R. Oil tanker disasters. (*Environment* Vol.19, no.9; Dec 1977: pp.16–27).
Gupta, Arvind. Geopolitical implications of Arctic meltdown. (*Strategic Analysis* Vol.33, no.2; Mar 2009: pp.174–177) – (GUP)
Gupta, Joyeeta & Ahlers, Rhodante. South Asia. (*South Asian Journal* no.28; Apr–Jun 2010: pp.6–13) – (GUP)
Hale, Thomas. Climate coalition of the willing. (*Washington Quarterly* Vol.34, no.1; Win 2011: pp.89–101) – (HAL)
Hallding, Karl, Han, Guoyi & Olsson, Marie. China's climate-and energy-security dilemma. (*China Aktuell Monatszeitschist* Vol.38, no.3; 2009: pp.119–134) – (HAL)
Hambro, E. The human environment: Stockholm and after. (*Yearbook of World Affairs*; 1974: pp.200–210).
Hamilton, Lawrence S. What are the impacts of Himalayan deforestation on the Ganges-Brahmaputra low lands and deltas? Assumption and facts. (*Mountain Research and Development* (Kathmandu) Vol.7, no.3; Aug 1987).
Hardy, Michael. The United Nations Environment Programme (*Natural Resources Journal*; Apr 13, 1973: pp.235–255).
Hare, F. Kenneth. Climate, drought and desertification. (*Nature and Resources* Vol.20, no.1; Jan–Mar 1984).
Harrison, Stephan. Climate change and regional security. (*Rusi Journal* Vol.153; Jun 2008: pp.88–93) – (HAR)
Hassan, Shaukat. *Problems of Internal Stability in South Asia*. PSIS Occasional Paper no.1, 1988. Geneva: The Graduate Institute of International. Studies, Jun 1988.
Hazarika, Sanjoy. Bangladesh and Assam: Land pressures, migration and ethnic conflict. Occasional Paper Series of the Project on Environmental Change Acute Conflict, A Joint Project of the University of Toronto and the American Academy of Arts and Sciences, no.3, Mar 1993.
Hazarika, Sanjoy. India admits failure to cut Bangladesh influx. *New York Limes*, Dec 16, 1992.
Henderson, Nancy. Securing our future. (*Foreign Policy* no.174; Sep–Oct 2009: pp.139–147) – (HEN)
Hilton, Isabel. Reality of global warming. (*World Policy Journal* Vol.25, no.1; Spring 2008: pp.1–8) – (HIL)
Hirono, Ryokichi. Japan's environmental cooperation with China during the last two decades. (*Asia Pacific Review* Vol.14, no.2; Nov 2007: pp.1–16) – (HIR)
Hodder, Patrick & Martin, Brian. No doomsday. (*Himal* Vol.22, no.10–11; Oct–Nov 2009: pp.74–77) – (HOD)
Holdden, Susan O. Statues and standards for pollution control in India. (*EPW* Vol.XXII, no.16; Apr 18, 1987).
Holroyd, Carin. National mobilization and global engagement. (*Asian Perspectives* Vol.33, NO.2; 2009: pp.73–96) – (HOL)

Holzer, Constantin & Zhang, Haibin. Potentials and limits of China-EU cooperation on climate change and energy security. (*Asia Europe Journal* Vol.6, no.2; Jul 2008: pp.217–227) – (HOL)

Hong, Zhao. Energy security concerns of China and ASEAN. (*Asia Europe Journal* Vol.8, no.3; Nov 2010: pp.413–426) – (HON)

Hongyuan, Yu. Challenge of climate change and China's action. (*Foreign Affairs Journal* no.93; Aug 2009: pp.77–88) – (HON)

Hovi, Jon, Bang, Guri & Froyn, Camilla Bretteville. Enforcing the Kyoto protocol. (*Review of International Studies* Vol.33, no.3; Jul 2007: pp.435–449) – (HOV)

Howell John M. An approach to the development of international jurisdiction to deal with environmental problems. (*Asian Perspectives* Vol.5, no.2; Fall–Winter 2018: pp.218–246).

Huang, Jing. Leadership of twenty (L20) within the UNFCCC. (*Global Governance* Vol.15, no.4; Oct–Dec 2009: pp.435–441) – (HUA)

Huchet, Jean-Francois & Marechal, Jean-Paul. Ethics and models of development. (*China Perspectives* no.1; 2007: pp.6–17) – (HUC)

Hughes, Larry & Chaudhry, Nikhil. Challenge of meeting Canada's greenhouse gas reduction targets. (*Energy Policy* Vol.39, no.3; Mar 2011: pp.1352–1362) – (HU)

Hultman, Nathan E. Can the world wean itself from fossil fuels? (*Current History* Vol.106, no.703; Nov 2007: pp.376–383) – (HUL)

Humphrey, Mathew. Rational irrationality and simulation in environmental politics. (*Government and Opposition* Vol.44, no.2; Apr 2009: pp.146–166) – (HUM)

Hurris, A. Radical economics and natural resources. (*International Journal of Environmental Studies* Vol.21; 1983: pp.43–53).

Islam, Muinul. Natural calamities and environment refugees in Bangladesh. (*Refugee* Vol.12, no.1; Jun 1992).

Istomin, Igor. Nothing new on the climate front. (*Russia in Global Affairs* Vol.8, no.4; Oct–Dec 2010: pp.112–123) – (IST)

Ivanchenko, A. Russia and the world after Copenhagen. (*International Affairs* (Moscow) Vol.56, no.6; 2010: pp.235–258) – (AVE)

Jam, Renu. The road from Rio. (*Development Alternatives* Vol.3, no.1; 1993: pp.1, 3).

Jamal, Amir H. The Socio-Economic Impact of New Biotechnologies in the Third World. In Cary Fowler et at. (eds.), *The Laws of Life: Another Development and the New Bio-Technologies*. Development Dialogue. New York: Dag Hammarskjold Foundation, 1998, Issues 1–2, pp. 5–8.

Jasparro, Christopher & Taylor, Jonathan. Climate change and regional vulnerability to transnational security threats in Southeast Asia. (*Geopolitics* Vol.13, no.2; Jun 2008: pp.232–256) – (JAS)

Johnson, Eric P. Air-source heat pump carbon footprints. (*Energy Policy* Vol.39, no.3; Mar 2011: pp.1369–1381) – (JO)

Karlsson-Vinkhuyzen, Sylvia I. United Nations and global energy governance. (*Global Change Peace and Security* Vol.22, no.1; Jun 2010: pp.175–195) – (KAR)

Kazuhiro, Ueta. Climate change and economic strategy. (*Japan Echo* Vol.34, no.4; Aug 2007: pp.45–47) – (KAZ)

Keirstead, James & Schulz, Niels B. London and beyond. (*Energy Policy* Vol.38, no.9; Sep 2010: pp.4870–4879) – (KE)

Kennedy, Christopher, Steinberger, Julia, Gasson, Barrie & Hansen, Yvonne. Methodology for inventorying greenhouse gas emissions from global cities. (*Energy Policy* Vol.38, no.9; Sep 2010: pp.4828–4837) – (KE)

Kenny, R., Law, C. & Pearce, J.M. Towards real energy economics. (*Energy Policy* Vol.38, no.4; Apr 2010: pp.1969–1978) – (KEN)

Keqiang, Li. Speech at the forum on China – EU strategic partnership. (*Foreign Affairs Journal Special Issue*. Nov 2009: pp.1–8) – (KEQ)

Khadka, Navin Singh. South Asia's vicious climatic cycle. (*South Asian Journal* no.28; Apr–Jun 2010: pp.22–26) – (KHA)

Khanna, G.N. Environmental protection act: A critical analysis. (*Indian CI Journal of Social Work* Vol.LI, no.1; 1990: pp.181–194).

Khanna, G.N. Problems of human environment: People's participation and legal solution. (*Social Action* Vol.39; Jul–Sept 1989: pp.275–297).

Khanna, K.P. & Modi, Ashok. Environment and development. (*EPWP* Vol.18; 1983: pp.638–642).

Khanom, Sufia. Gender issues in climate change. (*BIISS Journal* Vol.30, no.4; Oct 2009: pp.449–469) – (KHA)

Khatur, Renu. Organizational response to the environmental crisis in India. (*Indian Journal of Political Science* Vol.49, no.1; Jan–Mar 1989: pp.20–30).

Kilian, Bertil & Elgstrom, Ole. Still a green leader? the European Union's role in international climate negotiations. (*Cooperation and Conflict* Vol.45, no.3; Sep 2010: pp.255–273) – (KIL)

Kirton, John. Consequences of the 2008 US elections for America's climate. (*International Journal* Vol.64, no.1; Winter 2008–09: pp.153–162) – (KIR)

Kittikhoun, Anoulak & Weiss, Thomas G. Myth of scholarly irrelevance for the United Nations. (*International Studies Review* Vol.13, no.1; Mar 2011: pp.18–23) – (KI)

Klare, Michael T. Global warming battlefields. (*Current History* Vol.106, no.703; Nov 2007: pp.355–361) – (KLA)

Kokeyev, M. Kyoto. (*International Affairs* (Moscow) Vol.51, no.1; 2005: pp.118–126) – (KOK)

Kokorin, A. Climate, weather, ecology, and the end of the world. (*International Affairs* (Moscow) Vol.55, no.4; 2009: pp.133–138) – (KOK)

Kopytko, Natalie & Perkins, John. Climate change, nuclear power, and the adaptation-mitigation dilemma. (*Energy Policy* Vol.39, no.1; Jan 2011: pp.318–333) – (KOP)

Koser, Khalid. Why migration matters. (*Current History* Vol.108, no.717; Apr 2009: pp.147–153) – (KOS)

Kosnik, Lea. Potential for small scale hydropower development in the US. (*Energy Policy* Vol.38, no.10; Oct 2010: pp.5512–5519) – (KOS)

Kothari, Ashish. *India's Forests: Can Money Save Them?* Hindu Survey of the Environment, The Hindu, Chennai: Kasturi & Co, 2000.

Kothari, Ashish. Politics of biodiversity convention. (*Economic and Political Weekly* Vol.XXVII, Nos. 15–16; Apr 11–18, 1992: pp.749–755).

Kothari, Smith. Ecology versus development: The struggle for survival. (*Social Action* Vol.35; Oct–Dec 4, 1985: pp.146–153).

Kraemer, Andreas. Learning from Europe's mistakes. (*International Politic* Vol.10, no.1; Spring 2009: pp.52–55) – (KRA)

Kroll, Stephan & Shogren, Jason F. Domestic politics and climate change. (*Cambridge Review of International Affairs* Vol.21, no.4; Dec 2008: pp.563–583) – (KR)

Kshatriya, Gautam K. & Subramanium, P. Impact of climate change on human health. (*World Focus* Vol.31, no.3; Mar 2010: pp.103–105) – (KSH)

Kuik, Onno & Hofkes, Marjan. Border adjustment for European emissions trading. (*Energy Policy* Vol.38, no.4; Apr 2010: pp.1741–1748) – (KUI)

Kulshrestha, UC. & Gupta, Gyan Prakash. Climate change. (*World Focus* Vol.31, no.3; Mar 2010: pp.97–120) – (KUL)

Kulz, Helmut. Further water disputes between India and Paldstan. (*International and Comparative Law Quarterly* Vol.18; Jul 1969: pp.718–738).

Kumar, Manoj. Climate change and national security. (*Air Power* Vol.5, no.1; Jan–Mar 2010: pp.137–152) – (KUM)

Kumar, Vivek & Kumar, Atul. Emerging Asian giants in the energy and climate debate. (*World Affairs* Vol.12, no.1; Spring 2008: pp.32–47) – (KUM)

Kurtzman, Joel. Low-carbon diet. (*Foreign Affairs* Vol.88, no.5; Sep–Oct 2009: pp.114–122) – (KUR)

Kuwali, Dan. From the west to the rest. (*African Security Review* Vol.17, no.3; Sep 2008: pp.18–38) – (KUW)

Kydd, Andrew H. Learning together, growing apart. (*Energy Policy* Vol.38, no.6; Jun 2010: pp.2675–2680) – (KYD)

Lahiry, Sujit. Environment, sustainable development and climate change. (*Journal of Peace Studies* Vol.17, no.2–3; Apr–Sep 2010: pp.77–86) – (LAH)

Lal, Umpanu, Heikkila, Tanya, Brown, Casey & Siegfried, Tobias. Water in the 21st century. (*Journal of International Affairs* Vol.61, no.2; Spring/Summer 2008: pp.1–20) – (LAL)

Lee, Bernice. Managing the interlocking climate and resource challenges. (*International Affairs* Vol.85, no.6; Nov 2009: pp.1101–1116) – (LEE)

Lee, Jae-Seung. Energy security and cooperation in Northeast Asia. (*Korean Journal of Defense Analysis* Vol.22, no.2; Jun 2010: pp.217–233) – (LEE)

Leggett, Jeremy. Global warming: A Greenpeace view. In Jeremy Leggett (ed.), *Global Warming: The Greenpeace Report*. Oxford: Oxford University Press, 1990, pp.457–480.

Levi, Michael A. Copenhagen's inconvenient truth. (*Foreign Affairs* Vol.88, no.5; Sep–Oct 2009: pp.92–104) – (LEV)

Lewis, Joanna I. China's strategic priorities in international climate change negotiations. (*Washington Quarterly* Vol.31, no.1; Winter 2008: pp.155–174) – (LEW)

Lewis, Joanna I. Climate change and security. (*International Affairs* Vol.85, no.6; Nov 2009: pp.1195–1213) – (LEW)

Lewis, Joanna I. Evolving role of carbon finance in promoting renewable energy development in China. (*Energy Policy* Vol.38, no.6; Jun 2010: pp.2875–2886) – (LE)

Liao, Shu-Yi, Tseng, Wei-Chun & Chen, Chi-Chung. Eliciting public preference for nuclear energy against the backdrop of global warming. (*Energy Policy* Vol.38, no.11; Nov 2010: pp.7054–7069) – (LI)

Macey, Adrian. Climate change. (*Global Governance* Vol.15, no.4; Oct–Dec 2009: pp.443–449) – (MAC)

Macgill, Iain, Outhred, Hugh & Nolles, Karel. Some design lessons from market-based greenhouse gas regulation in the restructured Australian electricity industry. (*Energy Policy* Vol.34, no.1; Jan 2006: pp.11–25) – (MAC)

Macintosh, Andrew. Keeping warming within the 2 °C limit after Copenhagen. (*Energy Policy* Vol.38, no.6; Jun 2010: pp.2964–2975) – (MA)

Macintosh, Andrew & Wallace, Lailey. International aviation emissions to 2025. (*Energy Policy* Vol.37, no.1; Jan 2009: pp.264–273) – (MAC)

Madhukar, Uday. Battle royal. *India Today*, Jan 31, 1991: pp.66–68.

Mallapaty, Smriti. Glaciers take the heat. (*Himal* Vol.22, no.10–11; Oct–Nov 2009: pp.34–40) – (MAL)

Manzer, L.E. The CFC-Ozone issue: Progress on the development of alternatives to CFCs. (*Science* Vol.249; Jul 6, 1990: pp.31–35).

Marks, Danny. China's climate change policy process. (*Journal of Contemporary China* Vol.19, no.67; Dec 2010: pp.971–986) – (MAR)

Martin, Susan. Climate change, migration, and governance. (*Global Governance* Vol.16, no.3; Jul–Sep 2010: pp.397–414) – (MA)

Martin-Amo roux, Jean-Marine. Chinese coal and sustainable development. (*China Perspectives* no.1; 2007: pp.40–49) – (MAR)

Mathur, Deepa. Some thoughts on women, environment and development. (*Man and Development* Vol.XIX, no.4; Dec 1997: pp.79–83).

Matthew, Richard A. & Hammill, Anne. Sustainable development and climate change. (*International Affairs* Vol.85, no.6; Nov 2009: pp.1117–1128) – (MAT)

Matthews, Adam. Role for legislators. (*Global Governance* Vol.15, no.4; Oct–Dec 2009: pp.451–455) – (MAT)

May, Bernhard. Energy security and climate change. (*South Asian Survey* Vol.17, no.1; Mar 2010: pp.19–30) – (MAY)

McConnell, Grant. The conservation movement: Past and Present. *Western Political Quarterly* Vol.7, no.3; Sept 1954: pp.463–478.

McGee, Jeffrey & Taplin, Ros. Asia-Pacific partnership on clean development and climate. (*Global Change Peace and Security* Vol.18, no.3; Oct 2006: pp.173–192) – (MCG)

McNeely, Jeffrey A. Biodiversity: The economics of conservation and management. In James T. Winpenny (ed.), *Development Research: The Environmental Challenge*. London: ODI, 1991, pp.25–28.

Mehta, M.C. *Environmental Cases: What the Judiciary Can do*. The Hindu Survey of the Environment, The Hindu, Chennai: Kasturi & Co, 1992, pp.161–163.

Meidan, Michal. China in a post-Kyoto architecture. (*China Perspectives* no.1; 2007: pp.65–71) – (MEI)

Mideksa, Torben K. & Kallbekken, Steffen. Impact of climate change on the electricity market. (*Energy Policy* Vol.38, no.7; Jul 2010: pp.3579–3585) – (MID)

Mignone, Bryan K. International cooperation in a post-Kyoto world. (*Current History* Vol.106, no.703; Nov 2007: pp.362–368) – (MIG)

Miller, Kathleen A. Climate change and water resources. (*Journal of International Affairs* Vol.61, no.2; Spring/Summer 2008: pp.35–50) – (MIL)

Mitra, Abhijit, Zaman, Sufia & Banerjee, Kakoil. Effects of climate change on marine and estuarine ecosystems with special reference to Gangetic Delta. (*Journal of Indian Ocean Studies* Vol.16, no.3; Dec 2008: pp.195–201) – (MIT)

Moli, G. Poyya. Climate change and agriculture. (*World Focus* Vol.31, no.3; Mar 2010: pp.79–84) – (MOL)

Mooney, Chris. An inconvenient assessment. (*Bulletin of the Atomic Scientists* Vol.63, no.6; Nov–Dec 2007: pp.40–47) – (MOO)

Moriarty, Patrick & Honnery, Damon. What energy levels can the Earth sustain? (*Energy Policy* Vol.37, no.7; Jul 2009: pp.2469–2474) – (MOR)

Morton, Katherine. Climate change and security at the third pole. (*Survival: the IISS Quarterly* Vol.53, no.1; Feb–Mar 2011: pp.121–132) – (MO)

Morton, Katherine. China and environmental security in the age of consequences. (*Asia Pacific Review* Vol.15, no.2; Nov 2008: pp.52–67) – (MOR)

Mukhim, Patricia. Food security versus contract farming. (*Dialogue* Vol.10, no.3; Jan–Mar 2009: pp.15–23) – (MUK)

Munslow, Barry & O'Dempsey, Tim. From war on terror to war on weather. (*Third World Quarterly* Vol.31, no.8; 2010: pp.1223–1235) – (MUN)

Munslow, Barry & O'Dempsey, Tim. Globalization and climate change in Asia. (*Third World Quarterly* Vol.31, no.8; 2010: pp.1339–1356) – (MUN)

Murgatroyd, Clive. Defense in a changed climate. (*Rusi Journal* Vol.153, no.5; Oct 2008: pp.28–33) – (MUR)

Murphy, Ann Arie. Toward a US–Indonesia comprehensive partnership. (*Indonesian Quarterly* Vol.37, no.3; 2009: pp.265–282) – (MUR)

Murthy, C.S.R., *'Reforming the UN' in the World Focus*. New Delhi: United Nations at Fifty: Needs Restructuring, Sept 1995, Vol.16, no.9.

Murti, C.R. Krishna. Environmental challenges in developing society. (*Gandhi Mar 8* Vol.1; 1979: pp.142–148).

Mustafa, Malik Qasim. Climate change. (*Strategic Studies* Vol.27, no.3; Autumn 2007: pp.91–143) – (MUS)

Myers, Norman. Environment and security. (*Foreign Policy* no.74; Spring 1989: pp.23–41).

Myers, Norman. *Ultimate Security: The Environmental Basis of Political Stability*. New York: W.W. Norton, 1993.

Nabhi, Uma S. Environmental movements in India. (*India Quarterly* Vol.62, no.1; Jan–Mar 2006: pp.123–145) – (NAB)

Nandan, Deoki, Joon Vinod & Jaiswal, Vaishali. Global warming and the challenges posed by climate change. (*World Focus* Vol.30, no.9; Sep 2009: pp.362–367) – (NAN)

Naz, Antonia Corinthia C. & Naz, Mario Tuscan N. Ecological solid waste management in suburban municipalities. (*ASEAN Economic Bulletin* Vol.25, no.1; Apr 2008: pp.70–84) – (NAZ)

Nazareth, Samir. Need for old wine. (*Himal* Vol.22, no.10–11; Oct–Nov 2009: pp.70–71) – (NAZ)

Nishimura, Mutsuyoshi. Climate change diplomacy and the way forward for Japan. (*Asia Pacific Review* Vol.15, no.1; May 2008: pp.9–24) – (NIS)

Nishioka, Shuzo. Japanese climate policy based on science, equity and cooperation. (*Asia Pacific Review* Vol.15, no.1; May 2008: pp.25–35) – (NIS)

Nye, Joseph S. China's rise doesn't mean war . . . (*Foreign Policy* no.184; Jan–Feb 2011: pp.66–66) – (NYE)

Oberheitmann, Andreas & Sternfeld, Eva. Climate change in China. (*China Aktuell Monatszeitschist* Vol.38, no.3; 2009: pp.135–164) – (OBE)

Pachauri, R.K. Climate change and global warming. (*IIC Quarterly* Vol.33, no.2; Autumn 2006: pp.108–114) – (PAC)

Panda, Abanish & Jain, Arvind. Global warming. (*World Focus* Vol.31, no.3; Mar 2010: pp.90–96) – (PAN)

Papadakis, Elim. The green party in contemporary west German politics. (*Political Quarterly* Vol.54, no.3; Jul–Sept 1983: pp.302–307).

Parikh, Jyoti & Parikh, Kirti. Role of unsustainable consumption patterns and population in global environmental stress. (*Sustainable Development* Vol.1, no.1; Oct. 1991: pp.108–118).

Parks, Bradley C. & Roberts, J. Timmons. Inequality and the global climate regime. (*Cambridge Review of International Affairs* Vol.21, no.4; Dec 2008: pp.621–648) – (PA)

Paskal, Cleo. From constants to variables. (*International Affairs* Vol.85, no.6; Nov 2009: pp.1143–1156) – (PAS)

Peacock, A.D., Jenkins, D.P. & Kane, D. Investigating the potential of overheating in UK dwellings as a consequences of extant climate change. (*Energy Policy* Vol.38, no.7; Jul 2010: pp.3277–3288) – (PEA)

Peake, Stephen. Turbulence in the climate regime. (*Current History* Vol.109, no.730; Nov 2010: pp.349–354) – (PEA)

Pearce Fred. War over water. (*Down to Earth* Vol.2, no.11; Oct 31, 1993: pp.25–28).

Permitta, John C. Impacts of climate change and sea level rise on small Island States. (*Global Environmental Change*) Vol.2, no.1; 1992: pp.19–31.

Plant, Glen. Institutional and legal responses to global environmental change. In Ian H. Rowlands & Malroy Greene (eds.), *Global Environmental Change and International Relations*. London: Macmillan Academic and Professional Ltd., 1992, pp.122–144.

Porfiryev, Boris. Climate change and economy. (*Russia in Global Affairs* Vol.8, no.2; Apr–Jun 2010: pp.155–168) – (POR)

Prashad, Vijay. This frog won't leap. (*Himal* Vol.22, no.10–11; Oct–Nov 2009: pp.48–53) – (PRA)
Prescott, Matt & Taylor, Matthew. Every citizen a carbon trader. (*World Policy Journal* Vol.25, no.1; Spring 2008: pp.19–28) – (PRE)
Purdy, Margaret & Smythe, Leanne. Obscurity to action. (*International Journal* Vol.65, no.2; Spr 2010: pp.411–433) – (PUR)
Qingtai, Yu. Copenhagen climate change conference and China's positive contribution. (*Foreign Affairs Journal* Vol.95; Spr 2010: pp.1–20) – (QIN)
Qingtai, Yu. Copenhagen climate change. (*Foreign Affairs Journal* no.95; 2010: pp.1–8) – (QIN)
Qingtai, Yu. Tackling climate change. (*Foreign Affairs Journal* no.88; Summer 2008: pp.33–43) – (QIN)
Rachman, Gideon. Think again. (*Foreign Policy* no.184; Jan–Feb 2011: pp.59–63) – (RAC)
Rahman, I.A. '*South Asian Perspective on Human Rights and Environment*' *Perspectives on South Asia*, edited by V.A. Pal Panandiker & Navnita Chadha Behera. New Delhi: Centre for Policy Research, Konark Publishers, 2000, pp.417–423.
Rajamani, Lavanya. India and climate change. (*India Review* Vol.8, no.3; Jul–Sep 2009: pp.340–374) – (RAJ)
Raleigh, Clionadh. Political marginalization, climate change, and conflict in African Sahel states. (*International Studies Review* Vol.12, no.1; Mar 2010: pp.69–86) – (RAL)
Ramakrishna, Kilaparti. The emergence of environmental law in the developing countries: A case study of India. (*Ecology Law Quarterly* Vol.12, no.4; 1985: pp.907–935).
Ramakrishna, Kilaparti. Third world countries in the policy response to global environmental change. In Jeremy Leggett (ed.), *Global Warming the Greenpeace Report*. Oxford: Oxford University Press, 1990, pp.421–437.
Ramakrishnan, P.S. Tropical forests: Exploitation, conservation and management. (*Environment and Development* no.166; 1993).
Rangarajan, Mahesh. Ecology: Dilemmas and problems. (*EPW* Vol.XXXIII, no.1, 2; Jan. 1998: pp.23–25).
Ranjan, Rajiv. Copenhagen accord. (*World Focus* Vol.31, no.9; Sep 2010: pp.360–365) – (RAN)
Rao, J. Mohan. Economic reform and ecological refurbishment: A strategy for India. (*EPW* Vol.XXX, no.28; Jul 15, 1975).
Rao, P.K. *The Economics of Global Climatic Change*. Armonk: M.E. Sharpe, 2000.
Rao, R. Rama. *Environment. Problems of Developed and Developing Countries*. New Delhi, Economic and Scientific Research Foundation, 1976.
Red Clift, Michael. Sustainable development and global environmental change. (*Global Environmental Change* Vol.2, no.1; Mar 1992: pp.32–42).
Rhodes, Steven L. Climate change management strategies. (*Global Environmental Change* Vol.2, no.3; 1992: pp.205–214).
Robock, Alan. 20 reasons why geoengineering may be a bad idea. (*Bulletin of the Atomic Scientists* Vol.64, no.2; May–Jun 2008: pp.14–18) – (ROB)
Rosa, Fiona De. Preservationism and the place of people in the environmental movement. (*Social Alternatives* Vol.17, no.1; Jan 1998: pp.21–25).
Rothkopf, David. Is a green world a safer world? (*Foreign Policy* no.174; Sep–Oct 2009: pp.134–137) – (ROT)
Rothsteen, Robert L. Epitaph for a monument to a failed protest? A North-South retrospective. (*International Organization* Vol.42, no.4; Autumn 1988: pp.725–748).
Rowe, Elana Wilson. Who is to blame? Agency, causality, responsibility and the role of experts in Russian framings of global climate change. (*Europe-Asia Studies* Vol.61, no.4; Jun 2009: pp.593–619) – (ROW)

Rowlands, Ian H. The international politics of environment and development: The post-UNCED Agenda. In *The International Library of Politics and Comparative Government*. Dartmouth: The United Nations, Ashgate, 2000, Vol.11, pp.209–224.

Ru, Guo, Xiaojing, Cao, Xinyu, Yang & Yankuan, Li. Strategy of energy-related carbon emission reduction in Shanghai. (*Energy Policy* Vol.38, no.1; Jan 2010: pp.633–638) – (RU)

Rudig, Wolfgang & Lowe, Philip. The withered greening of British politics: A study of the ecology party. (*Political Studies* Vol.34, no.2; Jun 1986: pp.262–284).

Russell, James A. Environmental security and regional stability in the Persian Gulf. (*Middle East Policy* Vol.16, no.4; Winter 2010: pp.90–101) – (RUS)

Salander, Henrik. Principles and process. (*Arms Control Today* Vol.40, no.3; Apr 2010: pp.14–16) – (SAL)

Salehyan, Idean. From climate change to conflict? No consensus yet. (*Journal of Peace Research* Vol.45, no.3; May 2008: pp.315–326) – (SAL)

Samans, Richard, Schwab, Klaus & Malloch-Brown, Mark. Running the world, after the crash. (*Foreign Policy*, no.184; Jan–Feb 2011: pp.80–83) – (SAM)

Saran, Shyam. Climate change: Impact on economic development. (*Journal of Indian Ocean Studies* Vol.16, no.1–2; Apr–Aug 2008: pp.61–68) – (SAR)

Saran, Shyam. Global governance and climate change. (*Global Governance* Vol.15, no.4; Oct–Dec 2009: pp.457–460) – (SAR)

Saran, Shyam. Need for a South Asian perspective. (*Himal* Vol.22, no.10–11; Oct–Nov 2009: pp.30–33) – (SAR)

Sassoon, David. Weaker and worse. (*Himal* Vol.22, no.10–11; Oct–Nov 2009: pp.44–45) – (SAS)

Saunders, Clare. Stop climate chaos coalition. (*Third World Quarterly* Vol.29, no.8; 2008: pp.1509–1526) – (SAU)

Saward, M. Green theory. (*Environmental Politics* Vol.2, no.3; 1993: pp.509–512).

Saxena, K.D. *Environmental Planning, Policies and Programs in India*. New Delhi: Stupa, 1993, pp.34–48.

Saxena, K.P. Period of tribulations. (*World Focus*, New Delhi Vol.16, no.9; Sept 1995: pp.8–13).

Saxena, K.P. United nations and co-operation in development. In *The International Library of Politics and Comparative Government*. Dartmouth: The United Nations, Ashgate, 2000, Vol.11.

Schaffer, Teresita C. United States, India, and global governance. (*Washington Quarterly* Vol.32, no.3; Jul 2009: pp.71–87) – (SCH)

Scheffran, Jürgen. Climate change and security. (*Bulletin of the Atomic Scientists* Vol.64, no.2; May–Jun 2008: pp.19–24) – (SCH)

Schroeder, Miriam. Construction of China's climate politics. (*Cambridge Review of International Affairs* Vol.21, no.4; Dec 2008: pp.505–525) – (SC)

Scott, David. Environmental issues as a 'strategic' key in EU-China relations. (*Asia Europe Journal* Vol.7, no.2; Jun 2009: pp.211–224) – (SCO)

Scott, Shirley V. Securitizing climate change. (*Cambridge Review of International Affairs* Vol.21, no.4; Dec 2008: pp.603–619) – (SC)

Seddon, David. Insecure environment. (*Jane's Intelligence Review* Vol.19, no.5; May 2007: pp.6–13) – (SED)

Sen, Amartya. How is India doing? In Iqbal Khan(ed.), *Fresh Perspectives on India and Pakistan*. Oxford: Bougainvillea Books, 1985, pp.86–96.

Sharma, Manish. Climate change. (*World Focus* Vol.31, no.3; Mar 2010: pp.111–115) – (SHA)

Shea, Kevin P. A celebration of silent spring. (*Environment* Vol.15, no.1; Jan–Feb 1973: pp.4–5).

Shelton, David. Aborigines, environment and waste: A post-colonial perspective. (*Social Alternatives* Vol.17, no.1; Jan 1998: pp.7–10).

Sheth, Pravin. *Environmentalism, Politics, Ecology and Development.* Jaipur: Rawat Publications, 1997, p 55–68.

Sheth, Pravin. Gandhi: Eco-world view and its reinterpretation. (*The Fourth World*, no.5; April 1995: pp.57–72).

Sheth, Pravin. *Narmada Project: Politics of Eco-Development.* New Delhi: Har-Anand, 1994, pp.22–26.

Siddiqi, Toufiq. China and India. (*Journal of International Affairs* Vol.64, no.2; Spr/Sum 2011: pp.73–90) – (SID)

Siddiqui, Rehana. Environment issues and policy response in Pakistan. (*South Asian Journal* no.28; Apr–Jun 2010: pp.74–81) – (SID)

Sills, David L. The environmental movement and its critics. (*Human Ecology* Vol.3, no.1; 1975: pp.1–41).

Silori, C.S. Climate change. (*South Asian Journal*, no.21; Jul–Sep 2008: pp.8–23) – (SIL)

Singh, Abhoy. Our ecosystem in the 21st century. (*Vision* Vol.XVI, no.3–4; Jan–Jun 1997: pp.1–8).

Singh, Ajay. India's role in the new world order. (*Indian Defense Review* Vol.26, no.1; Jan–Mar 2011: pp.175–179) – (SIN)

Singh, Akshay K. Mitigating existential threat. (*World Focus* Vol.30, no.11–12; Nov–Dec 2009: pp.505–512) – (SIN)

Singh, Bhupendra Kumar. Ensuring India's energy security. (*World Focus* Vol.30, no.11–12; Nov–Dec 2009: pp.521–524) – (SIN)

Sinha, Uttam Kumar. Climate change. (*Strategic Analysis* Vol.33, no.2; Mar 2009: pp.169–173) – (SIN)

Sinha, Uttam Kumar. Climate change. (*Strategic Analysis* Vol.34, no.6; Nov 2010: pp.858–871) – (SIN)

Sinha, Uttam Kumar. Climate change and foreign policy. (*Strategic Analysis* Vol.34, no.3; May 2010: pp.397–408) – (SI)

Sinha, Uttam Kumar. Climate change and the road to Copenhagen. (*Strategic Analysis* Vol.33, no.5; Sep 2009: pp.634–636) – (SIN)

Sinha, Uttam Kumar. Climate summit at Copenhagen. (*Strategic Analysis* Vol.33, no.6; Nov 2009: pp.795–799) – (SIN)

Sinha, Uttam Kumar. Geopolitics of climate change and India's position. (*Indian Foreign Affairs* Journals Vol.4, no.2; Apr–Jun 2009: pp.97–112) – (SIN)

Situmeang, Hardiv Harris. Towards an equitable climate change regime. (*Indonesian Quarterly* Vol.38, no.1; 2010: pp.41–53) – (SIT)

Smith, D. & Blowers, A. Passing the buck-hazardous waste disposal as an international problem. (*Talking Politics* Vol.4, no.1; 1991: pp.4–9).

Smith, Heather A. Political parties and Canadian climate change policy. (*International Journal* Vol.64, no.1; Winter 2008–09: pp.47–66) – (SMI)

Smith, Julianne & Mix, Derek. Transatlantic climate change challenge. (*Washington Quarterly* Vol.31, no.1; Winter 2008: pp.139–154) – (SMI)

Smith, Paul J. Climate change, mass migration and the military response. (*Orbis* Vol.51, no.4; Fall 2007: pp.617–634) – (SMI)

Smith, Tony. Changing configurations of power in North-South relations. Since 1945. *International Organization* Vol.32, no.l; 1977: pp.1–27).

Socolow, Robert H & Glaser, Alexander. Balancing risks. (*Daedalus* Vol.138, no.4; Fall 2009: pp.31–44) – (SOC)

Sohn, Louis B. The Stockholm declaration on the human environment. (*The Harvard International Law Journal* Vol.14, no.3; 1973: pp.423–515).

Sorensen, Alan. Global progress report, 2010. (*Current History* Vol.109, no.723; Jan 2010: pp.3–30) – (SOR)
Sova cool, Benjamin K. & Brown, Marilyn A. Twelve metropolitan carbon footprints. (*Energy Policy* Vol.38, no.9; Sep 2010: pp.4856–4869) – (SO)
Speth, James Gustave. A post-Rio compact. (*Foreign Policy*, no.88; Fall l992: pp.l45–16l).
Steffen, Will. Just another environmental problem? (*Current History* Vol.106, no.703; Nov 2007: pp.369–375) – (STE)
Stein, Jana von. International law and politics of climate change. (*Journal of Conflict Resolution* Vol.52, no.2; Apr 2008: pp.243–268) – (STE)
Stepp, Matthew D. & Wine Brake, James J. Greenhouse gas mitigation policies and the transportation sector. (*Energy Policy* Vol.37, no.7; Jul 2009: pp.2774–2787) – (STE)
Stern, David I. & Jotzo, Frank. How ambitious are China and India's emissions intensity targets? (*Energy Policy* Vol.38, no.11; Nov 2010: pp.6776–6783) – (ST)
Stern, Todd & Antholis, William. Changing climate. (*Washington Quarterly* Vol.31, no.1; Winter 2008: pp.155–174) – (STE)
Stevenson, Hayley. Cheating on climate change? Australia's challenge to global warming norms. (*Australian Journal of International Affairs* Vol.63, no.2; Jun 2009: pp.165–186) – (STE)
Stix, Gary. Climate repair manual. (*Scientific American* Vol.1, no.16; Sep 2006: pp.16–19) – (STI)
Strong Maurice, F. Facing down Armageddon. (*World Policy Journal* Vol.26, no.2; Sum 2009: pp.25–32) – (STR)
Strong Maurice, F. The international community and the environment. (*Environmental Conservation* Vol.4, no.3; Autumn 1977: pp.165–172).
Strong Maurice, F. The way ahead. (*Our Planet* Vol.8, no.5; 1997: pp.6–8).
Swaminathan, M.S. *Biodiversity: Equity in Benefit Sharing*. Hindu Survey of the Environment, The Hindu. Chennai: Kasturi & Co, 2000.
Swaminathan, M.S. Biodiversity: Promoting efficiency in conservation and equity in utilization. RGICS (Rajiv Gandhi Institute for Contemporary Studies) Paper no.21, 1995.
Tenneson, Stein. Case for a proactive Indian and Chinese approach to climate change and energy security. (*Strategic Analysis* Vol.31, no.3; May 2007: pp.417–445) – (TON)
Thavasi, V. & Ramakrishna, S. Asia energy mixes from socio-economic and environmental perspectives. (*Energy Policy* Vol.37, no.11; Nov 2009: pp.4240–4250) – (TH)
Theisen, Ole Magnus. Blood and soil? Resource scarcity and internal armed conflict revisited. (*Journal of Peace Research* Vol.45, no.6; Nov 2008: pp.801–818) – (THE)
Thomas, C. Beyond UNCED: An introduction. (*Environmental Politics* Vol.2, no.4; 1993: pp.1–27).
Thompson, Alexander. Rational design in motion. (*European Journal of International Relations* Vol.16, no.2; Jun 2010: pp.269–296) – (TH)
Tolba, Mostafa F. Redefining UNEP. (*Our Planet* Vol.8, no.5; 1997: pp.9–11).
Tolba, Mostafa Kamal. *Development Without Destruction: Evolving Environmental Perception*. Dublin: Tycooly International, 1982.
Trombetta, Maria Julia. Environmental security and climate change. (*Cambridge Review of International Affairs* Vol.21, no.4; Dec 2008: pp.585–602) – (TR)
United Nations Environment Programme. Two decades of achievement and challenge. (*Our Planet* Vol.4, no.5; 1992).
Uyanaev, Sergei. End of the third round of the Russia-India-China academic dialogue. (*Far Eastern Affairs* Vol.38, no.2; 2010: pp.1–9) – (UYA)
Valentine, Scott, Sova cool, Benjamin K. & Matsuura, Masahiro. Empowered? evaluating Japan's national energy strategy under the DPJ administration. (*Energy Policy* Vol.39, no.3; Mar 2011: pp.1865–1876) – (VA)

Varghese, B G: Water conflict. (*USI Journal* Vol.140, no.580; Apr–Jun 2010: pp.160–167) – (VER)

Varma, Navarun, K., Sarangi, Gopal & Mishra, Arabinda. Impact of climate change on a hydro–geographic "region of conflict". (*World Affairs* Vol.13, no.2; Sum 209: pp.40–64) – (VAR)

Verbruggen, Aviel & Marchohi, Mohamed Al. Views on peak oil and its relation to climate change policy. (*Energy Policy* Vol.38, no.10; Oct 2010: pp.5572–5581) – (VER)

Vincent, A. The character of ecology. (*Environmental Politics* Vol.2, no.2; 1993: pp.248–276).

Visvanathan, Shiva. *Health: Semiotics of Waste.* Survey of the Environment, The Hindu, Chennai: Kasturi and Co, 2000.

Vogler, John. Climate change and EU foreign policy. (*International Politics* Vol.46, no.4; Jul 2009: pp.469–490) – (VOG)

Vogler, John. European contribution to global environmental governance. (*International Affairs* Vol.81, no.4; Jul 2005: pp.835–850) – (VOG)

Vogler, John. European Union as a protagonist to the United States on climate change. (*International Studies Perspectives* Vol.7, no.1; Feb 2006: pp.1–22) – (VOG)

Waid, 'S. Thinking global, acting local? British local authorities and their environmental plans. (*Environmental Politics* Vol.2, no.3; 1993: pp.453–478).

Walker, Martin. Southern supermen. (*World Policy Journal* Vol.25, no.4; Winter 2008: pp.133–143) – (WA)

Wallack, Jessica Seddom & Ramanathan, Veera Bhadran. Other climate changers. (*Foreign Affairs* Vol.88, no.5; Sep–Oct 2009: pp.105–113) – (WAL)

Weale, A. & Williams, A. Between economy and ecology? The single market and the integration of environmental policy. (*Environmental Politics* Vol.1, no.4; 1992: pp.45–64).

Webster, Mort, Paltsev, Sergey & Reilly, John. Hedge value of international emissions trading under uncertainty. (*Energy Policy* Vol.38, no.4; Apr 2010: pp.1787–1796) – (WEB)

Weiya, Huo. Save or splurge. (*Himal* Vol.22, no.10–11; Oct–Nov 2009: pp.66–67) – (WEI)

Westing, A.H. The atmosphere as a common heritage of humankind: Its role in environmental security. (*Scientific World* (London) Vol.34, no.4; 1990).

Wewerinke, Margreet & Doebbler, Curtis F.J. Exploring the legal basis of a human rights approach to climate change. (*Chinese Journal of International Law* Vol.10, no.1; Mar 2011: pp.141–160) – (WE)

Wickremasinghe, Anoja. Environmental deterioration in the hill country of Sri Lanka. (*Malaysian Journal of Tropical Geography* Vol.19; Jun 1989).

Will Catton. Dynamic carbon caps. Splitting the bill. (*Energy Policy* Vol.37, no.12; Dec 2009: pp.5636–5649) – (WIL)

Williams, Marc. Re-articulating the third world coalition: The role of the environmental agenda. (*Third World Quarterly* Vol.14, no.1; 1993: pp.7–29).

Winkler, Harald, Hughes, Alison & Haw, Mary. Technology learning for renewable energy. (*Energy Policy* Vol.37, no.11; Nov 2009: pp.4987–4996) – (WIN)

Wissen Bach, Uwe. Climate change as a human-security threat or a developmental issue. (*Korean Journal of Defense Analysis* Vol.22, no.1; Mar 2010: pp.29–41) – (WIS)

Witsenburg, Karen M. & Adano, Wario R. Of rain and raids. (*Civil Wars* Vol.11, no.4; Dec 2009: pp.514–538) – (WIT)

Wood William B. Tropic deforestation: Balancing regional development demands and global environmental concerns. (*Global Environmental Change* Vol.1, no.1; Dec 1990: pp.23–41).

World Bank Conference on Development Economics Proceedings, edited by Boris Pleskovic & Joseph E. Stiglitz Washington, DC: World Bank, 1998.

Wyss, David. Cost of keeping our cool. (*Current History* Vol.106, no.703; Nov 2007: pp.384–385) – (WYS)

Yang, Ming. Climate change and energy policies, coal and coalmine methane in China. (*Energy Policy* Vol.37, no.8; Aug 2009: pp.2858–2869) – (YAN)

Yap, Nomita. NGOs and sustainable development. (*International Journal* Vol. XLV, no.1; Winter 1989–90: pp.75–105).

Young, Oran R. Future of the arctic. (*International Affairs* Vol.87, no.1; Jan 2011: pp.185–193) – (YOU)

Young, Oran R. Global environmental change and international governance. In Ian H. Rowlands & Malory Greene (eds.), *Global Environmental Change and International Politics*. London: Macmillan Academic and Professional Ltd., 1992, pp.6–18.

Yumkella, Kandeh K. UN as a coalition of all talents. (*International Affairs* (Moscow) Vol.56, no.4; 2010: pp.21–35) – (YUM)

Yunfeng, Yan & Laike, Yang. China's foreign trade and climate change. (*Energy Policy* Vol.38, no.1; Jan 2010: pp.350–356) – (YU)

Zanni, Alberto M. & Bristow, Abigail L. Emissions of CO_2 from road freight transport in London. (*Energy Policy* Vol.38, no.4; Apr 2010: pp.1774–1786) – (ZAN)

Zhang, Zhong Xiang. Multilateral trade measures in a post-2012 climate change regime. (*Energy Policy* Vol.37, no.11; Dec 2009: pp.5105–5112) – (ZHA)

Zhuang, Guiyang. How will China move towards becoming a low carbon economy. (*China and World Economy* Vol.16, no.3; May–Jun 2008: pp.93–105) – (ZHU)

Zonglei, Wei. Prospects of climate change negotiation. (*Contemporary International Relations* Vol.20, no.6; Nov–Dec 2010: pp.62–75) – (ZON)

Zonglei, Wei. Tackling climate change. (*Contemporary International Relations* Vol.19, no.2; Mar 2009: pp.68–88) – (ZON)

Journals

Down to Earth
Front Line
Hindu Survey of Environment
India Today
Our Planet
Seminar
World Focus

Newspapers, Clippings and Newsletters

Newspapers

Amrit Bazar Patrika
The Bangladesh Observer
Dawn
Deccan Herald

Earth Summit Times
Guardian
The Hindu (Madras)
The Hindustan Times

The Independent
Indian Express

New York Times
The Statesman (Calcutta)
Times of India (New Delhi)
The Washington Post

Clippings

'Green File India' Vol.153, Sept. 2000.
'Green File India' Vol.155, Nov. 2000.
Green File India' A selection of clippings on the environment'. Vol.154, Oct. 2000.
Green File: 'South Asia: A selection of clippings on the environment'. Vol.5, no.12, July, 2000.
Green File 'South Asia: A selection of clippings on the environment', Vol.6, no.1. August, 2000.

Newsletters

IIPA: Newsletter
UN Newsletters.
UNEP Newsletter
WWF-India Newsletter.

INDEX

Agenda 21 58
air pollution 95, 96
American environmentalism 27, 28, 29, 30, 31
aquatic ecosystem 93, 94
assessment of UNEP 171–173

Bretton Woods Institutions 12, 13, 14, 194
British environmentalism 25, 26, 27
Brundtland, Gro Harlem xi, 2; Brundtland Commission 2, 9; Brundtland Report 7, 11, 12, 193

Carson, Rachel 31
Central Pollution Control Board 94
challenges of implementation 174, 175, 176
Convention of Biodiversity 44, 45

deforestation 89, 90, 91
desertification 86, 87, 88, 89
developed and developing countries views 35, 36
Dominant Social Paradigm 24

Earth Summit/UNCED 42
economic development 4, 5
economic and ecological globalization 179–182
environment 5, 6, 14, 15
environmental governance 105, 196
environmental issues and concerns 174, 175; environmental challenges viii, ix
environmental movements in India 78–85

environment assembly xiv, 39, 65, 165, 166, 167
environment in India 76, 77, 78

findings xv, xvi
forestry agreement 48
Founex Report 36

Ganga Action Plan 94, 95
Global Environmental Outlook (GEO) 53, 54, 63, 185, 186
global environment facility 46, 47, 48, 177
Governing Council 164, 165

Human Development Report 3, 6
human health and welfare 99, 100

International Union for Conservation of Nature 41

main and secondary hypothesis ix, x, 192
major achievements (Stockholm) 36, 37, 38, 49
Malmo Declaration 182, 183
Millennium Development Goals 15, 16
Multipurpose Valley Project 91, 92

National Conservation Strategy (Bangladesh) 112–117
NCS: Nepal 118–125
NCS: Pakistan 126–136

NCS: Sri Lanka 137–141
new environmentalism 32, 33
North-South divide xiv

Outcome of the Conference 43, 44, 45, 46, 47, 48, 49

Rio Conference xi, xii, 2, 42, 43
role of global environmental movements 23, 24

SACEP xiii, xiv, 141–154
South Asian environmental perspectives 105–111
state of environment report 185, 186
Stockholm Conference 33, 34, 35, 36, 37

sustainable development 7, 8, 9, 10, 11, 193–194
Sustainable Development Goals 16, 17; and challenges 18
system wide strategy of UNEP 91

treaty on climate change 45, 46

UN Environment Programme 38, 39, 40, 41, 178; critical problems 167–169, 170, 171; role of UNEP 40, 41, 42, 50–65
urbanization 99

WCED 9, 11
World Conservation Strategy (WCS) 8, 9
World Summit on Sustainable Development (2002) 58

Printed in the United States
by Baker & Taylor Publisher Services